Recent Advances in Lichenology

Dalip Kumar Upreti · Pradeep K. Divakar
Vertika Shukla · Rajesh Bajpai
Editors

Recent Advances in Lichenology

Modern Methods and Approaches in Biomonitoring and Bioprospection, Volume 1

Springer

Editors
Dalip Kumar Upreti
Lichenology Laboratory
CSIR-National Botanical Research Institute
Lucknow, Uttar Pradesh
India

Vertika Shukla
Lichenology Laboratory
CSIR-National Botanical Research Institute
Lucknow, Uttar Pradesh
India

Pradeep K. Divakar
Facultad de Farmacia, Departamento de Biologia Vegetal II
Universidad Complutense de Madrid
Madrid, Madrid
Spain

Rajesh Bajpai
Lichenology Laboratory
CSIR-National Botanical Research Institute
Lucknow, Uttar Pradesh
India

ISBN 978-81-322-2180-7 ISBN 978-81-322-2181-4 (eBook)
DOI 10.1007/978-81-322-2181-4

Library of Congress Control Number: 2014957397

Springer New Delhi Heidelberg New York Dordrecht London

Printed on acid-free paper

Springer (India) Pvt. Ltd. is part of Springer Science+Business Media (www.springer.com)

Dedicated to

Dr. D.D. Awasthi
Father of Indian Lichenology

Foreword

सीएसआईआर-राष्ट्रीय वनस्पति अनुसंधान संस्थान
CSIR-National Botanical Research Institute
(वैज्ञानिक तथा औद्योगिक अनुसंधान परिषद, नई दिल्ली)
राणा प्रताप मार्ग, लखनऊ – 226 001, उ.प्र., भारत
(Council of Scientific & Industrial Research, New Delhi)
Rana Pratap Marg, Lucknow - 226 001, U.P., India

डॉ. चन्द्र शेखर नौटियाल
टांटा इनोवेशन फैलो, एफएनए, एफएनएएससी, एफएनएएएस
निदेशक
Dr. Chandra Shekhar Nautiyal
TATA Innovation Fellow, FNA, FNASc, FNAAS
Director

FOREWORD

Lichens are unique organisms which require multidimensional approach to explore its potential in various fields of environment, botany and chemistry. Modern techniques especially molecular, culture, remote sensing technique has considerably contributed in the field of lichens. Multivariate analyses together with GIS approaches have established lichens as an ideal and reliable indicator of air pollution. Advanced culture techniques have increased the pharmacological application which was earlier restricted due to meager biomass of lichens. Advent of sophisticated analytical instrumentations facilitated isolation and characterization of bioactive constituents of lichens even in lower concentration for their bioprospection.

I am delighted to see that the editors made an effort to compile the advances in the field of lichenology contributed by experts of various fields of lichenology from around the globe, which will not only help to introduce various multidisciplinary approaches and techniques in the field of lichenology but also create interest among researchers to take up research on these unique organisms.

I am glad that this work has been accomplished in the Lichenology Laboratory of CSIR-National Botanical Research Institute, Lucknow.

I congratulate the editors for their endeavour in composing a comprehensive and valuable contribution about advances in the study of lichens worldwide and hope this book will be immensely helpful for environmentalists, botanist and phytochemists.

Date: 15th July 2014

(C.S. Nautiyal)

Phones : (Off.) (0522) 2205848, EPABX Phone : (0522) 2297800 to 2297999 Fax : (0522) 2205839, Post Box No. 436, Lucknow
e-mail : director@nbri.res.in, csnnbri@yahoo.com, Website: http://www.nbri.res.in

Preface

Lichens present an excellent example of symbiotic association. The unique composition of alga and fungi in lichens not only results in conferring differential sensitivity towards a range of environmental factors, but also makes them physiologically adapted and chemically diversed to combat various abiotic and biotic environmental stresses.

This book volume covers two very important aspects of lichens, i.e. biomonitoring and bioprospection. In order to understand the role of lichens as biomonitors, it is important to know the various factors which influence the growth of lichens in natural conditions.

Lichens have been known for long as bioindicators of air pollution and various studies have explored the biomonitoring potential of different lichen species. Recently, more standardised protocols have been adopted in Europe and America, which ensures the authenticity and reproducibility of lichen biomonitoring data.

Standardised methods for lichen diversity assessment not only authenticate classical approaches of lichen diversity changes related to ecological studies and air quality, but also provide an opportunity to extend to the studies related to climate change and land use changes. Lichen diversity studies coupled with remote sensing data can be used for mapping lichen species, which provide vital information regarding the present scenario and predict the future model of climate change or land use change. Applicability of lichen diversity-remote sensing studies further extends to estimating and mapping influence of pollutant in unsampled location based on sampled locations, which help to track the long-range transport of semi-volatile pollutants to high altitude ecosystems.

Lichens also play a major role in the functioning of the ecosystem by the process of soil formation (Lichens have the ability to grow on barren rock), but this characteristic is disadvantageous for conservation of ancient monuments as lichen growth results in biodeterioration. Studies on restricting lichen growth and preserving monuments are a subject of research and needs implementation of strategies for conservation of monuments.

Another aspect of lichens, which is gaining importance in recent climate change studies, is lichenometric studies, to observe glacier retreat phenomenon. As any change in shape and size of glacier provides vital information regarding climate change, therefore age of lichens growing on exposed moraines is a possible clue to retreat of glacier.

Apart from being excellent biomonitors, another not much explored aspect of lichens is their chemistry. For lichens their unique chemistry supplements physiological adaptation for sustaining in extreme climates, but for mankind lichens are a treasure house of pharmacologically important bioactive constituents capable of curing simple cough and cold to HIV and cancer.

The therapeutic potential of lichen secondary metabolites has been known since medieval times mainly through folklore and ethnobotanical uses. The main constraint which restricts its commercial utilisation is slow growth rate and low biomass of lichens. Recent advances in analytical instrumentation techniques have resulted in high throughput techniques for isolation, while highly sensitive detection techniques allow characterisation of bioactive compound even in very low quantity. Structural characterisation of bioactive compound paves the way for its synthesis in laboratory and further structural modifications.

Therefore, the book volume intends to introduce researchers to advancements in the field of lichenology with an aim to involve more active participation of multidisciplinary research in the study of lichens.

Contents

Editors and Contributors

About the Editors

Dalip Kumar Upreti FNASc, Head of the Lichenology Laboratory, CSIR-National Botanical Research Institute (CSIR-NBRI), Lucknow, has a Ph.D. (1983) in Botany on *Studies on the lichen family Cladoniaceae from India and Nepal* from Lucknow University under the guidance of Dr. D.D. Awasthi. Dr. Upreti has over 35 years of research experience and has published more than 300 research papers in peer-reviewed journals and co-authored eight books.

Apart from taxonomy, Dr. Upreti has also carried out extensive research on ecology, lichen chemistry, pollution monitoring, in vitro culture and biodeterioration studies and bioprospection of Indian lichens. Dr. Upreti was in Antarctica in 1991–1992. He is the Indian corresponding member to the British Lichen Society and International Association for Lichenology.

Pradeep K. Divakar has an M.Sc. in Botany and earned his Ph.D. (2002) on *Revisionary studies on the lichen genus Parmelia sensu lato in India* from Lucknow University. Dr. Divakar is Professor at the Department of Plant Sciences, Faculty of Pharmacy, Complutense University Madrid, Spain. Dr. Divakar has published two books and more than 70 papers in national and international journals. Currently, he is editor of *Biodiversity and Conservation*, *Mycokeys*, *Organisms Diversity and Evolution*, and *Lazaroa* (Iberian Journal of Botany).

His research activities centre on molecular systematics, genetic diversity, evolution, speciation events, gene mapping, gene flow, TCS estimation of gene genealogies and phenotypic character mapping, ancestral character state and genomics of fungi that form mutualistic symbioses with algae or cyanobacteria and mycobiont cultures.

Vertika Shukla holds an M.Sc. in Organic Chemistry and a Ph.D. (2003) on the *Chemical Study of Macrolichens of Garhwal Himalayas* from H.N.B. Garhwal University, Srinagar (Garhwal). She is currently working in the Lichenology Laboratory of CSIR-NBRI as a DST scientist and to date she has published more than 50 scientific articles.

Dr. Shukla has contributed to secondary metabolite chemistry, spatio-temporal behaviour of pollutants in high altitude ecosystems, lichen responses to climate change and the role of lichens in bioremediation of atmospheric fallouts.

Rajesh Bajpai who has an M.Sc. in Environmental Science and received his Ph.D. in 2009 from Babasaheb Bhimrao Ambedkar (Central) University, Lucknow on *Studies on lichens of some monuments of Madhya Pradesh with reference to Biodeterioration and Biomonitoring*, is working as a DST scientist in Lichenology Laboratory of CSIR-NBRI, Lucknow. Dr. Bajpai has published 30 research papers in various national and international journals.

Dr. Bajpai has carried out extensive research work on the interaction of lichens with the environment, climate-change modelling, arsenic pollution and biodeterioration.

Contributors

Richard A. Armstrong Department of Vision Sciences, Aston University, Birmingham, UK

Joël Boustie Faculté des Sciences Pharmaceutiques et Biologiques, Institut des Sciences Chimiques de Rennes, PNSCM-UMR 6226, Université Européenne de Bretagne, Rennes Cedex, France

Giorgio Brunialti Terra Data Environmetrics, Spin Off of the University of Siena, Monterotondo Marittimo, Italy

Pulak Das Department of Environmental Science, Indira Gandhi National Tribal University (IGNTU), Amarkantak, Madhya Pradesh, India

J. Garty Department of Molecular Biology and Ecology of Plants, Tel Aviv University, Tel Aviv, Israel

R.L. Garty-Spitz, Ganei Am, Israel

Paolo Giordani Botanic Centre Hanbury, DISTAV, University of Genova, Genova, Italy

Gaëtan Herbette Spectropole, Aix-Marseille Université, Marseille Cedex 20, France

Santosh Joshi Lichenology Laboratory, Plant Biodiversity and Conservation Biology Division, National Botanical Research Institute, CSIR, Lucknow, Uttar Pradesh, India

Pierre Le Pogam Faculté des Sciences Pharmaceutiques et Biologiques, Institut des Sciences Chimiques de Rennes, PNSCM-UMR 6226, Université Européenne de Bretagne, Rennes Cedex, France

Steven D. Leavitt Committee on Evolutionary Biology, University of Chicago, Chicago, IL, USA

C. Prabakaran EHD/BPSG/EPSA, Space Applications Centre (ISRO), Ahmedabad, Gujarat, India

Nupoor Prasad GIS Consultant, JPS Consultants Pvt. Ltd., New Delhi, India

Jayashree Rout Department of Ecology and Environmental Science, Assam University, Silchar, Assam, India

P.S. Roy Center for Earth and Space Sciences, University of Hyderabad, Hyderabad, India

M.R.D. Seaward Department of Archaeological, Geographical and Environmental Sciences (AGES), University of Bradford, Bradford, UK

Manoj Semwal ICT Department, CSIR—Central Institute of Medicinal and Aromatic Plants, Lucknow, India

C.P. Singh EHD/BPSG/EPSA, Space Applications Centre (ISRO), Ahmedabad, Gujarat, India

Larry L. St. Clair Department of Biology and M.L. Bean Life Science Museum, Brigham Young University, Provo, UT, USA

Gintarė Sujetovienė Department of Environmental Sciences, Vytautas Magnus University, Kaunas, Lithuania

1 The Influence of Environmental Factors on the Growth of Lichens in the Field

Richard A. Armstrong

Contents

R.A. Armstrong (✉)
Department of Vision Sciences, Aston University, Birmingham, B4 7ET, UK
e-mail: r.a.armstrong@aston.ac.uk

D.K. Upreti et al. (eds.), *Recent Advances in Lichenology*,
DOI 10.1007/978-81-322-2181-4_1

Abstract

The majority of studies of the effects of environmental factors on lichen growth have been carried out in the field. Growth of lichens in the field has been measured as absolute growth rate (e.g. length growth, radial growth, diameter growth, area growth, or dry weight gain per unit of time) or as a relative growth rate, expressed per unit of thallus area or weight, e.g. thallus specific weight. Seasonal fluctuations in growth in the field often correlate best with changes in average or total rainfall or frequency of rain events through the year. In some regions of the world, temperature is also an important climatic factor influencing growth. Interactions between microclimatic factors such as light intensity, temperature, and moisture are particularly important in determining local differences in growth especially in relation to aspect and slope of rock surface, or height on a tree. Factors associated with the substratum including type, chemistry, texture, and porosity can all influence growth. In addition, growth can be influenced by the degree of nutrient enrichment of the substratum associated with bird droppings, nitrogen, phosphate, salinity, or pollution. Effects of environmental factors on growth can act directly to restrict species distribution or indirectly by altering the competitive balance among different species in a community.

Keywords
Lichen · Radial growth rate · Dry weight gain · Relative growth rate · Climate · Substratum · Nutrient enrichment · Competition

1.1 Introduction

Lichens are influenced by many environmental factors including the long-term effects of climate, local changes in microclimate, and a variety of factors associated with the substratum such as type of rock, bark, or soil, substrate chemistry, and the degree of nutrient enrichment by birds, salinity, or pollution (Armstrong 1974; James et al. 1977). Studies of the effects of environmental factors on lichen growth are important for several reasons. First is to understand the factors determining the distribution of rare or interesting species and the structure of important lichen communities. Hence, a change in the environment to a less than optimal condition may inhibit lichen growth to such an extent that thalli rapidly fragment and disappear from a substratum. In addition, the environment can act more indirectly, subtle changes altering the competitive balance among different species in a community (Armstrong and Welch 2007). Second, knowledge of environmental growth effects is important in 'lichenometry', a technique which uses lichens to estimate the surface age of a substratum (Locke et al. 1979; Innes 1985; Matthews 1994; Benedict 2009). Third, it may be important to separate out growth effects due to natural factors, such as climate or microclimate, from those attributable to human activity, such as grazing by herbivores, pollution, or woodland management.

Two major problems have been encountered in studying lichen growth. First, studies have been limited by the relatively few facilities available for culturing whole lichen thalli in the laboratory for long periods of time. Hence, few 'factorial-type' experiments have been undertaken, in which the synergistic effects of different variables have been studied (Jones and Platt 1969; Armstrong 1994). As a result, most studies have been carried out in the field and therefore lack the control of the environment possible in a laboratory. Second, lichens are morphologically diverse consisting of crustose, foliose, placodiod, and fruticose forms, and they also grow on different substrata with contrasting properties such as rock, tree bark, and soil. Hence, different methods have been devised to measure growth, making it difficult to compare the results of different studies. This review considers various aspects of the influence of the environment on the growth of lichens as measured in the field including: (1) methods of measuring and expressing growth, (2) the influence of climate, microclimate, substratum factors, and nutrient enrichment on growth, and (3) the importance of the influence of growth on competition as a factor determining distribution and community structure (Armstrong and Welch 2007).

1.2 Measuring Lichen Growth

Different methods have been employed to measure lichen growth in the field depending on growth form and substratum type. Hence, foliose and crustose lichens both exhibit a flat dorsiventral thallus and when growing on relatively flat substrata such as rock and tree bark, similar methods can be used to measure their growth (Armstrong and Bradwell 2011). Current techniques enable measurements of foliose lichen growth to be made over relatively short-time intervals such as weeks and months (Hale 1970; Armstrong 1973) and are based on the original method described by Hale (1970, 1973). This involves measuring the advance of the thallus perimeter, either the tip of a lobe in foliose species or the edge of the hypothallus in many crustose species, with reference to fixed markers on the rock. Digital photography (Hooker 1980) and an image analysis system (Image-J) (Armstrong 2013) can then be used to measure growth increments. Measurements can be made over various time scales, viz. one month for faster-growing foliose (Armstrong and Bradwell 2011) and three months for slower-growing crustose species (Armstrong and Bradwell 2010).

Many studies have employed an absolute measure of growth such as radial growth, diameter growth, area growth, or dry weight gain. The most popular measure in many studies of foliose lichens has been rate of radial extension or radial growth rate (e.g. mm $year^{-1}$). A limitation of radial growth rate, however, is that it is strongly correlated with thallus size and exhibits a high degree of variability both within and between thalli (Armstrong and Bradwell 2010, 2011). As a consequence, measurement of relative growth rate has been used to compensate for these variations in size, growth being scaled to an existing thallus measure such as increase in area per unit of area in unit time (mm^2 mm^{-2}, $time^{-1}$) (Armstrong 1973, 1976a).

Fruticose lichens have a much more complex-branched morphology and different methods necessary to study their growth. Methods of quantifying growth of fruticose species have usually involved measuring length, internode length, diameter, thallus dry weight, or biomass (Karenlampi 1971). Karenlampi (1970) measured growth of thalli of *Cladonia alpestris* (L.) Rabenh. by dividing up thalli and weighing the various parts. The major growth zone was at the tip of the podetia, and relative growth values for length, diameter, and dry weight suggesting significantly higher values for younger than older portions of thalli. More recently, a non-destructive method of estimating biomass of *Usnea* species by harvesting a sample of sites and measuring dry weight gain, ratio of chlorophyll 'a' to 'b', and thallus area have been described, regression methods then being applied to calibrate biomass against the non-destructive measures (Bohuslavova et al. 2012).

1.3 The Influence of Climate on Growth

Beschel's detailed lichenometric studies in the 1950s and 1960s were some of the first to demonstrate the important role played by climate on lichen growth (Beschel 1958, 1961). Beschel found that the same species of the crustose genus *Rhizocarpon* grew considerably more slowly in Greenland than in the Swiss Alps (Beschel 1958, 1961). Re-measurement at Beschel's sites, 12 years after the original study (Tenbrink 1973), confirmed that growth of *Rhizocarpon geographicum* (L.) DC. was more than 50 % slower at those sites in Greenland furthest from the sea (6–7 mm/100 year compared with 17–18 mm/100 year). Subsequently, many lichenometric studies from the northern hemisphere have reported similar trends. Hence, Innes (1983) found a general decrease in growth rate of thalli of *Rhizocarpon* section *Rhizocarpon* from west to east in Scotland, UK. In addition, Bradwell (2001a) and Mathews (2005) found growth rate gradients across Iceland and southern Norway, respectively, probably reflecting reduced moisture availability associated with a diminishing maritime influence. Not all such studies, however, have found significant growth effects. For example, growth of transplanted specimens of *Dolichousnea longissima* (Ach.) Articus was studied in four contrasting habitats (Keon and Myer 2002), length growth or dry weight gain being measured after one year. Growth was similar at all sites studied including those in which the habitat was not predicted to be suitable for growth of this species. Similarly, dry weight gain of *Cladonia portentosa* (Dufour) Coem. was studied at five heathland sites differing in precipitation acidity, nitrogen deposition, and annual rainfall (Hyvarinen and Crittenden 1998), but no regional differences in growth were evident.

1.3.1 Rainfall

Evidence for the influence of climate on lichen growth has usually been obtained from studies of

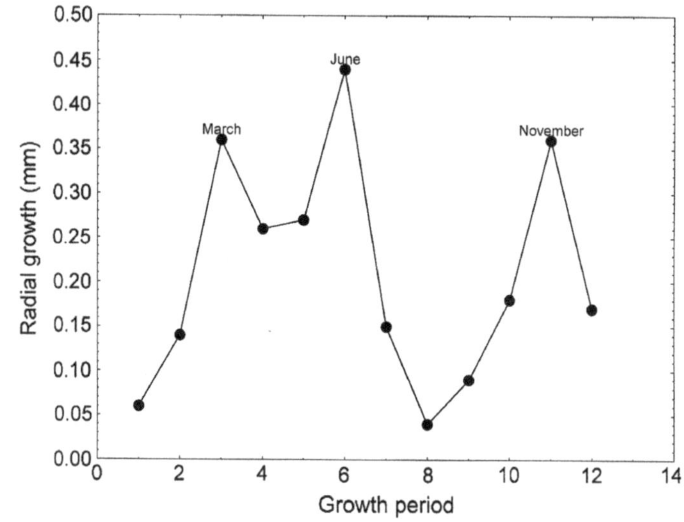

Fig. 1.1 Radial growth (RaG, mm) of the foliose lichen *Melanelia fuliginosa* ssp. *fuliginosa* (Fr. ex Duby) Essl. at a site in north Wales, UK, over 12 months. A large peak of growth occurred in June with smaller growth peaks in March and November corresponding to periods of maximum rainfall (data from Armstrong 1973)

annual or seasonal variations in growth (Rydzak 1961; Hale 1970; Armstrong 1973, 1993a; Showman 1976; Lawrey and Hale 1977; Fisher and Proctor 1978; Moxham 1981; Benedict 1990a). Hence, monthly relative growth rate of *Melanelia fuliginosa* ssp. *fuliginosa* (Fr. ex Duby) Essl. was measured over a period of one year in north Wales (Armstrong 1973, 1993a) (Fig. 1.1). A large peak of growth occurred in June with smaller growth peaks in March and November corresponding to periods of maximum rainfall. In general, monthly fluctuations in growth correlate best with average or total rainfall (Karenlampi 1971; Armstrong 1973; Golm et al. 1993), but linear regressions fitted to radial growth increments measured per month against total rainfall account for relatively small amounts of the total variance in growth (usually <40 %) (Armstrong 1988). For example, growth of *Xanthoparmelia conspersa* (Ehrh. Ex Ach.) Hale in each of 22 successive months in north Wales in relation to total monthly rainfall was studied by Armstrong (1988). A significant positive correlation between growth and rainfall over this period was evident ($r = 0.61$, $P < 0.001$), but the proportion of the variance accounted for by rainfall alone was low ($r^2 = 0.37$). In a study of *Cetraria* species, Paterson et al. (1983) found that moisture was the most important factor influencing growth, assimilation gains achieved during rainy days being sufficient to offset any carbon losses accrued over five dry days. In addition, mean daily rainfall in summer was an important determiner of growth in *Flavocetraria nivalis* (L.) Kärnefelt & A. Thell. although temperature was also considered an important factor (Karenlampi 1971). Similarly, in the high Arctic, the frequency of watering was the most important factor determining the relative growth rate of *Cetraria* species (Cooper et al. 2001). The effect of climate on growth on tree bark has frequently been studied using reciprocal transplants (Gaio-Oliveira et al. 2004). Hence, the relationship between thallus size and growth was studied in three successional forest stands across three boreal climate zones using transplants of *Lobaria pulmonaria* (L.) Hoffm. (Gauslaa et al. 2009). Stand-specific water availability was the most important factor increasing thallus area.

Because of significantly lower radial growth rates, fewer studies of the seasonal growth of crustose species have been carried out. Nevertheless, seasonal growth over three-month growth periods was studied in *R. reductum* Th. Fr. and *Lecidea tumida* Massal. growing on rocks in north Wales (Armstrong 1973). Both showed maximum growth between May and July, relatively little growth in November to January, and slow growth in August to October suggesting that a combination of higher rainfall and lower temperatures was not conducive to growth of these species.

One of the first lichenologists to study the relationship between growth and climate in fruticose lichens was Karenlampi (1971). At sites in Finland, relative growth rate of a number of fruticose species, measured as increase in weight or length, was correlated with mean daily rainfall in summer while temperature was inversely related to growth. In addition, growth of *Ramalina menziesii* (Taylor) was studied in California at coastal and inland sites (Boucher and Nash 1990), the primary determinant of growth appearing to be rainfall. However, the growth of a second species, viz. *R. farinacea*, (L.) Ach. was not be explained by any of the climatic variables.

1.3.2 Temperature

Although rainfall is the predominant variable in many regions, other climatic factors have also been identified as affecting lichen growth. Hence, Brodo (1965) studied annual variations in radial growth of the corticolous lichens *Lecanora chlarotera* Nyl. and *L. caesiorubella* Ach. and observed significantly greater growth during 1959–1961 compared with 1961–1962 largely attributable to differences in mean temperature. In addition, diameter growth of *R. geographicum* was studied at 47 sites covering 18 glaciers in south Norway. Annual growth was correlated with mean temperature and winter mean temperature but not with summer temperature and less correlation with rainfall was evident compared with other studies (Trenbirth and Mathews 2010).

In a study of the seasonal growth of *R. geographicum* (Armstrong 2006), radial growth rate of a sample of 20 thalli was measured in situ on a southeast-facing rock surface at three-month intervals over 51 months (Fig. 1.2). There were five periods of significant growth: July–September of 1993, 1994, and 1995, in January–March of 1996, and in April–June of 1997. In four of these periods, growth coincided with a mean temperature maximum over a three-month period exceeding 15 °C and three of the maxima also coincided with greater than 450 sunshine hours. In addition, two of the growth maxima coincided with three-month periods of total rainfall exceeding 300 mm and one with greater than 50 rain days.

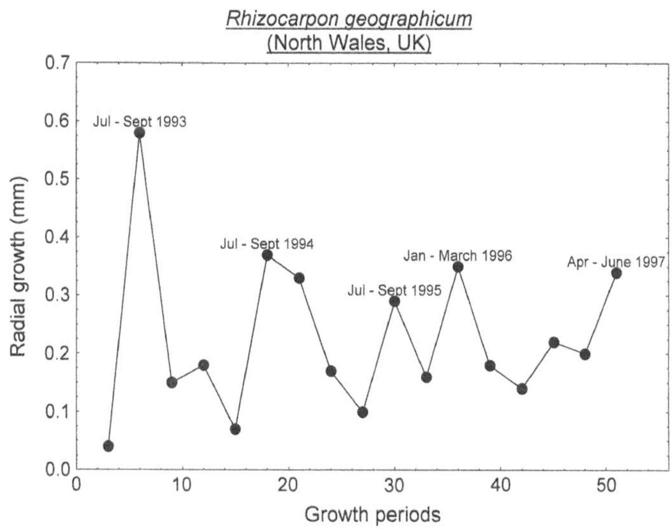

Fig. 1.2 Mean radial growth of thalli of *Rhizocarpon geographicum* measured in 17 three-month periods from 1993 to 1997 in south Gwynedd, Wales, UK. There were five periods of significant growth: July–September of 1993, 1994, and 1995, in January–March of 1996, and in April–June of 1997. Analysis of variance (two-way), between thalli, $F = 0.03$ ($P > 0.05$), between growth periods, $F = 3.86$ ($P < 0.001$) (data from Armstrong 2006)

There were no significant linear correlations between growth and individual climatic variables but significant nonlinear relationships were evident between growth and maximum temperature, mean temperature minimum, total number of air and ground frosts, and number of rain days, maximum temperature being the most significant (Armstrong 2006). Hence, in north Wales, maximum growth of *R. geographicum* may occur in any season, although July–September appeared the most favourable. Coxon and Kershaw (1983) studied changes in photosynthetic capacity in *R. superficiale* (Schaer.) Vain. through the season. This species showed a broad response to temperature and a high resistance to heat stress, characteristic of species living in 'boundary-layer' habitats, but there were no significant seasonal changes in photosynthetic capacity.

1.3.3 Frost Injury and Snowfall

At higher altitudes, the degree of winter frost injury and snowfall are important factors. Hence, Innes (1985) found that thalli of *Rhizocarpon* were smaller close to snow patches due to reduced growing season and ground instability. At some sites, however, the effects of increasing moisture balanced the negative influence of snow. In the Colorado Front Range (Benedict 1990b), frost damage, including 'spalling' of the upper cortex and algal layers, was evident especially in foliose species. Crustose species such as *R. riparium* Räsänen were less affected but some spalling did occur affecting the surface of areolae. Benedict (1990b) also demonstrated that, under snow, photosynthesis was limited due to impaired light penetration but thalli did continue to respire. To demonstrate this effect, lines of lichen-covered rocks were installed in two semi-permanent snow patches in the Colorado Front Range. Thalli of *R. geographicum* survived 5–8 years after transplanting when average duration of snow cover in the year exceeded 40–43 weeks. By contrast, *R. riparium* thalli survived the duration of the study and were clearly more snow tolerant than *R. geographicum*. The effect of snow on long-term lichen growth at high elevations has also been studied by measuring the size of *R. geographicum* along a transect traversing snow-free and snow-accumulation areas (Benedict 1991). No differences in maximum thallus diameter were observed despite large differences in duration of snow cover. Hence, zoned lichen communities, which are often observed adjacent to snow patches, are more likely to be attributable to episodic snow kill rather than to slow growth under seasonal cover.

1.3.4 Light Intensity

The influence of light on lichen growth in the field has been studied most frequently in woodland. Hence, in a transplant experiment, in which the influence of canopy structure on *L. pulmonaria* was studied, growth was strongly correlated with canopy light transmission, dry weight gain over 2 years being less than 5 % under canopies and rising to 20 % on branches associated with canopy gaps (Coxson and Stevenson 2007). In a further transplant experiment, the influence of canopy cover on growth was studied along a regional forest gradient in the boreal forest (Gauslaa et al. 2007). Dry weight gain was greatest in the Atlantic rainforest (36–38 %) but was reduced by low-light levels even in old forests and in most semi-exposed clear-cuts. The relationship between light intensity and growth of five macrolichens was studied in Sweden (Palmqvist and Sundberg 2000), a strong correlation being observed when thalli were moist. Hence, clear-cutting, which often creates an abrupt edge to forest stands, may affect foliose lichen growth. The effect of 'hard edges' and a less abrupt edge to the forest on growth was studied in British Columbia using transplants of *Lobaria retigera* (Bory) Trevis. (Stevenson and Coxson 2008). There was a high mortality of thalli and loss of biomass over 3 years at 'hard' edge sites but a significantly reduced loss when a less abrupt edge was present, suggesting that residual trees at the margin could reduce the impact of forest clearing on lichen diversity.

1.3.5 Interactions Between Factors

Factorial experimental designs have been relatively little used to study the synergistic effects of climatic and other factors on lichen growth. However, Jones and Platt (1969) studied the growth of *X. conspersa* using a three-factor experimental design incorporating four levels of radiation exposure, two contrasting climates, and two nutrient addition conditions. All factors significantly influenced radial growth, but climate had the most substantial effect. There were also significant interactions between climate, radiation exposure, and nutrient levels.

Stepwise multiple regression has also been used to analyse the various climatic factors determining seasonal trends in growth. Hence, Lawrey and Hale (1977) found that a multiple linear regression, which included maximum temperature, total rainfall, and per cent cloudy days as independent variables, explained 22 % of the variation in the growth of *Flavoparmelia baltimorensis* (Gyeln. & Fóriss) Hale. This method was also used to study the monthly radial growth of single lobes of *X. conspersa* in North Wales over 22 months (Armstrong 2009). Of eight climatic variables studied, the frequency of rain days, ground frosts per month, and average wind speed were selected as significant variables in total accounting for 49 % of the variance in the data.

1.4 The Influence of Microclimate on Growth

Investigation of lichen physiology suggests that interactions between light intensity, temperature, and moisture are likely to be important in determining local growth rates (Farrar 1973). Hence, the growth of the fruticose species *Cetrariella delisei* (Bory ex Schaer.) Kärnefelt & A. Thell. was measured in the high Arctic (Uchida et al. 2006). Positive net photosynthesis was recorded when thallus water contents were high, photosynthetic rates being lower on clearer days due to low thallus water contents. Net assimilation rate is influenced by the moisture content of the thallus (Harris 1971a, b, 1972; Kershaw 1972), reaching a peak between 65 and 90 % saturation, and falling at water contents near to saturation. In addition, in some species of temperate lichens, wetting the thallus results in losses of carbon by a release of carbon dioxide gas and re-saturation respiration (Smith and Molesworth 1973; Farrar 1973, 1976). Hence, to grow, thalli have to be wet sufficiently long in the light to overcome the carbon lost on rewetting, but not to be continuously saturated. An experimental demonstration of this effect was reported by Armstrong (1976b) who wetted thalli of three foliose species at different frequencies (Fig. 1.3). The radial growth of *M. fuliginosa* ssp. *fuliginosa* and *Phaeophyscia orbicularis* thalli (Neck.) Moberg declined with increased wetting frequency, while growth of *X. conspersa* increased with wetting frequency until 10 experimental wettings per month, declining at 15 wettings per month. Brown and Webster (1997) studied the growth of *Peltigera canina* (L.) Willd. transplanted onto garden soil. They found considerable linear growth of the transplants, especially when the soil was kept permanently watered, suggesting that a more continuous moisture regime may be beneficial in some species. Similarly, intact lichens grew most rapidly during periods of sustained moisture in the high Arctic (Cooper et al. 2001).

1.4.1 Aspect and Slope

Light, temperature, and rate of drying of the thallus may all vary with aspect (Armstrong 1975), slope (Sletvold and Hestmark 1998), rock and bark texture (Moxham 1981; Armstrong 1993b), and vertical location on a substratum (Harris 1971a; Armstrong 1978). The influence of aspect on the radial growth of *M. fuliginosa* ssp. *fuliginosa* at a site in north Wales was investigated by Armstrong (1975). In the periods September–October and March–July, growth of thalli on a northwest-facing rock surface was greater than on a south-facing rock surface. However, in the period January/February, growth on the south face was greater than on the northwest face. A physiological model which involved interactions between the levels of light, wetting

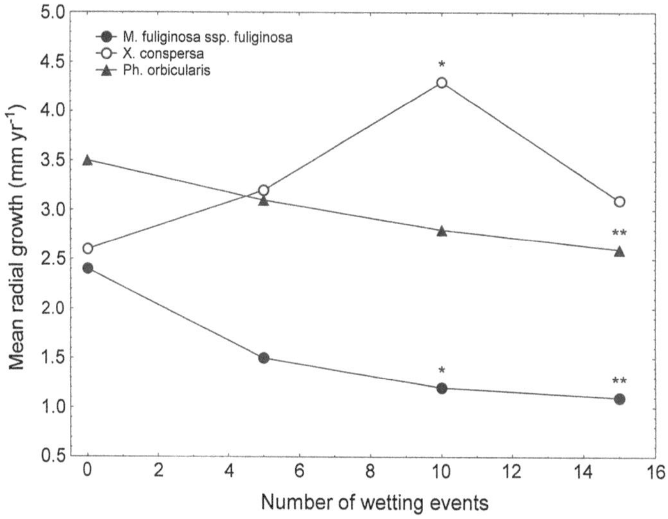

Fig. 1.3 The effect of frequency of wetting events on the radial growth (*RaG*) of three foliose lichens in south Gwynedd, Wales, UK. The radial growth of *Melanelia fuliginosa* ssp. *fuliginosa* and *Phaeophyscia orbicularis* thalli declined with increased wetting frequency while growth of *Xanthoparmelia conspersa* increased with wetting frequency until 10 experimental wettings per month, declining at 15 wettings per month. *Asterisk* indicates treatments significantly different from control (0) (data from Armstrong 1975)

regime, and rates of drying of the thalli was necessary to explain growth on the two surfaces (Armstrong 1975).

A number of studies have reported the effects of aspect on growth of crustose species. Hence, in the Brooks Range, Alaska, for example, radial growth of *R. geographicum* at sites with high to moderate light intensities was twice those at shaded locations (Haworth et al. 1986). Similarly, Bradwell (2001b) found that in Iceland, the largest thalli of *R. geographicum* were generally found on south-facing surfaces of boulders and are possibly a consequence of the longer thermal operating period and higher temperatures on southern surfaces. In the Cascade Mountains, Washington State, radial growth of *R. geographicum* on boulders was significantly greater on east-southeast and south-southeast facets compared with north-facing facets suggesting that at sites in the northern hemisphere, well-lit, southern facing surfaces are optimal for the growth of *R. geographicum* (Armstrong 2005) (Fig. 1.4). By contrast, variation in the slope of the facet appeared to have less influence on growth.

1.4.2 Vertical Height on Trees

Microclimatic differences also influence lichen growth with height on trees thus influencing vertical zonation. Hence, *Parmelia olivacea* (L.) Ach. em. Nyl. and *Parmeliopsis ambigua* (Wulf.) Nyl. differ in abundance and in vertical distribution especially on birch trees (Sonesson et al. 2011). During the cold season, *P. ambigua*, which was largely distributed on the lower trunk, was covered much of the time by snow, whereas *P. olivacea* occurred well above the snowline. The radial growth of *P. olivacea* was approximately twice that of *P. ambigua* but *P. ambigua* avoided competition by being adapted to the environment at the base of trees where snow accumulated and where *P. olivacea* could not survive. In addition, Harris (1971a, b, 1972) constructed a mathematical model to test the hypothesis that the vertical distribution of foliose lichens on trees in south Devon, UK, was a function of light intensity and water availability. Predicted vertical net carbon assimilation rates compared not unfavourably with the observed vertical distribution of the lichens (Harris 1972).

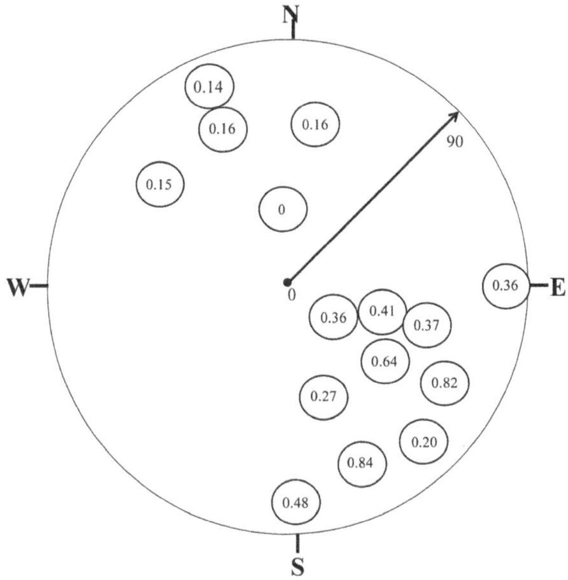

Fig. 1.4 Polar graph showing the influence of aspect and slope of the boulder facet on the radial growth (mm in 6 years) of *Rhizocarpon geographicum* in the north cascades, Washington State. Radial growth was significantly greater on east-southeast and south-southeast facets compared with north-facing facets suggesting that at sites in the northern hemisphere, well-lit, southern facing surfaces are optimal for growth. By contrast, variation in the slope of the facet appeared to have less influence on growth (data from Armstrong 2005)

1.5 The Influence of the Substratum on Growth

1.5.1 Physical Properties

The substratum may have several different effects on lichen growth and many of these factors could also interact with climate and microclimate. Hence, the relationship between the mechanical, chemical, and mineralogical properties of soil and the types of substratum chosen by terricolous lichens was studied by Garty et al. (1974). Species of *Squamarina* were found to be located on soils with the lowest shrinkage rates. When the shrinkage rate was above 5 %, however, the lichens grew on rock and moss instead suggesting that interactions between texture, porosity, rate of drying, and the physical attributes of the substratum on growth could have been responsible. In addition, *Xanthoria parietina* (L.) Th. Fr. can be affected by the texture of the substratum, thalli on rock, and smooth-barked surfaces having a higher radial growth than those on rough bark (Moxham 1981). Moreover, a combination of high water absorbing power and a high base content of the substratum favoured growth of lichens on roofs in suburban London (Brightman 1959). In addition, many lichens grow on moss which creates a very specific environment. In *Peltigera rufescens* (Weiss) Humb., for example, moss-associated thalli have a different morphology, have a 66 % higher net photosynthetic rate, and have a carbon dioxide concentration measured beneath the thalli more than 25 % greater than thalli growing on non-moss substrata (Colosie et al. 2012).

Relatively few studies of the growth of crustose lichens have been carried out on different types of substrata. *R. geographicum*, however, appears to be able to grow on a wide variety of siliceous rocks (Innes 1985). In addition, observations suggest *R. geographicum* may exhibit faster radial growth on sandstone than gneiss (Belloni 1973) and on andesite compared with granodiorite (Porter 1981).

The growth of the fruticose *Cladonia alpestris* (L.) Rabenh. and *C. rangiferina* (L.) Web. was studied by Kershaw and Rouse (Kershaw and

Rouse 1971). A strong correlation was observed between podetial length, diameter, internode length, and soil moisture; the level of soil moisture essentially integrating the effects of net radiation and drying rates of the lichen canopy as well as the relative metabolic activity of the lichen.

1.5.2 Rock Chemistry

The lichen flora of lime-rich rocks substantially differs from that of lime-poor, siliceous rocks and relatively few species are able to grow successfully on both types of substrata (James et al. 1977). To test the hypothesis that rock chemistry has a substantial effect on growth, fragments were cut from the perimeters of foliose lichens and glued onto different substrata, slate, granite, cement, and asbestos (Armstrong 1993b). The data for two foliose species are shown in Fig. 1.5. Radial growth of *X. conspersa* was inhibited on the calcium-rich compared with the siliceous substrates, while *X. parietina* would only grow successfully on a calcium-rich substratum, the fragments becoming discoloured and fragmented on lime-poor substrates.

A requirement for calcium may explain many aspects of the distribution of *Xanthoria parietina* in the UK. Hence, this species occurs on limestone rocks and walls, wall mortar, asbestos, tree bark where stemflow is enriched with calcium, and siliceous rocks near the sea (Armstrong 1990). By contrast, *Parmelia saxatilis* (L.) Ach. is virtually restricted to acid substrata, thalli of this species adopting a 'crescent-shaped' form after treatment with calcium carbonate, after which the thalli fragment and disappear (Armstrong 1990). Treatment with a 0.250 mM solution of calcium as calcium chloride, however, had little effect on growth of *P. saxatilis* (Fig. 1.5) suggesting that this species may tolerate mildly alkaline conditions. Gilbert (1971) found, for example, that *P. saxatilis* could occur at more alkaline sites in polluted environments. Zinc, copper, and mercury in rocks may also influence lichen growth as they have been shown to affect the chlorophyll content of lichen algae (Backor and Djubai 2004).

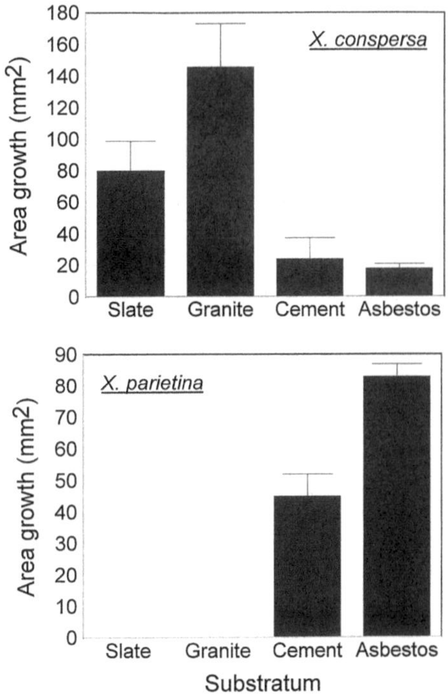

Fig. 1.5 The influence of the substratum on the radial growth (RaG mm year^{-1}) of two foliose saxicolous lichens in north Wales, UK. Analysis of variance (*ANOVA*): Radial growth of *Xanthoparmelia conspersa* was inhibited on the calcium-rich compared with the siliceous substrates while *Xanthoria parietina* would only grow successfully on a calcium-rich substratum. Between substrates $F = 11.09$***, *post hoc*: Asbestos = cement, Slate/Granite > Asbestos/Cement; *Xanthoria parietina*, Asbestos versus Cement $t = 4.71$**, ***$P < 0.001$, **$P < 0.01$ (data from Armstrong 1993b)

1.6 The Effect of Nutrient Enrichment on Growth

1.6.1 Bird Droppings

Several studies suggest that the degree of nutrient enrichment of a substratum especially by bird droppings can have a significant influence on lichen growth and community composition (Hale 1967). Bird droppings may influence growth by smothering the thalli, altering pH, or adding inhibitory and stimulatory compounds (Armstrong 1984). Hence, droppings from a variety of birds were applied as a thick paste (12 applications in one year) and as a suspension in deionized water (24 applications in one year) to several foliose lichens with differing distributions on and off bird perching stones (Armstrong 1984). Treatment with bird droppings increased the radial growth of *X. conspersa*, a species common on well-lit nutrient-enriched rocks. The paste and to a lesser extent the suspension inhibited growth of *P. saxatilis* (L.) Ach. a species which is relatively rare on bird perching stones. Uric acid, the most abundant nitrogenous component of bird droppings, did not influence growth when applied as a suspension to either species (Armstrong 1984). Hence, growth response to bird droppings was likely to be attributable to either increased pH or inorganic chemicals in the bird droppings.

Few direct studies of the influence of nutrient enrichment on growth of crustose species have been carried out. In a long-term study at Signy Island (Maritime Antarctic), however, growth of lichens at sites enriched by nitrogenous compounds derived from seabirds was rapid, especially that of *Acarospora macrocyclos* Vain. and species of *Buellia* and *Caloplaca* (Smith 1995). In addition, Wootton (1991) compared different maritime lichen communities with and without overhead seabird colonies. When guano was present, the 'orange' zone dominated by *Caloplaca marina* Wedd. was elevated further up the shore, and the grey *Physcia* zone eliminated in the supralittoral zone. Moreover, in the splash zone, *Verrucaria maura* Wahlenb. Ex Ach. decreased as a result of enhanced competition with the green alga *Prasiola*.

1.6.2 Nitrogen and Phosphorus

The response of *X. parietina* to varying nitrogen concentrations was studied by Gaio-Oliveira et al. (2005). Thallus specific weight was similar in all thalli without any significant effect of added nitrogen suggesting that this species may respond more to changes in pH than to nitrogen consistent with previous data (Armstrong 1990). The relationship between nitrogen concentration and growth was also studied in *Nephroma arcticum* (L.) Torss. and *Peltigera aphthosa* (L.) Willd. (Sundberg et al. 2001). After 4-month growth, control and nitrogen-fertilised thalli of *Nephroma arcticum* increased in area by 0.2 $m^2\ m^{-2}$ (mean relative growth rate = 3.8 mg $gm^{-1}\ d^{-1}$) and *Peltigera aphthosa* by 0.4 $m^2\ m^{-2}$ (mean relative growth rate = 8.4 mg $gm^{-1}\ d^{-1}$). Slower growth in *Nephroma arcticum* was explained by lower nitrogen and chlorophyll a concentration and, as a consequence, substantially lower light-energy conversion efficiency. The interaction between nitrogen availability and light exposure was also studied in *Platismatia glauca* (L.) WL Culb. & CF Culb. (Palmqvist and Dahlman 2006). Dry weight gain was significantly enhanced by nitrogen supply, variations in growth being most significantly accounted for by chlorophyll a concentration. Hence, *Platismatia glauca* may respond to increasing nitrogen concentration by increasing growth rate and carbon assimilation capacity through encouraging the production of algal cells. Phosphorus may also be important in some cyanobacteria lichens such as *Lobaria pulmonaria* (McCune and Caldwell 2009). Hence, this species doubled its annual biomass after a single 20-min immersion in a phosphorus solution suggesting that phosphorus is an important stimulant to growth.

Growth of the fruticose lichen *Cladonia stellaris* (Opiz) Brodo was studied in relation to levels of fertilisation, and growth not responding to nutrient additions suggesting that some fruticose species may not respond to short-term changes in the availability of nutrients (Hyvarinen et al. 2003). In a further study of *C. stellaris*, Makkonen et al. (2007) found that both nitrogen and phosphate could limit the growth of fruticose lichens with phosphate having a greater effect than nitrogen. Moreover, nitrogen and phosphorus enrichment may have differential effects on the algal and fungal partners (Makkonen et al. 2007). In addition, nitrogen was added to two lichen communities in south-west Norway dominated by species of *Cladonia* and *Cetraria*, respectively, (Fremstad et al. 2005). There was a decrease in lichen size and overall cover, and later, thalli became discoloured, the most sensitive species being *Alectoria nigricans* (Ach.) Nyl. and *Cetraria ericitorum*.

1.6.3 Salinity

Additional chemical factors may be important in determining growth at maritime sites (Fletcher 1976). Although calcium was shown experimentally to be an important ion in these environments, salinity exerted little effect, the species responding more to periods of immersion (Fletcher 1976). By contrast, Ramkaer (1978) found that the response of four different lichen fungi to salinity correlated well with the zonation on maritime rocks. No studies appear to have been published on the influence of chemical treatments on the radial growth of species in situ on maritime rocks. The radial growth of *X. parietina*, a common species of the submesic zone of the suppralittoral (Fletcher 1976) and many nutrient-enriched sites inland (Brodo 1973), however, was inhibited when transplanted to an inland site (Armstrong 1990). In addition, the transplanted thalli grew successfully inland when calcium carbonate was added as a paste to the thalli at intervals over a year, treatment with a 0.250 mM solution of calcium chloride at similar intervals having less effect. A relatively constant supply of calcium may therefore be necessary for successful growth of this species. This hypothesis is also supported by experiments showing that *X. parietina* thalli lost potassium ions when treated with distilled water and an application of a 0.250 mM solution of calcium to the medium prevented this loss in the light (Fletcher 1976). By contrast, addition of calcium carbonate to the calcifuge species *P. saxatilis* inhibited its growth which may explain its absence from lime-rich rocks.

1.6.4 Pollution

Despite the reinvasion of many urban sites by lichens in the last 20 years, polluted sites have significantly lower lichen diversity than rural sites, a testament to the powerful effect of air pollution on growth. Samples of foliose and crustose species were transplanted on rock fragments from north Wales to a city centre site in Birmingham, UK; no measurable growth was recorded, all thalli exhibiting evidence of fragmentation within 7 days of transplantation and all had disappeared completely within one month (Armstrong unpublished data). Gilbert (1971) directly measured the effects of air pollution on the growth of *Parmelia saxatilis*. As the level of sulphur dioxide increased, thalli became crescent-shaped, the thalli persisting for several years before flaking off the substratum. During this period, however, actively growing lobes were found on the outside margin and radial growth did not appear to be affected. In addition, both *P. saxatilis* and *X. parietina* developed an intense orange or green colour within the medulla, respectively, representing the breakdown of lichen acids (Gilbert 1971). In the West Riding of Yorkshire, UK, there was greater uniformity in radial growth of *Lecanora muralis* (Schreb.) Rabenh. as measured from year to year, than observed in rural environments (Seaward 1976). In addition, at nine sites within the WNW to NNE quadrant of Leeds, radial growth varied from 2.84 to 6.5 mm year^{-1} increasing with distance from the city centre. The study also suggested that urban environmental factors have a greater detrimental effect on growth during winter months suggesting that toxicity, due to air pollution, rather than drought was the most important factor.

1.7 Ecological Significance of Growth Effects

Ecological consequences may result from direct or indirect action of environmental factors on growth. A direct effect results from a significant change in an environmental factor which has such a detrimental effect on lichen physiology so as to completely inhibit growth. In such a circumstance, lichen thalli often exhibit the characteristic signs of degeneration, i.e. they become discoloured, crescent-shaped, or exhibit fragmentation of the thallus centre, ultimately disappearing from a substratum (Armstrong and Smith 1997). In many situations, however, growth effects are more subtle and indirect, changes in growth being manifest as alterations in the competitive balance among different species in a complex community (Armstrong and Welch 2007).

Few studies have examined the interaction between growth and competition in determining the composition of lichen communities (Armstrong and Welch 2007). In north Wales, for example, a complex relationship exists between growth and aspect of *R. geographicum* (Armstrong 1974, 2002). Hence, *R. geographicum* is abundant on south-facing surfaces at these sites but exhibits significantly greater growth on some northwest-facing surfaces (Armstrong 2002). The size frequency distributions of the thalli also differed between these two habitats, a higher percentage of thalli in the smallest size classes and a more restricted range of size being observed on the north-west surfaces. The rate of mortality of the thalli also appeared to be greater on north-west surfaces. It was concluded that the radial growth of *R. geographicum* did not predict its aspect distribution in north Wales but that the differing competitive environments on the two surfaces could explain its aspect distribution rather than the direct effects of environmental factors. In addition, Bradwell (2010) studied the radial growth of *R. geographicum* in a maritime subpolar environment in northern Scotland. The fastest radial growth of thalli greater than 10 mm in diameter at the two sites was 0.67 and 0.29 mm year^{-1}, the difference being attributable to the level of competition.

Armstrong (1991) studied the influence of rock surface aspect on competition between four foliose lichen species in experimental plots. In south-facing plots, the growth of *X. conspersa* was not reduced by the presence of any of its competitors and, *X. conspersa* is frequently found on well-lit, south-facing rock surfaces at the site (Armstrong 1974). By contrast, in north-facing plots, the growth of *X. conspersa* was reduced by *P. saxatilis*, thus reducing its competitiveness and possibly allowing *M. fuliginosa* ssp. *fuliginosa* to coexist with *X. conspersa*.

In a further experiment (Armstrong 2000), four foliose species were grown in well-lit plots with and without nutrient enrichment by added bird droppings. The results of the experiment for one of the species, *M. fuliginosa* ssp. *fuliginosa*, are shown in Fig. 1.6. Without addition of bird droppings, area growth of *M. fuliginosa* ssp. *fuliginosa* was reduced by each of the other three species, most notably by *X. conspersa*. In addition, performance in the 3-species mixture was not predictable from the 2-species mixture. Hence, *M. fuliginosa* ssp. *fuliginosa* survived better in the *Ph orbicularis/P. saxatilis* and *X. conspersa/Ph. orbicularis* mixtures than predicted but was completely eliminated in the *X. conspersa/P. saxatilis* mixture. In addition, the performance of *M. fuliginosa* ssp. *fuliginosa* in the 4-species mixture was not predictable from the 3-species mixture, this species surviving better than expected. In plots with bird droppings added, however, *M. fuliginosa* ssp. *fuliginosa* was not significantly reduced by *P. saxatilis* but was substantially reduced by *Ph. orbicularis* and *X. conspersa* and eliminated completely in the *X. conspersa/P. saxatilis*, *X. conspersa/Ph. orbicularis*, and *X. conspersa/P. saxatilis/Ph orbicularis* mixtures. Hence, addition of bird droppings altered the competitive balance between the species, reducing the effectiveness of *P. saxatilis* as a competitor, but increasing the competitive effect of *X. conspersa*. Hence, *M. fuliginosa* ssp. *fuliginosa* survives better in more complex mixtures but only under conditions where there is little nutrient enrichment.

Addition of nutrients, by increasing the growth of some species and reducing the growth of others, could increase the chance of competitive exclusion of a species and therefore could reduce species diversity (Lawrey 1981; Vagts and Kinder 1999). By contrast, Welch et al. (2006) grew lichen fragments attached to roofing slate in various combinations and sprayed them at different frequencies with a nutrient solution. At medium application rates, there was increasing thallus growth and an increase in the number of 'overlaps' between adjacent thalli compared with controls. In addition, where nutrients were added, *Flavoparmelia caperata* (L.) Hale and *X. parietina* were equally strong competitors as also were *Parmelia reddenda* Stirt. and *F. caperata*. Where no nutrients were added, this balance between the pairs of species was lost with *F. caperata* outcompeting *X. parietina* and

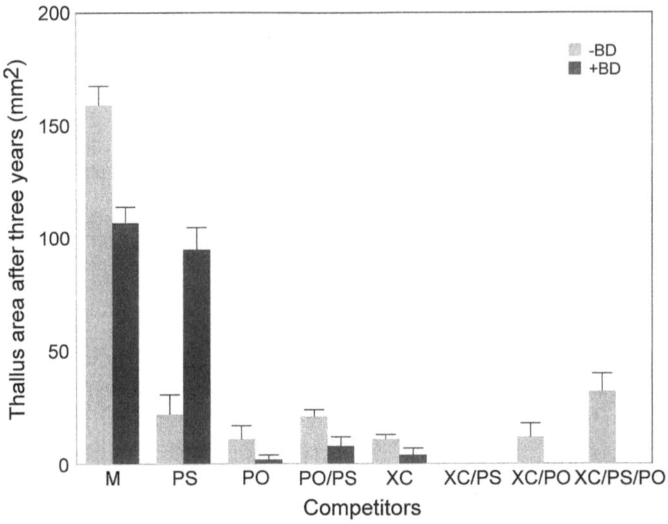

Fig. 1.6 The influence of nutrient enrichment by bird droppings (*+BD*) on the pattern of competition between *Melanelia fuliginosa* ssp. *fuliginosa* and three foliose saxicolous lichens (*M* growth of *M. fuliginosa* ssp. *fuliginosa* in monoculture, *PS Parmelia saxatilis*, *PO Phaeophyscia orbicularis*, *XC Xanthoparmelia conspersa*). Without addition of bird droppings, area growth of *M. fuliginosa* ssp. *fuliginosa* was reduced by each of the other three species, most notably by *X. conspersa*. In addition, performance in the 3-species mixtures was not predictable from the 2-species mixture. Hence, *M. fuliginosa* ssp. *fuliginosa* survived better in the *Ph orbicularis/P. saxatilis* and *X. conspersa/Ph. orbicularis* mixtures than predicted but was completely eliminated in the *X. conspersa/P. saxatilis* mixture. In addition, the performance of *M. fuliginosa* ssp. *fuliginosa* in the 4-species mixture was not predictable from the 3-species mixture, this species surviving better than expected. In plots with bird droppings added, however, *M. fuliginosa* ssp. *fuliginosa* was not significantly reduced by *P. saxatilis* but was substantially reduced by *Ph. orbicularis* and *X. conspersa* and eliminated completely in the *X. conspersa/P. saxatilis*, *X. conspersa/Ph. orbicularis*, and *X. conspersa/P. saxatilis/Ph orbicularis* mixtures (Data from Armstrong 2000)

P. reddenda outcompeting *F. caperata*. Hence, nutrient additions may not always reduce species diversity of a lichen community and further experiments of this type would be useful to understand the role that nutrient enrichment may play in determining the structure of lichen communities.

1.8 Conclusions

Despite the difficulties of growing lichens under controlled conditions and the challenges of field experimentation, there is a substantial literature on the effects of environmental factors on lichen growth. The major conclusions that can be drawn from these studies are as follows:

1. Seasonal fluctuations in radial growth often correlate best with average or total rainfall, the number of rain days, or rainfall in a specific season. Temperature may also an important climatic factor in some regions.
2. Interactions between microclimatic factors and especially light intensity, temperature, and moisture are important in determining local annual growth rates and differences in growth with aspect and slope of the substratum or with height on trees.
3. The physical and chemical nature of the substratum can affect growth especially the effects of texture, porosity, and rate of drying.
4. Whether the substratum is lime-rich or lime-poor and the degree of nutrient enrichment or maritime influence can have profound effects on growth both directly and by altering the competitive balance between species.
5. Effects of environmental factors on growth can alter the competitive balance on a substratum

ultimately influencing the composition of lichen communities.

6. Factorial experiments are needed to investigate the synergistic influence of different environmental factors and between the environment and competition and would lead to a more realistic understanding of the interactive processes influencing lichen growth in the field.

References

Armstrong RA (1973) Seasonal growth and growth rate-colony size relationships in six species of saxicolous lichens. New Phytol 72:1023–1030. doi:10.1111/j.1469-8137.1973.tb02078.x

Armstrong RA (1974) The descriptive ecology of saxicolous lichens in area of South Merionethshire, Wales. J Ecol 62:33–45. doi:10.2307/2258878

Armstrong RA (1975) The influence of aspect on the pattern of seasonal growth in the lichen *Parmelia glabratula* ssp. *fuliginosa*. New Phytol 75:245–251. doi:10.1111/j.1469-8137.1975.tb-01393.x

Armstrong RA (1976a) Studies on the growth rate of lichens. In: Brown DH, Bailey RH, Hawksworth DL (eds) Progress and problems in lichenology. Academic Press, London, pp 309–322

Armstrong RA (1976b) The influence of the frequency of wetting and drying on the radial growth of three saxicolous lichens in the field. New Phytol 77:719–724. doi:10.1111/j.1469-8137.1976.tb-04666.x

Armstrong RA (1978) The colonization of a slate rock surface by a lichen. New Phytol 81:85–88. doi:10.1111/j.1469-8137.1978.tb01606.x

Armstrong RA (1984) The influence of bird droppings and uric acid on the growth of five species of saxicolous lichens. Environ Exp Bot 4:95–99. doi:10.1016/0098-8472(84)90065-0

Armstrong RA (1988) Substrate colonization, growth and competition. In: Galun M (ed) Handbook of lichenology. CRC Press, Cleveland, pp 3–16

Armstrong RA (1990) The influence of calcium and magnesium on the growth of the lichens *Xanthoria parietina* and *Parmelia saxatilis*. Environ Exp Bot 30:51–57. doi:10.1016/0098-8472(90)90008-R

Armstrong RA (1991) Competitive interactions between four foliose lichens on north and south facing rock surfaces. Environ Exp Bot 31:51–58. doi:10.1016/0098-8472(91)90007-B

Armstrong RA (1993a) Seasonal growth of foliose lichens in successive years in South Gwynedd, Wales. Environ Exp Bot 33:225–232. doi:10.1016/0098-8472(93)90068-Q

Armstrong RA (1993b) The growth of six saxicolous lichens transplanted to lime-rich and lime-poor substrates in South Gwynedd, Wales. Symbiosis 15:257–267

Armstrong RA (1994) The influence of bird droppings on the growth of lichen fragments transplanted to slate and cement substrates. Symbiosis 17:75–86

Armstrong RA (2000) Competitive interactions between four foliose lichen species with and without nutrient enrichment. Symbiosis 28:323–335

Armstrong RA (2002) The effect of rock surface aspect on growth, size structure and competition in the lichen *Rhizocarpon geographicum*. Environ Exp Bot 48:187–194. doi:10.1016/S00098-8472(02)00040-0

Armstrong RA (2005) Radial growth of *Rhizocarpon* section *Rhizocarpon* lichen thalli over six years at Snoqualmie Pass in the Cascade Range, Washington State. Arct Antarct Alp Res 37:411–415. doi:10.1657/1523-0430(2005)037[0411:RGORSR]:RGORSRJ2.0.CO;2

Armstrong RA (2006) Seasonal growth of the crustose lichen *Rhizocarpon geographicum* (L.) DC. in south Gwynedd Wales. Symbiosis 41:97–102

Armstrong RA (2009) Monthly fluctuations in radial growth of individual lobes of the lichen *Parmelia conspersa* (Ehrh. ex Ach.) Ach. Symbiosis 47:9–16

Armstrong RA (2013) Development of areolae and growth of the peripheral prothallus in the crustose lichen *Rhizocarpon geographicum*: an image analysis study. Symbiosis 60:7–15. doi:10.1007/s13199-013-0234-2

Armstrong RA, Welch AR (2007) Competition in lichen communities. Symbiosis 43:1–12

Armstrong RA, Bradwell T (2010) Growth of crustose lichens: a review. Geogr Ann A Phys Geogr 92A:3–17

Armstrong RA, Bradwell T (2011) Growth of foliose lichens: a review. Symbiosis 53:1–16. doi:10.1007/s13199-011-0108-4

Armstrong RA, Smith SN (1997) Factors associated with degeneration of the thallus center in foliose lichens. Symbiosis 22:293–302

Backor M, Djubai A (2004) Short-term and chronic effects of copper, zinc and mercury on the chlorophyll content of four lichen photobionts and related algae. J Hattori Bot Lab 95:271–284

Belloni S (1973) Eicerche lichenometriche in Valfurva e nella Valle di Solda. Bull Com Glaciol Ital 21:19–33

Benedict JB (1990a) Experiments on lichen growth. I. Seasonal patterns and environmental controls. Arct Alp Res 22:244–254. doi:10.2307/1551587

Benedict JB (1990b) Winter frost injury to lichens: Colorado Front Range. Bryologist 93:423–426. doi:10.2307/3243606

Benedict JB (1991) Experiments on lichen growth II. Effects of a seasonal snow cover. Arct Alp Res 23:189–199. doi:10.2307/1551382

Benedict JB (2009) A review of lichenometric dating and its applications to archeology. Am Antiq 74:143–172

Beschel RE (1958) Lichenometrical studies in West Greenland. Arctic 11:254

Beschel RE (1961) Dating rock surfaces by lichen growth and its application to the glaciology and physiography (lichenometry). In: Raasch GO (ed) Geology of the

Arctic. University of Toronto Press, Toronto, pp 1044–1062

Bohuslavova O, Smilauer P, Elster J (2012) *Usnea* lichen community biomass estimation on volcanic mesas, James Ross Island, Antarctica. Polar Biol 35:1563–1572. doi:10.1007/S00300-012-1197-0

Boucher VL, Nash TH (1990) Growth patterns in *Ramalina menziesii* in California; Coastal vs. inland populations. Bryologist 93:295–302. doi:10.2307/3243516

Bradwell T (2001a) A new lichenometric dating curve for southeast Iceland. Geogr Ann 83A:91–101. doi:10.1111/j.0435-3676.2001.00146.x

Bradwell T (2001b) Glacier fluctuations, lichenometry and climatic change in Iceland. PhD thesis, University of Edinburgh, 365 pp

Bradwell T (2010) Studies on the growth of *Rhizocarpon geographicum* in northwest Scotland and some implications for lichenometry. Geogr Ann A Phys Geogr 92:41–52. doi:10.1111/j.1468-0459.2010.00376.x

Brightman FH (1959) Some factors influencing lichen growth in towns. Lichenologist 1:104–108

Brodo IM (1965) Studies on growth rates of corticolous lichens on Long Island, New York. Bryologist 68:451–456

Brodo IM (1973) Substrate ecology. In: Ahmadjian V, Hale ME (eds) The lichens. Academic Press, London

Colesie C, Scheu S, Green TGA, Weber B, Wirth R, Budel B (2012) The advantage of growing on moss: facilitative effects on photosynthetic performance and growth in the cyanobacterial lichen *Peltigera rufescens*. Oecologia 169:599–607. doi:10.1007/S00442-011-2224-5

Cooper EJ, Smith FM, Wookey PA (2001) Increased rainfall ameliorates the negative effect of trampling in the growth of high Arctic forage lichens. Symbiosis 31:153–171

Coxson DS, Kershaw KA (1983) The ecology of *Rhizocarpon superficiale* II. The seasonal response of net photosynthesis and respiration to temperature, moisture and light. Can J Bot 61:3019–3030

Coxson DS, Stevenson SK (2007) Growth rate responses of *Lobaria pulmonaria* to canopy structure in even-aged and old-growth cedar-hemlock forests of central-interior British Columbia, Canada. Forest Ecol Manage 242:5–16. doi:10.1016/j.foreco.2007.01.031

Farrar JF (1973) Lichen physiology. In: Ferry BW, Baddeley MS, Hawksworth DL (eds) Progress and problems in lichenology. Athlone Press, University of London, London, pp 238–282

Farrar JF (1976) The lichen as an ecosystem: observation and experiment. In: Brown DH, Bailey RH, Hawksworth DL (eds) Progress and problems in lichenology. Academic Press, London, pp 385–406

Fisher PJ, Proctor MCF (1978) Observations on a season's growth of *Parmelia caperata* and *P. sulcata* in South Devon. Lichenologist 10:81–89. doi:10.1017/S0024282978000092

Fletcher A (1976) Nutritional aspects of marine and maritime lichen ecology. In: Brown DH, Bailey RH, Hawksworth DL (eds) Progress and problems in lichenology. Academic Press, London, pp 359–384

Fremstad E, Paal J, Mois T (2005) Impacts of increased nitrogen supply on Norwegian lichen-rich alpine communities: 1 10-year experiment. J Ecol 93:471–481. doi:10.1111/j.1365-2745.2005.00995.x

Gaio-Oliveira G, Dahlman L, Maguas C, Palmqvist K (2004) Growth in relation to micoclimatic conditions and physiological of four *Lobaria pulmonaria* populations in two contrasting habitats. Ecography 27:13–28. doi:10.1111/j.0906-7590.2004.03577.x

Gaio-Oliveira G, Dahlman L, Palmqvist K, Maguas C (2005) Responses of the lichen *Xanthoria parietina* (L.) Th Fr. to varying nitrogen concentrations. Lichenologist 37:171–179. doi:10.1017/S0024282904014598

Garty J, Gal M, Galun M (1974) The relationship between physiochemical soil properties and substrate choice of 'multisubstrate' lichen species. Lichenologist 6:146–150

Gauslaa Y, Palmqvist K, Solhaug KA, Holien H, Hilmo O, Nybakken L, Myhre LC, Ohlson M (2007) Growth of epiphytic old forest lichens across climatic and successional gradients. Can J For Res—Rev 37:1832–1845. doi:10.1139/X07-048

Gauslaa Y, Palmqvist K, Solhaug KA, Hilmo O, Holien H, Nybakken L, Ohlson M (2009) Size-dependent growth of two old-growth associated macrolichen species. New Phytol 181:583–692. doi:10.1111/j.1469-8137.2008.02690.x

Gilbert OL (1971) Studies along the edge of a lichen desert. Lichenologist 5:11–17

Golm GT, Hill PS, Wells H (1993) Life expectancy in a Tulsa cemetery: growth and population structure of the lichen *Xanthoparmelia cumberlandia*. Am Mid Nat 129:373–383. doi:10.2307/2426518

Hale ME (1967) The biology of lichens. Edward Arnold, London

Hale ME (1970) Single-lobe growth-rate patterns in the lichen *Parmelia caperata*. Bryologist 73:72–81

Hale ME (1973) Growth. In: Ahmadjian V, Hale ME (eds) The lichens. Academic Press, New York, pp 473–492

Harris GP (1971a) The ecology of corticolous lichens. I. The zonation on oak and birch in south Devon. J Ecol 59:431–439

Harris GP (1971b) The ecology of corticolous lichens II. The relationship between physiology and the environment. J Ecol 59:441–452

Harris GP (1972) The ecology of corticolous lichens III. A simulation model of productivity as a function of light intensity and water availability. J Ecol 60:19–40. doi:10.2307/2258038

Haworth LA, Calkin PE, Ellis JM (1986) Direct measurement of lichen growth in the central Brooks Range, Alaska USA, and its application to lichenometric dating. Arct Alp Res 18:289–296. doi:10.2307/2258323

Hooker TN (1980) Lobe growth and marginal zonation in crustose lichens. Lichenologist 12:313–323. doi:10.2307/2258324

Hyvarinen M, Crittenden PD (1998) Growth of the cushion-forming lichen *Cladonia portentosa* at nitrogen-polluted and unpolluted heathland sites. Environ Exp Bot 40:67–76. doi:10.1016/S0098-8472(98)00022-7

Hyvarinen M, Walters B, Koopmann R (2003) Impact of fertilization on phenol content and growth rate of *Cladina stellaris*: a test of the carbon-nutrient balance hypothesis. Oecologia 134:176–181. doi:10.1007/S00442-002-1105-3

Innes JL (1983) Development of lichenometric dating curves for highland Scotland. Trans R Soc Edinb 74:23–32

Innes JL (1985) Lichenometry. Prog Phys Geog 9:187–254. doi:10.1177/030913338500900202

James PW, Hawksworth DL, Rose F (1977) In: Seaward MRD (ed) Lichen communities in the British Isles: a preliminary conspectus. Academic Press, London, pp 295–419

Jones JM, Platt RB (1969) Effects of ionizing radiation, climate and nutrition on growth and structure of a lichen *Parmelia conspersa* (Ach.) Ach. Radioecology Symp 2:111–119

Karenlampi L (1970) Morphological analysis of the growth and productivity of the lichen *Cladonia alpestris*. Rep Kevo Subarct Res 7:9–15

Karenlampi L (1971) Studies on the relative growth rate of some fruticose lichens. Rep Kevo Subarct Res 7:33–39

Keon DB, Muir PS (2002) Growth of *Usnea longissima* across a variety of habitats in the Oregon Coast range. Bryologist 105:233–242. doi:10.1639/0007-2745(2002)105[0233:GOULAA]2.0.CO;2

Kershaw KA (1972) The relationship between moisture content and net assimilation rate of lichen thalli and its ecological significance. Can J Bot 50:543–555

Kershaw KA, Rouse WR (1971) Studies on lichen-dominated systems II. The growth pattern of *Cladonia alpestris* and *Cladonia rangeriferina*. Can J Bot 49:1401–1410

Lawrey JD (1981) Evidence for competitive release in simplified saxicolous lichen communities. Am J Bot 68:1066–1973. doi:10.2307/2442716

Lawrey JD, Hale ME (1977) Studies on lichen growth rates at Plummers Island, Maryland. Proc Biol Soc Wash 90:698–725

Locke WW, Andrews JT, Webber PJ (1979) A manual for lichenometry. British Geomorphological Research Group, Technical Bulletin 26, pp 1–47

Makkonen S, Hurri RSK, Hyvarinen M (2007) Differential responses of lichen symbionts to enhanced nitrogen and phosphorus availability: an experiment with *Cladina stellaris*. Ann Bot 99:877–884. doi:10.1093/aob/mcm042

Mathews JA (1994) Lichenometric dating: a review with particular reference to 'Little Ice Age' moraines in southern Norway. In: Beck C (ed) Dating in surface context. New Mexico Press, Albuquerque, pp 185–212

Matthews JA (2005) Little Ice Age glacier variations in Jotunheimen, southern Norway: a study in regionally controlled dating of recessional moraines with implications for climate and lichen growth rates. Holocene 15:1–19

McCune B, Caldwell BA (2009) A single phosphorus treatment doubles growth of cyanobacterial lichen transplants. Ecology 90:567–570. doi:10.1890/08-0344.1

Moxham TH (1981) Growth rates of *Xanthoria parietina* and their relationship to substrate texture. Cryp Bryol Lichenol 2:171–180

Palmqvist K, Sundberg B (2000) Light use efficiency of dry matter gain in five macro-lichens: relative impact of microclimate conditions and species-specific traits. Plant Cell Env 23:1–14. doi:10.1046/j.1365-3040.2000.00529.x

Palmqvist K, Dahlman L (2006) Responses of the green algal foliose lichen *Platismatia glauca* to increased nitrogen supply. New Phytol 171:343–356. doi:10.1111/j.1469-8137.2006.01754.x

Paterson DR, Paterson EW, Kenworthy JB (1983) Physiological studies on temperate lichen species. I. A mathematical model to predict assimilation in the field based on laboratory responses. New Phytol 94:605–618. doi:10.1111/j.1469-8137.1983.tb04869.x

Porter SC (1981) Lichenometric studies in the Cascade range of Washington: establishment of *Rhizocarpon geographicum* growth curves at Mount Rainier. Arc Alp Res 13:11–23. doi:10.2307/1550622

Ramkaer K (1978) The influence of salinity on the establishment phase of rocky shore lichens. Bot Tidsskr 72:119–123

Rydzak J (1961) Investigations on the growth rate of lichens. Ann UMCS (Lublin, Poland) Sec C 16:1–15

Seaward MRD (1976) Performance of *Lacanora muralis* in an urban environment. In: Brown DH, Bailey RH, Hawksworth DL (eds) Progress and problems in lichenology. Academic Press, London, pp 323–358

Showman RE (1976) Seasonal growth of *Parmelia caperata*. Bryologist 79:360–363

Sletvold N, Hestmark G (1998) Thallus growth asymmetry and habitat inclination in *Lasallia pustulata*. Bryologist 100:465–466

Smith DC, Molesworth S (1973) Lichen physiology XIII. Effects of rewetting dry lichens. New Phytol 72:525–533. doi:10.1111/j.1469-8137.1973.tb04403.x

Smith RIL (1995) Colonization by lichens and the development of lichen-dominated communities in the maritime Antarctic. Lichenologist 27:473–483. doi:10.1016/S0024-2829(95)80007-7

Sonesson M, Grimberg A, Sveinbjornsson B, Carlsson BA (2011) Seasonal variation in concentrations of carbohydrates and lipids in two epiphytic lichens with contrasting, snow-depth related distribution on subarctic birch trees. Bryologist 114:443–452 doi:10.1639/0007-2745-114.3.443

Stevenson SK, Coxon DS (2008) Growth responses of *Lobaria retigera* to forest edge and canopy structure in the inland temperate rainforest, British Columbia.

Forest Ecol Manage 256:618–623. doi:10.1016/j.foreco.2008.05.025

Sundberg B, Nasholm T, Palmqvist K (2001) The effect of nitrogen and key thallus components in the two tripartite lichens *Nephroma arcticum* and *Peltigera aphthosa*. Plant Cell Environ 24:517–527. doi:10.1046/j.1365-3040.2001.00701.x

Tenbrink NW (1973) Lichen growth rates in West Greenland. Arct Alp Res 5:323–331

Trenbirth HE, Matthews JA (2010) Lichen growth rates on glacier forelands in southern Norway: preliminary results from a 25-year monitoring programme. Geogr Ann A Phys Geogr 92A:19–40

Uchida M, Nakatsubo T, Kanda H, Koizumi H (2006) Estimation of annual primary production of the lichen *Cetrariella delisei* in a glacier foreland in the high Arctic, Nylalesund, Svalbard. Polar Res 25:39–49. doi:10.1111/j.1751-8369.2006.tb00149.x

Vagts I, Kinder M (1999) The response of different *Cladonia* species after treatment with fertiliser or lime in heathland. Lichenologist 31:75–83

Webster M, Brown DH (1997) Preliminary observations on the growth of transplanted *Peltigera canina* under semi-natural conditions. Lichenologist 29:91–96

Welch AR, Gillman MP, John EA (2006) Effect of nutrient application on growth rate and competitive ability of three foliose lichen species. Lichenologist 38:177–186. doi:10.1017/S0024282906004555

Wootton JT (1991) Direct and indirect effects of nutrients on intertidal community structure: variable consequences of seabird guano. J Exp Mar Biol 151:139–153. doi:10.1016/0022-0981(91)90121-C

2 Sampling and Interpreting Lichen Diversity Data for Biomonitoring Purposes

Paolo Giordani and Giorgio Brunialti

Contents

Abstract

Although lichen diversity values are broadly used as bioindicators, mainly for air pollution, lichen communities can be substantially influenced by other ecological factors, such as tree species and forest structure, and microclimatic conditions. In particular, species composition may be a suitable indicator for climate and land-use effects as well. For effective utilization of lichen diversity data in biomonitoring studies including air pollution, ecosystem functioning, and forestry studies, standardized sampling procedure and avoiding sampling and non-sampling errors are the important aspects to be considered. Further interpretation of lichen diversity data requires careful data analysis for providing affirmative results related to ambient air quality. In any lichen biomonitoring program, expected deliverables are based on a hypothesis, which may be achieved by standardization of the sampling procedures based on the functional requirement of the dependent environmental variables. The chapter discusses the procedures and methodology for sampling and interpreting lichen diversity data for biomonitoring purposes.

Keywords

Air pollution · Ecosystem functioning · Quality assurance · LDV

P. Giordani (✉)
Botanic Centre Hanbury, DISTAV, University of Genova, Corso Dogali, 1M, 16136 Genova, Italy
e-mail: giordani@dipteris.unige.it

G. Brunialti
Terra Data Environmetrics, Spin Off of the University of Siena, Via Bardelloni 19, 58025 Monterotondo Marittimo, Italy
e-mail: brunialti@terradata.it

D.K. Upreti et al. (eds.), *Recent Advances in Lichenology*,
DOI 10.1007/978-81-322-2181-4_2

2.1 Introduction

Due to their physiology, lichens are sensitive to a series of environmental parameters (namely light, air humidity, UV-B radiation, temperature, and airborne chemicals such as SO_2 and NO_x making them useful indicators for air pollution and climate change (Poikolainen et al. 1998; Tarhanen et al. 2000; Cornelissen et al. 2001; Kricke and Loppi 2002; Castello and Skert 2005). Biomonitoring methods based on the diversity of epiphytic lichens are among the most used worldwide. Apart from assessing the effects of gaseous pollutants, these approaches were recently extended to a suite of other anthropogenic disturbances (Nimis et al. 2002). Several aspects of lichen diversity (e.g., species richness and abundance, species composition, indicator species, functional traits and groups) are usually considered, each of them for a particular reason (Nimis et al. 2002). In simplified terms, it is possible to identify three main purposes to perform lichen monitoring:

1. Air quality monitoring. The sensitivity of lichens to phytotoxic gases (mainly SO_2 and NO_x) underlies their use to assess and monitor the effects of atmospheric pollution (Hawksworth and Rose 1970; Richardson 1993). They have been used as long-term biomonitors of air pollution both for small-scale (Nimis et al. 1990; Loppi et al. 2004; Giordani 2007) and large-scale surveys (Nimis et al. 1991; Van Dobben and DeBakker 1996; Bennett and Wetmore 1999; Frati and Brunialti 2006; Giordani 2006). Lichen biomonitoring surveys are often used to integrate instrumental data of atmospheric pollution (Nimis et al. 1990; Giordani et al. 2002; Giordani 2007; Pinho et al. 2004) and for developing forecasts in connection with human health (Cislaghi and Nimis 1997).
2. Sustainable forestry. Epiphyte diversity is related to forest structure and dynamics. Forest management, directly or indirectly, affects several environmental factors relevant to dispersal, establishment, and maintenance of lichen species. Studies on lichen diversity clearly demonstrate dramatic losses of species caused by forest management in European temperate and boreal forests (Hauck et al. 2013). In general, higher lichen diversity is related to lower management intensity, even though in some cases non-intensively managed forests may provide better conditions for epiphytic lichens than recently abandoned forests. For instance, several studies suggest that selective cutting is less detrimental to forest lichens than the shelterwood system or clear cutting (Nascimbene et al. 2013b).
3. Ecosystem functioning. Epiphytic lichens play an important role in both the forest water cycle, increasing canopy interception of precipitations and forest–nutrient cycling, above all for N-fixing 'cyanolichens' (lichens with a cyanobacterial symbiont; for a review see Ellis 2012). In addition to nutrient cycling, lichens are a focal point in forest food webs. They influence the ecological success of forest-dwelling animals in a wide variety of ways, e.g., for macrofauna in nest building, and as forage (Ellis 2012). In some cases, indicator species have been proposed as proxies for lichen species richness more generally. As an example, forest stands with the flagship species *Lobaria pulmonaria* tended to have more red-listed species and more continuity indicator species than stands without this lichen (Nilsson et al. 1995), tentatively pointing to covariation among continuity indicators. Similarly, comparing stands with different abundance and fertility attributes for *L. pulmonaria*, these attributes were associated with higher tree and stand-level macrolichen species richness and with significant differences in species composition among *L. pulmonaria* site-types, i.e., cyanolichens and rare species skewed to the '*L. pulmonaria* stands' (Nascimbene et al. 2010).

In this chapter, we intend to explore a series of key issues on the topic in order to provide a framework of the subject with an overview of recently published researches.

In this regard, Fig. 2.1 provides a step-by-step flow diagram of operations required to address a lichen monitoring program. As a rule, these are

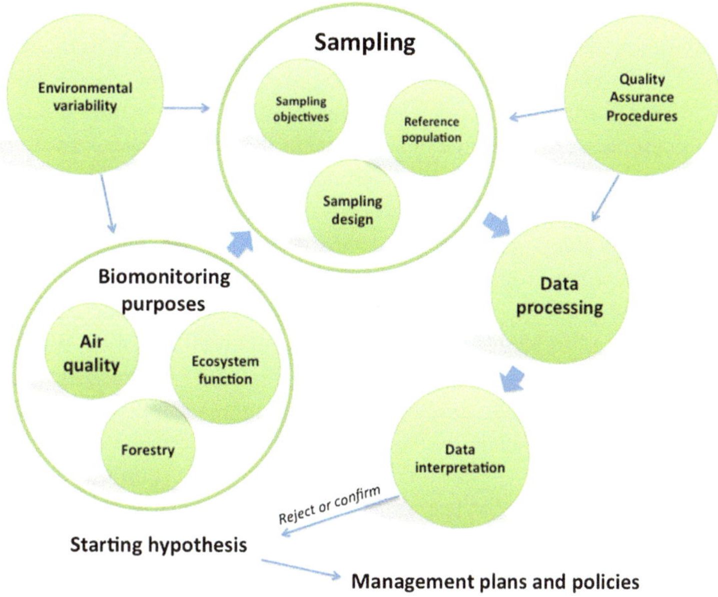

Fig. 2.1 Schematic diagram of the operations required to address a lichen monitoring program

the typical questions to be answered: does air quality affect lichen diversity in our study area? Or else, does forest management (or fragmentation) influence lichen communities in our investigated forests? Or even, do lichen diversity play a role in ecosystem function of our natural park?

The first step in the organization of a lichen monitoring program is to establish the initial null versus alternative hypotheses. For example, we may state the basic hypothesis that the expected lichen diversity is equal for impact and control sites (with respect to air pollution or to forest management). However, this is where environmental variability comes into play. Thus, the second step is to qualify, quantify, and control the main ecological factors that can affect our response variable, both including those of our interest and also the ones which may represent a source of noise (Brunialti et al. 2010a). Indeed, the effect of large-scale (such as climate) or local scale (as tree substrate) natural factors may represent relevant sources of variability and noise in relation to the phenomenon we are studying (Giordani et al. 2013; Ellis 2012). Also, some of these effects may be cumulative (Jovan and McCune 2004; Pinho et al. 2004; Giordani 2006; Caruso and Thor 2007). For instance, we should study air pollution effect on lichen diversity in forest sites with different forest management: We will have to set the study in order to disentangle the effect of individual factors and determine which of the two acts as main driver affecting lichen diversity or the presence/abundance of indicator species (Giordani 2007; Giordani et al. 2014a).

Our opinion is that this issue may have a central role in biomonitoring studies. For this reason, an entire Sect. 2.2.1 of the chapter focuses on these aspects, where we explore the main sources of variability affecting lichen diversity at different spatial scales. This variability being considered to adequately respond to our initial question, and the third step of our scheme concerns the adoption of an appropriate sampling design, from the selection of objectives and target population to the development of an adequate sampling strategy. We address this topic in Sect. 2.2.2, where we explore the main standardized methods for lichen diversity assessment, and analyze their suitability for various purposes.

Section 2.2.3 focuses on quality assurance procedures that are closely linked to sampling design and should be taken into account to obtain reproducible, reliable, and defensible data (Ferretti 2011). In this respect, we must remember that there is a strong link between data quality, information quality, and decision-making processes (Crumbling 2002).

The fourth step is then the phase of data processing. A broad range of statistical tools may be adopted to test the initial hypothesis. A subsequent interpretation of the results is essential to verify that the statistical significance of the test has a meaning even from an ecological point of view (Sect. 2.3). Starting from raw data, it is possible to adopt a series of interpretative devices that allow analyzing various aspects of the phenomenon under investigation. Only in this way, we may obtain useful information for management and decision making.

In this regard, the last step of our process involves the rejection or acceptance of the starting null hypothesis. On the basis of this result, we get essential information to proceed with the process of planning and management (policies to reduce air pollution, forest and landscape management, etc.). Indeed, without reliable information on changes to ecological systems and on the causes of those changes, decision making cannot deal effectively with these issues (Ferretti 2011).

The main objectives reported above are addressed in most cases with management and decision-making intent. In this perspective, if biomonitoring should be taken as a serious basis for decision making, it needs to produce robust, defensible data and documented quality (Brunialti et al. 2004). With this in mind, it is essential to approach the issue by means of a suitable sampling design and also appropriate statistical and interpretative tools that should be carefully selected to address properly each of these purposes (Ferretti and Erhardt 2002).

2.2 Sampling Lichen Diversity

There is no doubt that ecological monitoring has a fundamental role in providing baseline data on the status and trend of environmental resources. However, as underlined by Legg and Nagy (2006), in several cases, ecological conservation programs suffer from the lack of details of goal and hypothesis formulation, survey design, data quality and statistical power at the start. As a result, sometimes, they are likely to fail to reach the necessary standard of being capable of rejecting a false null hypothesis with reasonable power. This is mainly related to the adoption of an unsuitable sampling design, and this is basically the reason why sampling has frequently been considered as the weakest point of monitoring programs. This is unfortunate because failure to provide convincing evidence of the overall quality of environmental information can have serious consequences in environmental decision making. Indeed, we must consider that the defensibility of the data supporting the decision is fundamental (Ferretti 2011). To avoid this and to promote a correct process, the selection of a suitable sampling design represents the first step to reduce data variability due to sampling error. When selecting the proper design, the objectives (biomonitoring purposes, Fig. 2.1) of the survey and environmental variability should be taken into account carefully (Brunialti et al. 2004).

With respect to lichen biomonitoring, environmental variability acting at different spatial scales has been taken into account during the standardization process of the sampling protocols (Giordani et al. 2001; Brunialti and Giordani 2003; Nimis et al. 2002), thereby providing useful information for the selection of suitable sampling designs.

This section provides useful insights on several issues relevant to environmental variability, sampling design, and quality assurance procedures, which are the basis of a proper environmental monitoring program.

2.2.1 Variability of Lichen Diversity Across Scales

Variability is an inherent property of ecological systems and every attempt to measure and interpret the environment should consider it (Brunialti et al. 2004). Thus, in the assessment of environmental quality by means of biomonitors, it is important to understand the environmental processes driving the variability of biological data, since this may affect the forecasting precision of these techniques (Laskowky and Kutz 1998). Environmental factors such as geomorphology, climatic variables, and substrate could have a great impact on the ecosystem property being studied in order to assess environmental quality, such as the rate of indicator species, the biodiversity of a community, or the presence of injuries on organisms. In simple terms, environmental heterogeneity may control the variability of lichen diversity and the composition of lichen communities at least at three spatial scales: (1) tree level; (2) plot level, and (3) landscape level. In this section, we highlight the possible effect of these ecological factors on lichen diversity by referring to the main studies conducted so far.

2.2.1.1 Spatial Scale of Variation in Lichen Monitoring Data

Quantifying spatial and temporal variation is a key element for understanding mechanisms and processes that structure the species assemblages for assessing the environmental impact and for identifying an appropriate scale of sampling (Levin 1992; Underwood et al. 2000). Nonetheless, biodiversity assessments are strongly dependent on observation scales (Gaston 2000; Purvis and Hector 2000), and natural spatial variation or the spacing of sampling units may decisively affect results (Dungan et al. 2002).

Although several studies have investigated spatial variation in plants (see, e.g., Palmer 1990; Gunnar and Moen 1998), lichens are still poorly investigated from this point of view (Ferretti et al. 2004), even though descriptors of lichen diversity are widely used in applicative studies for assessing alterations of the environment (Nimis et al. 2002), covering areas ranging from some dozen to some thousands square kilometers (e.g., Giordani et al. 2002). In particular, bioindication techniques usually focus on between-site variability in epiphytic lichen diversity using a high sampling density and do not sufficiently consider possible within-site variability, probably because this would imply sampling more trees, thus increasing the cost of the survey (Ferretti et al. 2004; Brunialti et al. 2010a). Controlling the spatial variation at macroscales is necessary but not sufficient, if the variation at more detailed levels (e.g., considering areas <1 km^2) remains unknown.

A recent study explored the variability of lichen composition throughout the spatial levels using sound sampling protocols, along biogeographic, climatic, and pollution gradients (Giordani et al. 2013). The authors showed that the distribution of variation in lichen diversity was impressively similar across the spatial scales (Giordani et al. 2013). In particular, they found that it was fairly constant despite climatic variation, landscape complexity, air pollution, and general anthropogenic pressure. Moreover, the predictability of lichen bioindication methods, as estimated by the between-plot variability/total variability ratio, was limited to less than 40 %, thus calling for some refinements in sampling and interpretation phases (Fig. 2.2). This finding suggests the adoption of specific experimental designs, such as a stratified random sampling or a tree-based sampling to improve the detection of the sources of variation.

2.2.1.2 From Tree-level to Landscape-level Variability

Epiphytic lichens distribution depends on a complex set of environmental and substrate-related explanatory variables, acting from the tree to the landscape level (Nimis et al. 2002). Several tree-level environmental factors are important for the growth of epiphytic lichens and drive the diversity and composition of their communities (Ellis 2012). Figure 2.3 summarizes the main ecological factors acting at this level on a hypothetical tree with a different canopy in winter and summer periods or a deciduous and an evergreen tree, respectively.

We can distinguish at least two main categories of ecological factors that have been explored by many studies (see Ellis 2012): microclimatic variables (light availability, temperature, and relative humidity) and bark properties (pH, texture, exfoliation, water-holding capacity). In this respect, both substrate eutrophication, due to the grazing of cattle and wild animals and also to soil re-suspension (Loppi and De Dominicis 1996; Frati et al. 2007) and circumference, and

Fig. 2.2 Estimated components of variance (%) in lichen diversity values, calculated for each spatial scale (modified from Giordani et al. 2013)

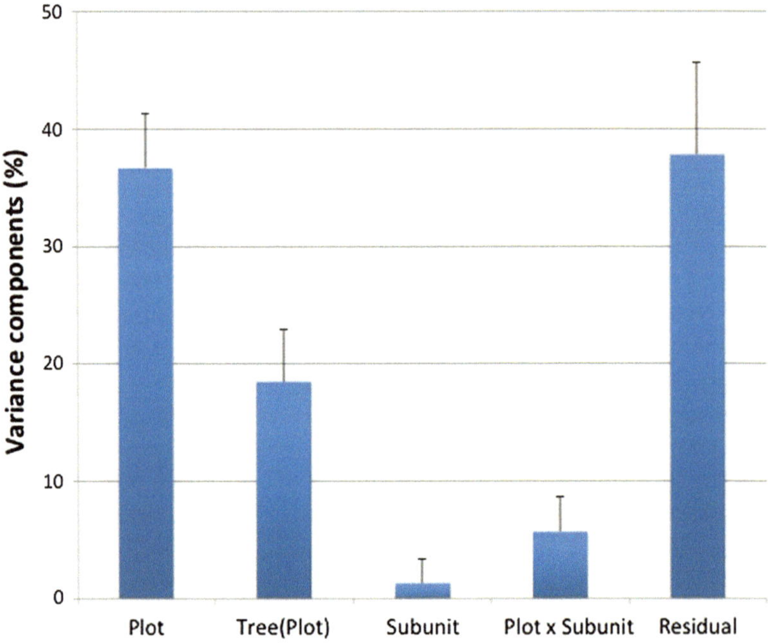

Fig. 2.3 Conceptual model illustrating the main ecological factors potentially affecting lichen communities under the canopy of a broad-leaved tree in winter and summer periods (see explanation in the text)

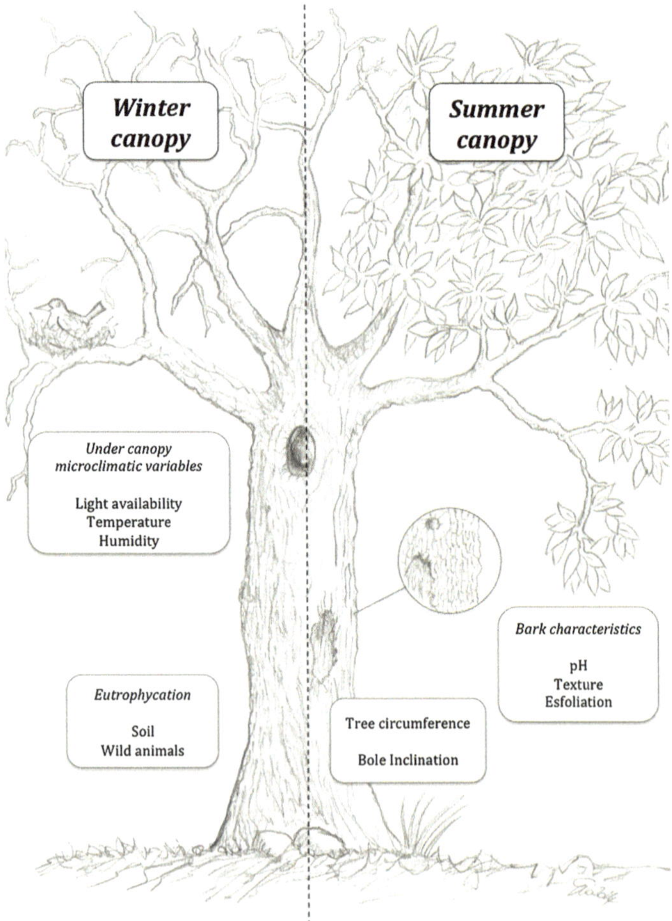

inclination of the bole may contribute indirectly to modify these ecological factors (see, e.g., Giordani et al. 2001).

Figure 2.4 summarizes the main ecological factors affecting lichen communities at plot and landscape levels. They are mainly related to the distribution and density of trees (e.g., the gradient from isolated trees to near-natural forest stands) and to the land-use and geomorphological variability (e.g., altitudinal range, agricultural, or forested landscapes). Light and humidity availability generally play the role of direct factors in this context.

Basically, all these aspects are very interconnected and show that the effects largely covariate. Canopy structure decisively controls light conditions at the lower bole (Moe and Botnen 1997; Fritz et al. 2009), and this may result as a negative effect of canopy shading on species richness, both at tree level (Loppi and Frati 2004) and at plot level (Brunialti et al. 2010b). While focusing on edge effects in fragmented boreal and Mediterranean forests, several authors pinpointed light as a main driving factor for lichen colonization, affecting lichen abundance, diversity, and composition at tree level (Esseen and Renhorn 1998; Belinchón et al. 2007; Boudreault et al. 2008; Brunialti et al. 2012b). In particular, higher lichen species richness was found at the side of the trunk facing the edge (Brunialti et al. 2012b), or in correspondence with the first light peak (Belinchón et al. 2007). Loppi and Frati (2004) found higher lichen diversity values on *Tilia platyphyllos* compared to *Quercus ilex.* These findings were explained by significantly higher winter light conditions and water-holding capacity of the bark for *Tilia* trees. Bark-pH was not found as discriminant parameter among the two tree species. On the contrary, many studies have demonstrated a strong epiphyte response to bark-pH, explaining epiphyte community variation compared among tree species in both boreal and temperate systems (Hyvärinen et al. 1992; Gauslaa 1995; Kuusinen 1996; Jüriado et al. 2009; Lewis and Ellis 2010; Leppik et al. 2011). Moreover, intraspecific studies, among boles of the same tree species, have indicated that bark-pH/nutrient status may be modified by an interaction with the soil environment (Gauslaa 1995; Gustafsson and Eriksson 1995; Kermit and

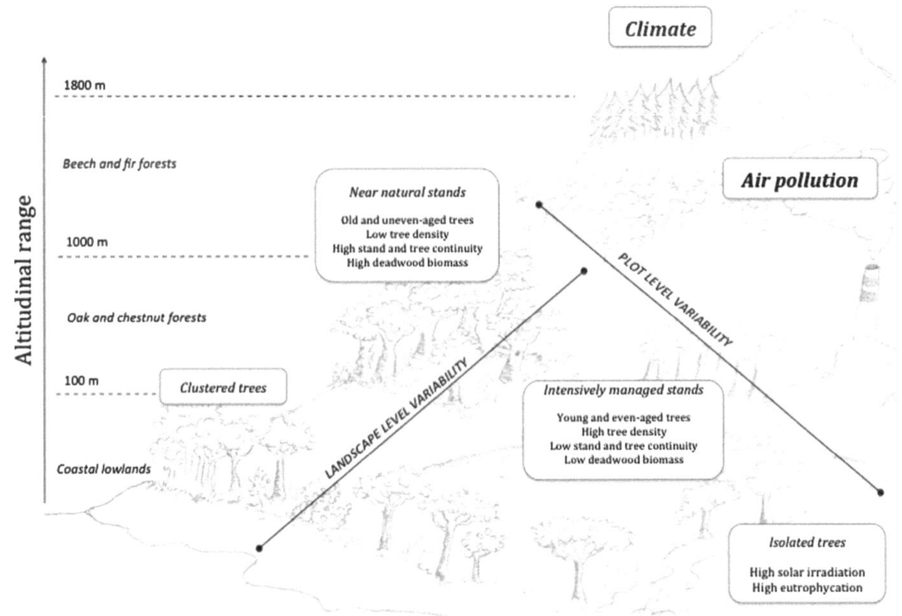

Fig. 2.4 Conceptual model illustrating the main ecological factors potentially affecting lichen communities at plot and landscape levels (see explanation in the text)

Gauslaa 2001) and with a subsequent effect on epiphyte community composition. Interestingly, differences in tree age/size (dbh) correlate with bark-pH among individual boles of the same species, though the direction of this relationship cannot be generalized. Bark-pH may decrease with increasing tree circumference (Bates 1992; Kuusinen 1994) and/or age (Ellis and Coppins 2007), or pH may increase with circumference (Jüriado et al. 2009) and/or tree age (Fritz et al. 2009). Also, bark texture was found to be an important tree species factor influencing lichen diversity (Bates 1992; Cácares et al. 2007; Ranius et al. 2008; Fritz et al. 2009). This is subject to an effect of tree age/size (dbh) and an interaction with vertical height on the trunk (Johansson et al. 2007; Ranius et al. 2008; Fritz 2009), and it is related with a species-specific relationship between bark roughness and tree age/size (Uliczka and Angelstam 1999).

The standardization process of air quality biomonitoring method took into account all these sources of variability acting on lichen diversity at tree level, making it possible to isolate environmental gradients in a simplified system (Asta et al. 2002; EN 16413 2014; see paragraph interpreting lichen diversity). For this reason, the standard features of the sampling trees (slope, pH of the bark, etc.) have been selected carefully (Table 2.1).

With respect to plot-level variability, Giordani (2006) carried out a hierarchical evaluation of the effects of substrate- and environmental-related variables on lichen diversity. Elevation, mean annual temperature, and latitude were the main factors influencing epiphytic lichen distribution, together with disturbances (such as atmospheric pollution, forest fires and agricultural practices) and habitat heterogeneity that covary along the strongest gradient of lichen community composition. Lichen species richness was positively associated with mean annual rainfall and longitude, negatively associated with harvesting, and positively associated, though weakly, with the occurrence of past forest fires (these burnt sites were characterized by recolonization processes).

In a study of epiphytic lichens of forest ecosystems of Tuscany, Loppi et al. (1999) concluded that habitat characteristics are more important than phorophyte properties and there is evidence that the epiphytic lichen vegetation of deciduous *Quercus* trees follows a distribution which is related to elevation and climate, with great differences in community structure along the altitudinal gradient (Loppi et al. 1997).

As for the aspects affecting lichen distribution at landscape level, many authors pointed out the role of climatic factors (McCune et al. 1997; Goward and Spribille 2005; Hauck and Spribille 2005; Giordani 2006, 2007). In this respect, Giordani and Incerti (2008) found that the distribution of more than 30 % of epiphytic species was associated to macroclimatic variables. A significant subset of epiphytic lichens in the study area has been proved to be efficient bioclimatic indicator for montane, humid sub-Mediterranean, and Mediterranean units.

Further, also land-use intensity drives the local variation of lichen diversity, both in Mediterranean (Giordani et al. 2010) and in boreal ecosystems (Holt et al. 2008). Land-use intensity

Table 2.1 List of features to of a standard tree (see EN 16413 2014; Asta et al. 2002)

Feature	Description
Suitable tree species	The sampling tree belongs to one of the groups with similar bark physicochemical properties (EN 16413 2014; Asta et al. 2002). Indicatively, species belonging to the same group can be used interchangeably
Trunk circumference	The sampling tree has a trunk circumference (at 130 cm from the ground level) between 50 and 250 cm
Trunk inclination	Each exposition (N, E, S, W) has an inclination (at the center of each grid) <20°
Bark damage	The area of the trunk that is unsuitable for recording (damage, decortication, branching, knots and/or other epiphytes or climbing plants such as ivy, preventing growth of lichens) within each of the 4 grids when summed <20 %

was detected as a good proxy for describing the lichen distribution and abundance under anthropogenic pressures. In particular, land-use categories, mainly based on vegetational features, differed in epiphytic and epilithic lichen communities, and the strongest differences were observed among the forested sites versus managed agroforestry lands. These latter were mainly characterized by common lichen vegetation (mainly xerophytic–nitrophytic species).

Similarly, in Scotland, lichen communities differed between natural ancient pine forests and those of trees within managed sites both in forest and agricultural areas (Wolseley et al. 2006). The alteration was mainly due to increasing nitrophytes, most evident for epiphytes, but also significant for saxicolous communities. According to Bergamini et al. (2005), the trend of modification of lichen communities under changing land use is nearly constant along large latitudinal gradient, ranging from northern Europe to the Mediterranean. Also, at this large scale, strong differences were detected between the forested and the more open land-use types, especially for epiphytic crustose lichens.

2.2.2 Standardizing Lichen Diversity Sampling

Differences between methods, difference in the application of the same method, measurement error, sampling and non-sampling error, and errors related to model applications are all terms of the whole error budget that inevitably affects environmental surveys (Gertner et al. 2002). In this perspective, the extent to which the objective of the survey is matched depends very much on the ability to manage the various sources of variability (Khol et al. 2000; Wagner 1995) by adopting suitable standard operating procedures (Brunialti et al. 2004, 2012b).

A recent example of standardization process relates to the sampling protocol to assess epyphitic lichen diversity for air quality. In fact, the European Committee for Standardization (CEN, Comité Européen de Normalization) has recently published an European standard, reporting the sampling protocol for lichen diversity assessment (EN 16413 2014). The process of standardization started in 2007 and took into account the previous European and national guidelines (Asta et al. 2002; VDI 2005; AFNOR 2008). In the meantime, some field tests have been performed to obtain information on the type and size of errors and the uncertainty of the methodologies under standardization (Brunialti et al. 2012a, Cristofolini et al. 2014). To fill this gap, the tests dealt with the entire process from survey design to field measurements. In particular, the comparative tests, consisting in multiple exercises, were run with the aim of comparing the results obtained by different, well-experienced operators faced with the same problem, at the same time, under the same field conditions and following the same standard operating procedures (SOPs).

A similar standardization sampling procedure has been adopted in forest monitoring. An example is represented by the sampling protocol of the EU project 'Forest BIOdiversity Test phase Assessments' (ForestBIOTA), carried out by 12 European countries in the framework of the ICP Forests Expert Panel on Biodiversity and Ground Vegetation Assessments (Fisher et al. 2009). In this context, Stofer et al. (2003, 2012) prepared a standardized sampling protocol for lichen diversity assessment that takes into account all the main steps of the sampling from the tree- to the stand level.

2.2.2.1 Sampling Objectives

A clear definition of objective is important; otherwise, any statement about data quality remains elusive (Ferretti 2011). As a consequence, although it is not a typical source of error per se, ambiguous monitoring objectives can be great promoters of errors: On the one hand, the monitoring design cannot be properly addressed if the objective is not clear, and incorrect design may jeopardize the whole monitoring. For this reason, objective must be explicit with respect to target population and its attributes, spatial and temporal domains, desired precision level and minimum detectable change, and type I and II acceptable error rates.

With respect to the CEN standard for biomonitoring ambient air, the sampling objective is to obtain an estimate of the parameter of the response variable (e.g., mean species richness or mean lichen diversity value, LDV) over the study domain with a given precision (EN 16413 2014). The precision level should be expressed in terms of confidence intervals for a defined probability level. It is required that the sampling objective is defined for each study. For example, we may want to obtain an estimate of the mean LDV for the study domain with a confidence interval ±10 % of the mean value at a probability (*P*) level of 95 %.

Since the computation of estimates and confidence intervals depends on the sampling design adopted, each study shall define precision and probability levels, taking into account the requirements of the study framework and considering the available resources (EN 16413 2014).

2.2.2.2 Target Population

In statistical terminology, the target population refers to a different concept than that associated with the community, population, individual, and genetic concept of biological systems. The target population is the collection of elements about which information is wanted (Cochran 1977). Both the 'target population' for which information is wanted and the 'elements' that make up the target population must be rigorously defined. As clearly stated, target population should be carefully selected before starting each monitoring program. It must be representative of the whole population in order to extend the results obtained by the selection of a sample of it to the entire population. In fact, we must consider that data obtained outside a formal sampling design cannot be considered representative for the entire population, and therefore, conclusions cannot be extended without making assumptions about it (Ferretti 2011). Similarly, models built upon non-representative data can be seriously biased. A classical example is the biodiversity data collected by taxonomists who are inclined to concentrate their efforts in the localities that guarantee success in the collection of as many species as possible, thus resulting in species-richness bias (Sastre and Lobo 2009).

For instance, a recent comparative test to identify critical issues in lichen biomonitoring demonstrated that different teams may select different target populations when planning the work (Brunialti et al. 2012a). Although all of them may be formally correct, these differences are a source of inconsistency in the results and could potentially lead to differing conclusions. It is therefore extremely important that every step of the design process is properly documented and reported in survey documents so that future repetitions may be made.

2.2.2.3 Sampling Design and Sampling Schemes

For broader applications (e.g., at the European level), a standard protocol should be flexible enough to preserve the representativeness of the data under different ecological conditions. In general, it is difficult, if not impossible, to find a single sampling design that is suitable for each situation. Different options of sampling schemes may be adopted, mainly in relation to the complexity of the survey area and with the distribution of the trees to be sampled (Elzinga et al. 2001; EN 16413 2014; Ferretti and Erhardt 2002). With respect to air monitoring pollution, the recent CEN standard proposes the adoption of different schemes in relation to ecologically homogeneous and heterogeneous areas. In particular, in the first case, we may have tree options:

1. When standard trees are abundantly and homogeneously distributed over the study domain, a simple random or systematic design is recommended. Plot sampling is recommended, with sample plots allocated according to a regular grid, with the starting point of the grid chosen at random (e.g., Giordani et al. 2002).
2. When standard trees are abundantly scattered in clusters over the study domain, tree-based cluster sampling or two stage sampling is recommended. A criterion to identify clusters should be initially set, then identify, and list

all the clusters and obtain a random sample of them. It should be noted that a two-stage sampling requests a subsampling of the sampling unit, and it introduces a further source of variability which may affect the quality of the data. It is important to take this into account when performing statistical analysis of the data (EN 16413 2014).

3. When standard trees are infrequently scattered over the study domain, a simple tree-based random sampling is recommended. It is possible to obtain a list of the individual trees on the basis of the aerial photo and to select randomly the sample trees.

Similar options may be selected also in ecologically heterogeneous areas. In this case, however, a stratified random sampling design is recommended both with standard trees homogeneously distributed and scattered in clusters over the study domain. A tree-based stratified random design is suitable when we have standard trees infrequently scattered over the study domain.

Notwithstanding this great variability of solutions, it is not so obvious that each of these is properly taken into account when a monitoring program is designed, since the experts in the field may be more influenced by examples from the past (seen as best practices) that by newly suggested options. The results of a comparative test carried out of operators on the same area and with the same standard procedures by five independent groups partly confirm this statement. In this study, in fact, most teams adopted a stratified random sampling (Brunialti et al. 2012a). Although they largely agreed on selection of the sampling scheme, considerable differences occurred in subsequent steps of the sampling design: For example, the number of selected land cover categories ranged from 2 to 8 and the sampling density 23–43. The authors observed that training courses on sampling design and basic statistics may be successful in reducing this source of variability in routine field studies.

2.2.2.4 Sampling Units

As reported above, both plot-based and tree-based sampling schemes can be adopted. According to the latter option, the sampling units are represented by the standard trees available in our survey area. Instead, when considering the plot-based sampling, the selection of the optimal shape and size of the plot remains an open question. In this case, the goal is to find an acceptable trade-off between a good representativeness of the study area and a cost-effective sampling effort.

Concerning the shape of sampling units, circular or square plots are used in most cases. In general, the former are suggested in air pollution monitoring programs (Giordani et al. 2002), while the latter are mainly adopted in forest surveys (Giordani 2006). The basis for a proper selection of the shape of sampling units relies both on probabilistic issues and on practical questions (Elzinga et al. 2001). For rectangular sampling units, the associated edge error is larger than for circular plots. Moreover, a circular plot requires only one measurement (radius) to be installed. In special cases, such as for the assessment of the effect of forest fragmentation, rectangular transects were taken into account, as they could fit better with the natural shape of forest fragments, which were further divided into subplots (Brunialti et al. 2012b).

As for the plot dimensions, McCune and Lesica (1992) found trade-offs between species capture and accuracy of cover estimates for three different within-site sample designs for inventory of macrolichen communities in forest plots. On average, whole-plot surveys captured a higher proportion of species than did multiple microplots, while giving less accurate cover estimates for species. The reverse was true for microplots, with lower species captures and much better cover estimates for common species. Belt transects fell between the over two sampling designs.

Ravera and Brunialti (2013) showed that a probabilistic sampling based on the selection of only three trees within small circular plots (14 m diameter) can be effective for assessing the occurrence of species of conservation concern in old growth forests in an Italian National Park. They found that most of the species were present in a few sampling sites and only a small group of species were common to more than 50 % of the plots. This may suggest that the adopted sampling design allowed not only to determine the most local common species but also to detect the presence of sporadic species (Ravera and Brunialti 2013). To confirm this, their lichen list reported several new and interesting species and represented 30 % of the indicators of forest continuity known from the region. This is a considerable result if we consider the nature of the adopted sampling and also the fact that regional floras mainly report records from preferential surveys. We must keep in mind, however, that floristic surveys, mainly carried out by means of a preferential sampling, are biased by the fact that they are primarily carried out to discover rare species in specific habitats or ecological niches. Humphrey et al. (2002) obtained similar results in a study carried out to compare lichen and bryophyte communities between planted and seminatural stands. A high percentage of species was recorded only once, and very few species were common to more than half the plots. This 'local rarity' phenomenon has been noted in other studies (Vitt et al. 1995; Collins and Glenn 1997; Qian et al. 1999; Humphrey et al. 2000) and is partly related to sampling area. The authors of that study observed that it is possible that a 1-ha sampling plot used is too small to capture a representative sample of lower plant diversity in forest stands. For example, Rose (1993) recommends a minimum sampling area of 1 km^2, but this depends on the objective of the survey. Hence, these findings suggest to adopt one of the following approaches for future investigations: (1) extend the surveys to plot with a broad area and, at the same time, detect a higher number of trees and/or substrates (rocks, soil, etc.). Although this approach could be very time consuming, it should ensure the finding of a large number of species; (2) improve the number of small plots in the study area. In this way, with a lower sampling effort at the individual plot, it is possible to obtain information on a wider territory. Moreover, this second approach should be mostly useful for preliminary lichen surveys in poorly studied wild areas (Ravera and Brunialti 2013).

2.2.2.5 Sampling Density

Optimal sampling density is an essential requirement of lichen biomonitoring surveys for obtaining precise and unbiased estimates of population parameters and maps of known reliability (Ferretti et al. 2004). This aspect is often a sore point of biomonitoring projects and should be carefully addressed to respond correctly to the sampling objective and to select a sample with a proper sample size to avoid jeopardizing the effectiveness of the investigation. Apart from catching a sufficient amount of variability, a decision on a proper sample size forcely considers the sampling effort and its costs and it often happens to reduce the number of sampling sites in order to limit the costs.

Both the ecological complexity of a given area and the desired level of precision (in terms of confidence interval) drive the sampling design. In heterogeneous areas, a greater number of sampling units and more detailed land cover stratification are needed to obtain accurate estimates. For broader application (e.g., at the European level), a standard protocol should be flexible enough to preserve the representativeness of the data under different ecological conditions (Brunialti et al. 2012a). In this respect, Ferretti et al. (2004), starting from the results of two large-scale surveys undertaken in Italy, carried out a study on the effects of different sampling densities (the number of sampling units in the study area) and on the reliability (in terms of confidence intervals and relative error in the mean values) of the estimates of the lichen diversity values of a given area and of lichen diversity maps. An iterative approach was taken into account to generate subsets with lower sampling density with respect to the original sampling units (ordinary kriging interpolations). Obviously, a higher sampling density may lead to a low error rate but may be financially unsustainable. A very low sampling density, on the other hand, may provide uncertain data as to be of no real use (Khöl et al. 1994; Ferretti and Erhardt 2002). Their findings suggested that a large-scale lichen diversity spatial pattern can be detected with a much more relaxed grid density than those originally applied (Ferretti et al. 2004). This is extremely important, since reducing the sampling effort can result in considerable savings in resources that can be diverted to additional, more detailed investigation with denser sampling in those areas identified as problematic by the study based on the low sampling density. These results were also confirmed by a similar study (Frati and Brunialti 2006), within a long-term monitoring program, that showed the possibility to reduce the sampling effort in future monitoring surveys, resulting in a considerable reduction of the sampling effort maintaining a good data quality.

2.2.2.6 Lichen Survey

As discussed in the previous sections, a series of micro- and macroenvironmental variables can affect the composition and diversity of epiphytic lichen communities. For this reason, a standardized sampling strategy to count lichen diversity is extremely important to ensure comparable and accurate results. The recently developed European standards for lichen biomonitoring (Stofer et al. 2012; EN 16413 2014) have been developed as the result of a standardization process carried out in the last few years starting from the Index of Atmospheric Purity (IAP) approach (Hawksworth and Rose 1970; Nimis et al. 1990, 1991) and upgrading it with previous guidelines (Asta et al. 2002; VDI 3957 2005). Furthermore, also field experience of several European researchers and the results of recent comparative tests (Brunialti et al. 2012a; Cristofolini et al. 2014) were useful to obtain the current sampling strategy for lichen survey. At the end of this process, to reduce the effect of several possible sources of error (e.g., different size of explored area on the trunk, subjectivity in the positioning of the sampling grid, etc.), several parameters have been standardized. Hence, the abundance of each lichen species is currently sampled by means of a sampling grid consisting of a 10 × 50 cm ladder divided into 5 10 × 10 cm quadrants. This ladder grid is placed systematically on the N, E, S, and W side of the bole of each tree (4 per tree), with the top edge 1.5 m above ground, following the standards suggested by Asta et al. (2002). Summary measurements of species richness and abundance are usually calculated for each plot: mean number of species per tree; total number of species within the plot; and mean lichen diversity value (LDV—Asta et al. 2002) calculated as the sum of the abundance of each species within the sampling grids on a tree, averaged for all trees within a plot.

2.2.3 Quality Assurance

Biomonitoring investigations are subjected to a variety of error sources that need to be acknowledged and documented in order to be managed properly (Brunialti et al. 2004). The quality of the data originating from biological measurements depends heavily on at least three factors (see, e.g., Kovacs 1992; Brunialti et al. 2004): (1) variability of the biomonitoring organisms (interactions between the organisms and environmental factors); (2) type of sampling (sampling design, density of sampling points); and (3) operators involved in assessing lichen diversity, which is a method requiring a relatively high taxonomic knowledge.

The first issue has already been discussed above (Sect. 2.2.1), and we have seen that this source of variability can be controlled with a good knowledge of the ecological characteristics of biological indicators considered and adopting proper sampling designs. Regarding the type of sampling, the sampling errors are closely associated with the sampling design adopted and its quality. As for the third factor, the effect of operators' subjectivity and expertise has been widely addressed within lichen biomonitoring so as to obtain useful information to further improve standardized protocols.

To respond to these issues and to take all the steps of the monitoring survey into account, the adoption of quality assurance procedures is strongly recommended. QA is an organized group of activities defining the way in which tasks are to be performed to ensure an expressed level of quality. The main benefit of a QA plan is the improved consistency, reliability, and cost-effectiveness of a program through time (Ferretti 1998). A QA plan is essential since it forces program managers to identify and evaluate most of the factors involved in the program. In addition, the assessment of data quality enables mathematical management of uncertainty due to the method used (Ferretti 1998; Cline and Burkman 1989).

It is therefore important that environmental biologists and field ecologists consider QA as a key attribute of their work in order to provide robust and defensible data to decision makers (Brunialti et al. 2004).

2.2.3.1 Sampling Errors

Sampling errors are generated by the nature of the sampling itself and by the degree of variability in the target population. As reported above, such kinds of errors always occur but can be controlled by appropriate sampling design (Cochran 1977; Köhl et al. 2000; Ferretti and Erhardt 2002).

A good sampling design is essential either to collect data amenable to statistical analyses and to control errors in relation to the costs (Brunialti et al. 2004). An important issue, in this respect, is selectivity, which seems particularly important in ecological measurement. A protocol is selective if the response provided as a measurement depends only on the intended ecosystem property (Olsen et al. 1999). Regarding this aspect, Yoccoz et al. (2001) suggest that quantitative state variables characterizing the system well should be privileged. For example, when defining management objectives in terms of changes of densities of indicator species, the program should incorporate tests to ensure that selected species are indeed indicators of the process and variables of interest (Yoccoz et al. 2001). For this reason, it is important to establish a priori the variables of interest in a sampling protocol. The criteria for this selection should be based on data quality, applicability, data collection, repeatability, data analysis and interpretation, and cost-effectiveness. In particular, the use of quantitative state variables is recommended in order to reduce the error in data collecting due to the subjectivity of the operators.

2.2.3.2 Non-sampling Errors

Non-sampling errors include measurement, classification, and observer errors, which are rooted in how the standard operating procedures (SOPs) are prepared and applied and how well-trained and skilled the field crews are (Ferretti 2009). In general, non-sampling errors can occur when the methodology is poorly standardized, when teams have insufficient skills or insufficient care is taken in applying the method, or when there are problems with instrument calibration (Ferretti and Erhardt 2002). Many papers have focused on these topics and have shown that non-sampling errors can be a significant source of variability in monitoring studies (Gertner and Köhl 1992; McCune et al. 1997; Giordani et al. 2009; Francini et al. 2009; Gottardini et al. 2009; Marchetto et al. 2009; Sastre and Lobo 2009).

In the specific case of epiphytic lichen assessment, non-sampling error may basically occur at two stages (Brunialti et al. 2012a): the identification of standard trees and the counting of lichen species. As for the former case, an imprecise definition of suitable trees in the SOPs might be one of the main reasons for differences in the number of suitable trees found by the different teams. As for the counting of lichen species, floristic knowledge is a crucial issue that needs to be addressed in lichen biomonitoring where the protocol is based on assessment of all lichen species, including groups of lichens which are hard to identify in the field, such as crustose lichens (Asta et al. 2002). Differences in the floristic skills of the teams can cause serious errors (Brunialti et al. 2002; Giordani et al. 2009). Brunialti et al. (2012a) confirmed that considerable underestimation of species richness may occur even when sampling within an a priori positioned grid. Floristic skill is even more important when assessing crustose-dominated communities, where poorly developed thalli often occur (Giordani et al. 2009). Undoubtedly, variability among crews could be reduced with intercalibration courses and harmonization procedures (McCune et al. 1997; Brunialti et al. 2002, 2004). In fact, there is evidence that operators often improve in accuracy during the same test and that their accuracy improves with taxonomic training and, above all, continuous fieldwork (Brunialti et al. 2002).

As far as precision is concerned, very high levels are usually registered among operators, suggesting a high reproducibility of the lichen diversity counts (Brunialti et al. 2002). This is extremely important for correct evaluation of time series in biomonitoring studies. However, it should be borne in mind that changes in operators in long-term monitoring of permanent plots can give misleading results (McCune et al. 1997) and should be carefully addressed.

2.3 Interpreting Lichen Diversity

As described in the previous paragraphs, a considerable effort has been made in the recent years for standardizing the sampling design and strategies of lichen biomonitoring. In the following paragraphs, we will describe some of the most used approaches for interpreting lichen diversity data in terms of effects of various anthropogenic disturbances. In general, the interpretation of geographic patterns and temporal trends of lichen diversity may be assisted by using ecological indicator values (Hawksworth and Rose 1970; Wirth 2010; Nimis and Martellos 2001, 2002), multivariate statistics, such as numerical analysis of matrices of species (Giordani et al. 2002; Giordani 2006), nonparametric models (McCune and Mefford 2004; Giordani 2007), or other statistical tools.

2.3.1 The Concepts of Biomonitoring

We can define monitoring as the process of gathering information about some system–state variables at different points in time for the purpose of assessing the state of the system and making inferences about changes in state over time (Yoccoz et al. 2001). If our focus is on the monitoring of biological diversity, the systems of interest to us are typically ecosystems or components of such systems (communities and populations), and the variables of interest include quantities such as species richness, species diversity, biomass, and population size.

2.3.1.1 Lichen Diversity Value, α-Diversity

According to the recently developed standards for lichen biomonitoring (Stofer et al. 2012; EN 16413 2014), the basic results of lichen diversity sampling are aggregated matrices of the species frequencies at nested spatial levels of sampling, i.e., a matrix of species at aspects of each tree; at trees of each sampling units; and, finally, at sampling unit level. Several diversity indices are calculated a posteriori basing on these basic matrices. This recommendation comes to the fact that several interpretative tools may be applied to basic data, which are used in a various manner in different countries.

As a simple but effective approach, α-diversity (number of species at plot level) is a robust parameter for interpreting patterns of epiphytic lichen communities along gradient of anthropogenic effects, such as pollution or forest management.

Among the possible descriptors obtained from the basic species × sampling unit matrix, the lichen diversity value (LDV) by Asta et al. (2002) is by far one of the most used in applicative biomonitoring surveys. The value for a given sampling unit is calculated as the sum of the frequencies of all lichen species found on each tree within the unit and averaged by the number of sampled trees. Relevant differences in lichen growth may be expected on different cardinal aspects of the trunks; therefore, it is suggested that frequencies are also summed separately for each aspect, and possibly, additional analysis might be carried out in this sense. LDV in its basic definition has been extensively used in applicative studies all over Europe (Paoli et al. 2006; Svoboda et al. 2010; Giordani et al. 2014a), and its relationships with pollution and other environmental factors had been analyzed in details. In the Mediterranean, together with pollution, the mean annual rainfall of the survey area seems to be the best predictor for LDV variability (Giordani 2006). However, the decisive variables affecting the lichen diversity are apparently different in urban versus forested areas (Giordani 2007). In these latter, harvesting and forest fires showed a predominant effect. Contrarily, in urban areas, air pollutants, mainly SO_2, are the main limiting factors, even if this relationship is lowering under ameliorating conditions of atmospheric pollution. Similarly, Svoboda et al. (2010) found that lichen diversity in Central Europe responded differently to environmental predictors depending on different human impact. These authors observed that in industrial regions, air pollution was the strongest factor affecting lichen diversity, whereas in agricultural to highly forested regions, LDV was mainly influenced by forest age and forest fragmentation.

Starting from the basic species × sampling unit matrix, further parameters can be derived from the data set of the species frequencies, including the relative LDVs of morphofunctional groups of lichens associated to particular sources of atmospheric pollution (e.g., nitrophilus versus acidophilus species—see Sect. 3.2.3).

2.3.1.2 Rapid Biodiversity Assessment (RBA) Based on Morphospecies

The application of biomonitoring methods based on high levels of taxonomic knowledge, such as the recently standardized lichen biomonitoring method (EN 16413 2014), requires an adequate number of specialists that are not always available, especially in large-scale biodiversity assessments (Wilkie et al. 2003). The use of guilds or morphological groups as indicators of changes in ecosystem function has been considered by several authors as a good compromise between the need for specialized knowledge and rapid field procedures employing non-specialist technicians, thus providing a possible shortcut in assessing total species richness (see, e.g., Pharo et al. 2000; Giordani et al. 2009). This issue has been considered in several ecological monitoring fields to explore the possibility of using surrogate species for estimate total biological diversity. As an example, several studies have been performed to assess the congruence among vascular plants, vertebrate, invertebrate, bryophytes, and lichens within large-scale biomonitoring surveys using simplified assessment methods (see Oliver et al. 1998; Pharo et al. 2000; Wilkie et al. 2003; Santi et al. 2010). However, some conflicting results obtained from the works listed above suggest that this is not always the ideal solution and several critical issues emerge: Among the others, there are sources of variation coming from the fact that the communities do not always behave in a linear and unambiguous manner. Also, the data quality in surveys involving non-specialist crews may vary at such a level which could drastically compromise the reliability of the results. In this respect, Giordani et al. (2009) carried out a study to compare data obtained by non-specialists through simplified methods based on morphospecies (RBA), with those collected by specialists using the lichen diversity value (LDV) method. They found that lichen diversity estimated by means of Rapid Biodiversity Assessments (i.e., based on morphospecies) showed good correlations with the results of a classical, systematic identification of species only when performed by operators with high taxonomic knowledge. Furthermore, the use of sampling lists based on highly simplified morphospecies did not lead to significant advantages in terms of time needed for fieldwork (Fig. 2.5).

In some other cases, Rapid Biodiversity Assessments of lichen biodiversity led to interesting results even at large scale. Recently, citizen science approaches have been applied to lichen biomonitoring of the effects of atmospheric pollution. These are voluntary schemes engaging members of the public in the collection of scientific information. The OPAL Air Survey in the UK (Davies et al. 2011) used presence and abundance data for 9 selected lichens, collected by more than 4,000 volunteers in a public survey of lichens on trees, to examine the response of individual species and groups of indicator lichens to air pollution and climate drivers on a national scale. The use of these macrolichen indicators has shown to have robust relationships with

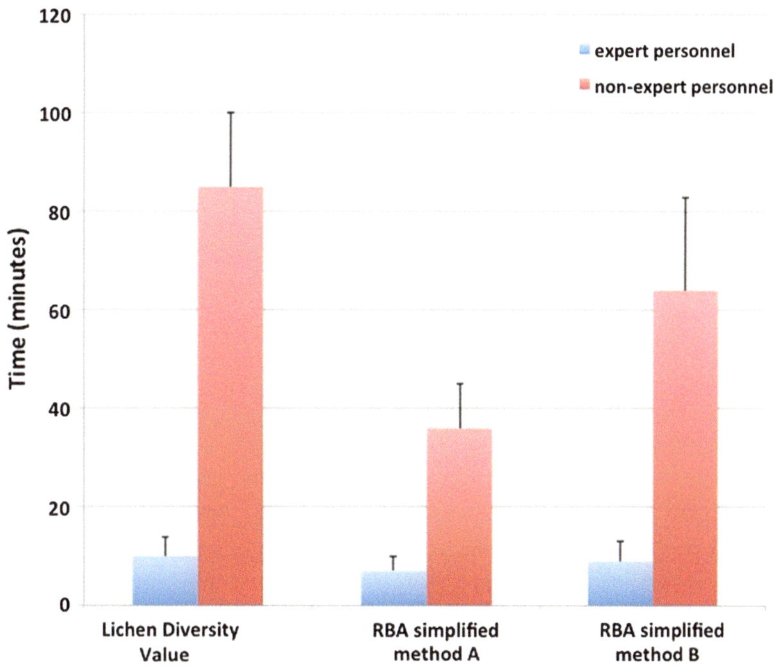

Fig. 2.5 Time (minutes) required for each lichen diversity assessment (tree level), using different sampling lists. Lichen diversity value (sampling of lichens at species level) was compared with two different simplification methods of rapid biodiversity assessment (*RBA*, simplified methods A and B). Data from Giordani et al. (2009)

modeled nitrogenous pollutants at the national scale (Seed et al. 2013). In the USA, Casanovas et al. (2014) proposed a citizen scientist-based survey methodology for macrolichen diversity in which parataxonomic units (PUs), as identified in lichen photographs, served as species surrogates to estimate lichen diversity. Although in most cases the authors showed that the observed and estimated cumulative richnesses from both techniques were not statistically significantly different from each other, the extensive use of these approaches in biomonitoring surveys should be carefully evaluated, as misidentifications of morphologically similar species could led to wrong interpretation of data, e.g., in terms of relative abundance of functional groups. At this regards, it has been shown that the use of electronic devices and identification tools may help to increase the quality of RBA. The project Dryades developed interactive identification keys in the form of applications for mobile devices. Keys were generated from databases of morphoanatomical characters. The applications were tested Europe wide during the project KeyToNature and have proved to be useful in education and in projects of citizen science (Nimis et al. 2012).

2.3.1.3 β-Diversity

Most biomonitoring applicative surveys and researches using epiphytic lichens have focused on analyzing plot-level species richness (e.g., alfa-diversity), abundance, and/or composition patterns along ecological gradients. Recently, it has been suggested that also analyses of β-diversity may provide insights into mechanisms and drivers influencing lichen communities, thus contributing to a better interpretation of the results. Beta-diversity has been interpreted mostly as the extent of change in community composition (Whittaker 1960). Basically, beta-diversity patterns are originated from two distinct processes, the replacement and the loss of species (Carvalho et al. 2012). With the aim of differentiating the relative influences of these components on beta-diversity, various measures have been proposed, which relied on an additive rather than multiplicative approach (Baselga 2010). As a first application to lichen communities, Nascimbene et al. (2013a) used the conceptual scheme by Podani and Schmera (2011) to evaluate the relative importance of β-diversity, nestedness, and agreement in species richness in presence–absence data matrices via partitioning pairwise gamma diversity into additive components. Podani and Schmera (2011) considered three complementary indices that measure similarity (S), relative species replacement (R), and relative richness difference (D) for given pairs of observations. In particular, according to these authors, β-diversity is defined as the additive result of R and D, whereas other descriptors, namely nestedness and richness agreement, result from the additive effects of similarity with the other two complementary components (S + D and S + R). By analyzing the diversity in *L. pulmonaria* communities in Italian forests, Nascimbene et al. (2013a) showed that both species replacement and similarity were generally associated with forest structure predictors, such as the number of trees in the plot and the distance between trees, while richness difference was mainly associated with geographic predictors, with special reference to longitude and altitude. Giordani et al. (2014b) coupled the analysis of β-diversity with the approach based on the relative abundance of functional groups. These authors explored the shift in functional groups for nitrogen tolerance along a gradient of increasing cattle load in epilithic lichen communities of alpine pasturelands. An increasing cattle load caused a decreasing replacement of oligotrophic species and consequently a decrease in β-diversity. Conversely, when considering a data set with only N-tolerant species, there was very high pairwise similarity among sampling plots, irrespectively by the cattle load gradient to which they were exposed.

2.3.1.4 Indicator Species

Detecting species that best characterize some set of sites is an important step in evaluating classifications in community ecology. With reference to lichen biomonitoring, this approach has been mostly applied for detecting the effects of forest management or to assess the ecological continuity of ancient woodlands. In some cases, lists of indicators species were compiled on the basis of expert assessments (e.g., Rose 1976). As for other biologically based approaches, also for biomonitoring, the use of proper methods for measuring the explanatory power of species is essential, whenever priority must be given to species that best reflect environmental quality (Dufrêne and Legendre 1997; Podani and Csányi 2010).

Whittet and Ellis (2013) tested 29 indicator species of forest continuity, as proposed by the current suite of British lichen indicators (Coppins and Coppins 2002) for different biogeographic regions in the UK. In accordance with previous studies (e.g., Sætersdal et al. 2005; Giordani and Incerti 2008), these authors confirmed that indicator species are likely to have a restricted geographic scope. Moreover, they suggested that only some of the studied taxa could be actually accurate indicators of ecological continuity through a dependency on the long-term persistence, whereas in other cases, several species may possibly be associated with specialist microhabitats under a sub-optimal climate, whereas they would not be significantly associated with ancient woodlands. Giordani (2012) compared the performance of four potential indicators for monitoring the effects of forest management on epiphytic lichens in broadleaved Mediterranean forests. Indicators included total lichen diversity (LDV) and the abundances of species associated with intensive management, species associated with aged coppiced woodlands and indicator species ratio (ISR). ISR was defined as the ratio between the difference between the species associated to aged coppice forests and those associated to intensively managed forests and the total abundance. At each of 50 sampling sites, the four indicators were calculated using indicator value analysis (Dufrêne and Legendre 1997) and compared through correspondence analysis. By balancing the partial information provided by both sensitive and resistant species, ISR was shown to be a more effective indicator, being independent of floristic composition and the occurrence of rare species. The main drawback of the indicator species approach is that evaluation of the effects of a given stress (e.g., forest management) is possibly biased due to the fact that lichen species are strongly threatened by several anthropogenic disturbances occurring at the same time (e.g., high levels of air pollutants or forest fires).

2.3.2 Interpretative Tools

2.3.2.1 Interpretative Scales of Alteration

Following the definition of Nimis (1999), biomonitoring techniques estimate the degree of alteration from normal conditions resulting from the effects of pollution on the reactive components (e.g., lichens) of the ecosystems. However, defining 'normal conditions' in ecology is extremely critical and can only rely on an operational basis. At this regard, some authors proposed an interpretation of lichen diversity data in terms of percentile deviations from an observed maximum diversity. Loppi et al. (2002a, b) sampled lichen diversity on about 3,000 trees in northern Tyrrhenian, Italy, and considered as 'natural' those values ≥ the 98° percentile of their frequency distribution. The average of these values was taken as an operational definition of 'naturality,' which represents a 'maximum potential lichen diversity' in a given area. These authors considered that a 25 % deviation from 'normal conditions' could still be regarded as 'natural,' owing to natural fluctuations of lichen diversity. Starting from this point, an interpretative scale of naturality/alteration was build basing the different degrees of naturality/alteration on progressive 25 % deviations from 'normal conditions.'

As lichen distribution strongly depends on macro- and mesoclimatic factors (see Ellis 2012), such interpretative scales of lichen diversity in terms of deviation from maximum potential conditions information should be referred to a regional level. Thus, direct comparisons between biomonitoring surveys carried out in different bioclimatic regions are often poorly informative and can also led to misleading interpretations. A similar approach of regionalization has been followed and debated in the case of other biomonitoring techniques (e.g., Moog et al. 2004). For example, the EU Water Framework Directive suggested qualitative reference for evaluating the ecological status of water bodies, assessing the highest potential quality, based on the composition of aquatic communities. It was possible to predict how the response of diatom communities to anthropogenic pressure in each French hydroecoregion was predicted (Tison et al. 2007), in relation to the topology of running waters and validated with benthic macroinvertebrates from reference sites (Wasson et al. 2002).

The delimitation of eco-regions or bioclimatic regions bases upon differences in the composition of their lichen flora. At this regards, Giordani and Incerti (2008) applied a nonparametric multiplicative regression model to the lichen flora of a climatically heterogenous area and detected 59 species which were significantly associated to macroclimatic variables (i.e., annual rainfall and temperature). A cluster analysis grouped the taxa into four subsets that were related to different climatic niches (warm–humid, cold–humid, mesothermic–humid, warm–dry) corresponding to distinct bioclimatic regions.

As an example of regionalization of interpretative scales, Castello and Skert (2005) provided evaluation scales of environmental alteration based on lichen diversity in the North Adriatic sub-Mediterranean bioclimatic region. These authors sampled deciduous oaks in 11 reference sites in open stands within natural woods or near very small isolated villages in rural or natural areas, far from large urban areas, industrial zones, and long-distance transport of air pollutants. As a result of relatively dry condition occurring in North Adriatic, the LDV threshold for naturality class was slightly lower than the one calculated for the more humid Thyrrenian Italy (Giordani 2004). Regionalized interpretations are also strongly recommended by the German guidelines (VDI 3957 2005), which states that a comparison of surveys of different regions is only possible if the surveyed areas have a similar climate and, therefore, a comparable lichen flora, as is the case with Central Europe (excepting the Alps) for which the evaluation scale was calibrated.

In some cases, ad hoc interpretative scales have been calculated for local situations. Due to the lack of an interpretative scale for semiarid Mediterranean bioclimatic region, Paoli et al. (2006) developed a calibrated scale for a small area (36 km^2) according to the protocol suggested by Loppi et al. (2002a, b) for the assessment of environmental deviation from natural conditions. Macroregional- and local-based interpretations of lichen diversity are not mutually exclusive and can provide integrate information on the actual environmental conditions.

Thought that nitrogen is becoming one of the most relevant limiting factors for lichen communities, especially in Central Europe, the German standard for lichen biomonitoring (VDI 3957 2005) proposed an interesting integrate approach for interpretation, which is basically based on the relative diversity of functional groups for nitrogen requirements. According to the German guidelines, nitrophytic species which respond positively and oligotrophic species which respond negatively to eutrophication (referred to as 'reference species') are calculated separately. The two partial LDVs of a sampling unit are combined to form the 'air quality index'. The LDV of the reference species is entered along the ordinate axis, whereas the LDV of the indicators of eutrophication is entered along the abscissa. The quality class assigned to the field in which the crossing point of the lichen diversity values comes to lie gives the evaluation of the air quality of the sampling unit. According to the German approach, the thresholds of the air quality classes in the matrix are chosen so that the class width is approximately equal to three times of standard deviation of the LDV in the study area. In this way, the lichen cover of sampling units belonging to different air quality classes is significantly different if the respective classes are separated by at least one other class.

The use of interpretative scales of lichen diversity based on deviation from maximum potential diversity also presents some weak points. From a theoretical point of view, it is well known that natural situations are not necessarily associated with maximum values of biodiversity. According to influential ecological theories, such as the intermediate disturbance hypothesis (Connell 1978), intermediate levels of disturbance, in terms of frequency and/or intensity of the phenomena, will maximize species diversity. The question has been shown to be more complex and less generalizable than this, with recent studies demonstrating that diversity could show monotonic, unimodal, or even flat response to disturbance, depending on the studied organisms and on the disturbance aspects considered (Hall et al. 2012). In an applicative perspective, the lacking of undisturbed reference situations in some eco- or bioclimatic regions (e.g., the Po Plain in Italy or many anthropized areas of Central Europe and North America) makes difficult to define properly the values associated to the highest classes of the scale, thus affecting the entire interpretation process.

2.3.2.2 Mapping Lichen Diversity

Mapping lichen diversity is an attractive approach, which allows an immediate representation of the results. Showing spatial distribution patterns of the studied descriptors, maps of lichen biodiversity, or abundance enable a quick and clear identification of areas with different levels of disturbance (Pinho et al. 2004; Asta et al. 2002). Spatial mapping of lichen diversity or associated measurements had been extensively used both in research and applicative lichen biomonitoring works (Giordani et al. 2002; Pinho et al. 2004, 2008a; Geiser et al. 2010). A large suite of GIS softwares provide tools for interpolating (i.e., estimating) the values of the response variable (e.g., lichen diversity) in non-measured parts of the survey area, basing on the data collected at sampling sites. In this chapter, we do not review the technical aspects of this approach. Basically, it follows the principle of geostatistical modeling theory originally developed for applied geology and more recently applied to ecology (see, e.g., Perry et al. 2002 for a detailed description). However, when applying these techniques, lichenologists should be aware that incorrect settings of the geostatistical model could lead to misleading results and interpretation and that the error associated to a poor spatial model might be also larger than those imputable to other sources of errors, such as sampling design or taxonomic misidentification.

Estimated values in lichen biodiversity maps are related to sampling densities (the number of sampling units in the study area), which in turn affect the reliability of the collected data in terms of confidence intervals and relative error in the mean values (Geiser and Neitlich 2007). By analyzing two large surveys of lichen mapping carried out in Italy (Nimis et al. 1991; Giordani et al. 2002), Ferretti et al. (2004) showed that even a considerable reduction (up to 50 %) of the original sampling effort led to a much smaller increase in mapping errors (<18 %). These data suggested that reducing the sampling effort ensures a considerable saving in resources that can be diverted to additional and/or more detailed investigation with denser sampling in those areas identified as problematic by the study based on the low sampling density.

Despite the fact that the interactions of pollution with other confounding factors in determining the distribution of lichens has been largely ascertained (Nimis et al. 2002), surprisingly, this source of variation was not always explicitly considered when building a spatial model for lichen biomonitoring data. A good example of integrate interpretation of lichen data, accounting the variability of both pollution and climate, was provided by Geiser and Neitlich (2007). Analyzing the effects of air quality and climate on epiphytic lichens in the US Pacific Northwest, they produced a kriged map of air scores, which were derived by the coordinates on the main axis of an NMS ordination of sampling plots located along gradients of pollution and climate. The estimated values of the air scores in non-sampled cells of the study area were calculated using the Gaussian semi-variogram model and a variable search radius including 20 sampled points. The spatial scale of analysis of a geostatistical model is often determined a priori. Pinho et al. (2008a) warned that, using this approach, some relationships between indicator and environment may be overlooked. Ribeiro et al. (2013) faced this problem investigating the relationships between ecological indicators and underlying environmental factors in Portugal. They used a multivariate geostatistical method, a linear model of coregionalization, to analyze the joint distribution of biodiversity variables in their study area. They were able to assess the strength of the relationship between each environmental factor and ecological indicator at several spatial scales. Basically, they related information on land cover and climatic variables with the abundance of fruticose lichen species, which were expected to be very sensitive to multiple environmental drivers. Their analysis implied the calculation of a nested variogram function to quantify the intensity and direction of correlations between the abundance of fruticose lichens and environmental factors at relevant spatial scales. These authors found that at medium scales (c. 15 km), open-space areas (considered as a proxy variable for particle emissions) were more important for shaping the abundance of this lichen group, whereas at larger scales (c. 45 km), open artificial areas (as a proxy for gaseous pollutants) and climate were preponderant.

2.3.2.3 The Use of Lichen Functional Groups to Detect Critical Loads and Critical Levels of Pollutants

'Functional groups' is the term used to describe sets of species exerting a comparable effect upon a particular process or responding in a similar manner to changes in their external constraints (Lévêque and Mounolou 2003). In particular, species' functional groups have been proved to be a valuable tool for comparing lichen diversity across diverse regions where high levels of floristic variation may occur (Giordani et al. 2012). Interestingly, recent revisions of the critical loads (CLOs) and critical levels (CLEs) for a large number of ecosystems are based on the response of lichens to main pollutants (e.g., Cape et al. 2009; Fenn et al. 2008). CLOs and CLEs have been defined to set sustainable thresholds for the protection of ecosystems from the effects of pollutants. The CLO is defined as "a quantitative estimate of deposition of one or more pollutants below which significant harmful effects on specified elements of the environment do not occur according to present knowledge" (Posthumus 1988), whereas the CLEs refer to the concentration of pollutants in the atmosphere above which adverse effects occur (Cape et al. 2009). Pinho et al. (2014) recently proposed a new tool to calculate CLEs by stratifying ammonia concentrations into classes and focusing on the highest diversity values. Based on the significant correlations between ammonia and biodiversity, the CLE of ammonia for Mediterranean evergreen woodlands was found to be 0.69 $\mu g\ m^{-3}$, below the currently accepted pan-European CLE of 1.0 $\mu g\ m^{-3}$.

Concerning the detection of CLOs, the occurrence of oligotrophic lichen species provided information on the actual impact of reduced nitrogen compounds (mainly ammonia) in the forest plots of the European network ICP Forests (Giordani et al. 2014a). The critical load causing a significant change of the expected composition of epiphytic lichen vegetation occurred at nitrogen deposition = 2.4 kg N $ha^{-1}\ year^{-1}$. Lichen functional groups for eutrophication and/or, more specifically, nitrogen tolerance have been extensively used to assess the critical level and critical load of nitrogen compounds in several forest ecosystems all over the world (Fenn et al. 2008; Geiser et al. 2010; Pinho et al. 2008b, 2011). Interestingly, the results for European plots were in accordance with those of other areas of the world. For example, in conifer forests of the Pacific North West of USA, Geiser et al. (2010) found, for wet deposition, a critical load ranging from 0.7 to 4.4 kg $ha^{-1}\ year^{-1}$, depending on the amount of precipitation. The concept of critical load for lichen communities has been also applied to other sources of disturbance. Giordani et al. (2014b) established the cattle critical load in alpine pasturelands, in terms of Adult Cattle Units (ACU) per hectare per year. These authors showed that the relative frequency of oligotrophic epilithic lichen species significantly decreased as ACU increased. The cattle critical load was set for ACU = 0.12 ACU $ha^{-1}\ year^{-1}$.

2.4 Open Questions

Despite the relevant number of researches in the field of lichen biomonitoring carried out in the last 25 years, there are still open points which have not yet been sufficiently addressed. In some cases, these questions are far from being minor and responses are urged in order to make the interpretation of lichen biomonitoring more robust. Among the others, the insufficient knowledge on the timescale of the response of epiphytic lichen communities to disturbances and the lacking of an adequate integration about lichen biomonitoring and other standard biomonitoring techniques particularly call for more attention by researchers and stakeholders. According to its technical definition, monitoring is the collection and analysis of repeated observations or measurements to evaluate changes in condition and progress toward a management objective (Elzinga et al. 2001). Thus, also for lichen biomonitoring studies, an accurate knowledge on the temporal variation of the observed phenomenon is crucial for getting reliable results. However, despite the huge literature on physiological effects of disturbances on lichen thalli, the time span between the disturbance (e.g., pollution) and the alteration on epiphytic lichen communities in terms of species loss, recover, or changes in species composition has not been fully explored. Such effects result from complex interactions between temporal trends of limiting factors (e.g., phytotoxic gases) and natural dynamics of the communities, which include re-colonization processes driven by dispersal, substrate availability, establishment of propagules, and intra- and interspecific competition (Werth et al. 2006).

Data from long observation periods showed clear trends in lichen diversity (Lisowska 2011). The diversity of epiphytic species in London has continued to increase from the 1970s to 2004 as a response to decreasing NO_x atmospheric concentrations (Hawksworth and Rose 1970; Davies 2007). In the urban area of Turin (N-Italy), contrasting trends of the numbers of both lichen species presence and abundance were observed over a period of 200 years as a result of changing pollution scenarios (Isocrono et al. 2007). A dramatic species loss was detected in seminatural broadleaved forests in northwestern Germany from the nineteenth century to date (Hauck et al. 2013). Up to 70 % of the species became rarer during the 100- to 150-year long observation period, and an extinction rate of 28 % was estimated.

Against these robust evidences of long-term variations, information on short-term trends of lichen diversity and composition are surprisingly scarce and urge more detailed researches. Loppi et al. (2004) carried out 5 repeated surveys of lichen from 1993 to 2000 with time spans ranging from 1 to 3 years. They showed that despite their slow growth rate, lichens respond rapidly to decreasing concentrations of air pollutants, allowing annual changes to be detected. Total species richness increased from 1993 to 1999 and then decreased again in 2000, while the mean number of species per station increased from 1993 to 1999 and remained constant in 2000. The β-diversity decreased linearly from 1993 to 1999, indicating that sampling stations became floristically more similar in time.

From an applicative perspective, when the aim is the evaluation of change between subsequent measurements, there are several implications related to the statistical analysis for detecting changes, which should carefully considered (EN 16413 2014). Among them, one should decide whether to make the sampling units temporary or permanent. When sampling units are temporary, the random sampling procedure is carried out independently at each sampling period (Elzinga et al. 2001). The principal advantage of using permanent instead of temporary sampling units is that the statistical tests for detecting change from one time period to the next in permanent sampling units are much more powerful than the tests used on temporary sampling units.

References

AFNOR (2008) Biosurveillance de l'environnement—Détermination d'un indice biologique de lichens épiphytes (IBLE). NF X43-903

Asta J, Erhardt W, Ferretti M, Fornasier F, Kirschbaum U, Nimis PL, Purvis OW, Pirintsos S, van Scheidegger C, Haluwyn C, Wirth V (2002) Mapping lichen diversity as an indicator of environmental quality. In: Nimis PL, Scheidegger C, Wolseley PA (eds) Monitoring with lichens—monitoring lichens. Kluwer Academic Publishers, Dordrecht, pp 273–279

Baselga A (2010) Partitioning the turnover and nestedness components of beta diversity. Glob Ecol Biogeogr 19:134–143

Bates JW (1992) Influence of chemical and site factors on Quercus and Fraxinus epiphytes at Loch Sunart, western Scotland: a multivariate analysis. J Ecol 80:163–179

Belinchón R, Martínez I, Escudero A, Aragón G, Valladares F (2007) Edge effects on epiphytic communities in a Mediterranean Quercus Pyrenaica forest. J Veg Sci 18:81–90

Bennett JP, Wetmore CM (1999) Changes in element contents of selected lichens over 11 years innorthern Minnesota, USA. Environ Exp Bot 41:75–82

Bergamini A, Scheidegger C, Stofer S, Carbalho P, Davey S, Dietrich M, Dubs F, Farkas E, Groner U, Kärkäinen K, Keller C, Lökös L, Lommi S, Máguas C, Mitchell R, Pinho P, Rico VJ, Aragón G, Truscott A-M, Wolseley P, Watt A (2005) Performance of macrolichens and lichen genera as indicators of lichen species richness and composition. Conserv Biol 19:1051–1062

Boudreault C, Bergeron Y, Drapeau P, Mascarúa López L (2008) Edge effects on epiphytic lichens in remnant stands of managed landscapes in the eastern boreal forest of Canada. Forest Ecol Manage 255:1461–1471

Brunialti G, Giordani P (2003) Variability of lichen diversity in a climatically heterogeneous area (Liguria, NW Italy). The Lichenologist 35:55–69

Brunialti G, Giordani P, Isocrono D, Loppi S (2002) Evaluation of data quality in lichen biomonitoring studies: the Italian experience. Environ Monit Assess 75:271–280

Brunialti G, Giordani P, Ferretti M (2004) Discriminating between the good and the bad: quality assurance is central in biomonitoring studies. Environ Monit 20:443–464

Brunialti G, Frati L, Incerti G, Rizzi G, Vinci M, Giordani P (2010a) Lichen biomonitoring of air pollution: issues for applications in complex environments. In: Romano GC, Conti AG (eds) Air quality in the 21st century. Nova Science Publishers Inc., New York

Brunialti G, Frati L, Aleffi M, Marignani M, Rosati L, Burrascano S, Ravera S (2010b) Lichens and bryophytes as indicators of old-growth features in Mediterranean forests. Plant Biosyst 114(1):221–233

Brunialti G, Frati L, Cristofolini F, Chiarucci A, Giordani P, Loppi S, Benesperi R, Cristofori A, Critelli P, Di Capua E, Genovesi V, Gottardini E, Innocenti G, Munzi S, Paoli L, Pisani T, Ravera S, Ferretti M (2012a) Can we compare lichen diversity data? A test with skilled teams. Ecol Ind 23:509–516

Brunialti G, Frati L, Loppi S (2012b) Fragmentation of Mediterranean oak forests affects the diversity of epiphytic lichens. Nova Hedwig 96(1–2):265–278

Cácares MES, Lücking R, Rambold G (2007) Photophyte specificity and environmental parameters versus stochasticity as determinants for species composition of corticolous crustose lichen communities in the Atlantic rain forest of north-eastern Brazil. Mycol Prog 6:117–136

Cape JN, van der Eerden LJ, Sheppard LJ, Leith ID, Sutton MA (2009) Evidence for changing the critical level for ammonia. Environ Pollut 157:1033–1037

Caruso A, Thor G (2007) Importance of different tree fractions for epiphytic lichen diversity on *Picea abies* and *Populus tremula* in mature managed boreonemoral Swedish forests. Scand J Forest Res 22:219–230

Carvalho JC, Cardoso P, Gomes P (2012) Determining the relative roles of species replacement and species richness differences in generating beta-diversity patterns. Glob Ecol Biogeogr 21:760–771

Castello M, Skert N (2005) Evaluation of lichen diversity as an indicator of environmental quality in the North Adriatic submediterranean region. Sci Total Environ 336:201–214

Casanovas P, Lynch HJ, Fagan WF (2014) Using citizen science to estimate lichen diversity.Biol Cons 171:1–8

Cislaghi C, Nimis PL (1997) Lichens, air pollution and lung cancer. Nature 387:463–464

Cline SP, Burkman WG (1989) The role of quality assurance in ecological programs. In: Bucher JB, Bucher-Wallin J (eds) Air pollution and forest decline. Birmensdorf, IUFRO, p 361

Cochran WG (1977) Sampling techniques, 3rd edn. Wiley, New York

Collins SL, Glenn SM (1997) Effects of organismal and distance scaling on analysis of species distribution and abundance. Ecol Appl 7:543–551

Connell JH (1978) Diversity in tropical rain forests and coral reefs. Science 199:1302–1310

Coppins AM, Coppins BJ (2002) Indices of ecological continuity for woodland epiphytic lichen habitats in the British Isles. British Lichen Society, London

Cornelissen JHC, Callaghan TV, Alatalo JM, Michelsen A, Graglia E, Hartley AE, Hik DS, Hobbie SE, Press MC, Robinson CH, Henry GHR, Shaver GR, Phoenix GK, Gwynn Jones D, Jonasson S, Chapin FS III, Molau U, Neill C, Lee JA, Melillo JM, Sveinbjörnsson B, Aerts R (2001) Global change and arctic ecosystems: is lichen decline a function of increases in vascular plant biomass? J Ecol 89:984–994

Cristofolini F, Brunialti G, Giordani P, Nascimbene J, Cristofori A, Gottardini E, Frati L, Matos P, Batič F, Caporale S, Fornasier MF, Marmor L, Merinero S, Nuñez Zapata J, Tõrra T, Wolseley P, Ferretti M (2014) Towards the adoption of an international standard for biomonitoring with lichens—consistency of assessment performed by experts from six European countries. Ecol Ind 45(2014):63–67

Crumbling D (2002) In search of representativeness: evolving the environmental data quality model. Qual Assur 9:179–190

Davies L, Bates JW, Bell JNB, James PW, Purvis WO (2007) Diversity and sensitivity of epiphytes to oxides of nitrogen in London. Environ Pollut 146:299–310

Davies L, Bell JNB, Bone J, Head M, Howard C, Hobbs SJ, Jones DT, Power SA, Rose N, Ryder C, Seed L, Stevens G, Toumi R, Voulvoulis N, White PCL (2011) Open air laboratories (OPAL): a community driven research programme. Environ Pollut 159(9): 2203–2210

de Hauck M, Bruyn U, Leuschner C (2013) Dramatic diversity losses in epiphytic lichens in temperate broad-leaved forests during the last 150 years. Biol Conserv 157:136–145

Dufrêne M, Legendre P (1997) Species assemblages and indicator species: the need for a flexible asymmetrical approach. Ecol Monogr 67:345–366

Dungan JL, Perry JN, Dale MRT, Legendre P, Citron-Pousty S, Fortin MJ et al (2002) A balanced view of scale in spatial statistical analysis. Ecography 25:626–640

Ellis CJ (2012) Lichen epiphyte diversity: a species, community and trait-based review. Perspect Plant Ecol, Evol Syst 14:131–152

Ellis CJ, Coppins BJ (2007) Reproductive strategy and the compositional dynamics of crustose lichen communities on aspen (*Populus tremula* L.) in Scotland. The Lichenologist 39:377–391

Elzinga CL, Salzer DW, Willoughby JW, Gibbs PJ (2001) Monitoring plant and animal populations. Blackwell Science, Malden 360 pp

EN 16413 (2014) Ambient air—biomonitoring with lichens—assessing epiphytic lichen diversity. Comité Européen de Normalisation, 33 pp

Esseen P-A, Renhorn K-E (1998) Edge effects on an epiphytic lichen in fragmented forests. Conserv Biol 12:1307–1317

Fenn ME, Jovan S, Yuan F, Geiser L, Meixner T, Gimeno BS (2008) Empirical and simulated critical loads for nitrogen deposition in California mixed conifer forests. Environ Pollut 155:492–511

Ferretti M (1998) Potential and limitations of visual indices of tree condition. Chemosphere 36:1031

Ferretti M (2009) Quality assurance in ecological monitoring—towards a unifying perspective. J Environ Monit 11:726–729

Ferretti M (2011) Quality assurance: a vital need in ecological monitoring. CAB Rev 6(2011). http://www.cabi.org/cabreviews

Ferretti M, Erhardt W (2002) Key issues in designing biomonitoring programmes. Monitoring scenarios, sampling strategies and quality assurance. In: Nimis PL, Scheidegger C, Wolseley P (eds) Monitoring with lichens-monitoring lichens. Kluwer Academic Publishers, Dordrecht, pp 111–139

Ferretti M, Brambilla E, Brunialti G, Fornasier F, Mazzali C, Giordani P, Nimis PL (2004) Reliability of different sampling densities for estimating and mapping lichen diversity as a bioindicator of environmental quality. Environ Pollut 127(2):249–256

Fischer R, Granke O, Chirici G, Meyer P, Seidling W, Stofer S, Corona P, Marchetti M, Travaglini D (2009) Background, main results and conclusions from a test phase for biodiversity assessments on intensive forest monitoring plots in Europe. iForest 67–74

Francini A, Pellegrini E, Lorenzini G, Nali C (2009) Non-sampling error in ozone biomonitoring: the role of operator training. J Environ Monit 11:736–744

Frati L, Brunialti G (2006) Long-term biomonitoring with lichens: comparing data from different sampling procedures. Environ Monit Assess 119:391–404

Frati L, Santoni S, Nicolardi V, Gaggi C, Brunialti G, Guttova A, Gaudino S, Pati A, Pirintsos SA, Loppi S (2007) Lichen biomonitoring of ammonia emission and nitrogen deposition around a pig stockfarm. Environ Pollut 146:311–316

Fritz O, Brunet J, Caldiz M (2009) Interacting effects of tree characteristics on the occurrence of rare epiphytes in a Swedish boreal forest. Bryologist 112:488–505

Gaston KJ (2000) Global patterns in biodiversity. Nature 405:220–227

Gauslaa Y (1995) The Lobarion, an epiphytic lichen community of ancient forests threatened by acid rain. The Lichenologist 27:59–76

Geiser LH, Neitlich PN (2007) Air pollution and climate gradients in western Oregon and Washington indicated by epiphytic macrolichens. Environ Pollut 145:203–218

Geiser LH, Jovan SE, Glavich DA, Porter MK (2010) Lichen-based critical loads for atmospheric nitrogen deposition in Western Oregon and Washington Forests, USA. Environ Pollut 158:2412–2421

Gertner G, Köhl M (1992) An assessment of some non-sampling errors in a national survey using an error budget. Forest Sci 41:758

Gertner G et al (2002) Mapping and uncertainty of predictions based on multiple primary variables from joint co-simulation with Landsat TM image and polynomial regression. Remote Sens Environ 83:498

Giordani P (2004) Licheni epifiti come biomonitors dell'alterazione ambientale. Influenza delle variabili ecologiche sulla diversità lichenica. Tesi di dottorato. Università di Trieste

Giordani P (2006) Variables influencing the distribution of epiphytic lichens in heterogeneous areas: a case study for Liguria, NW Italy. J Veg Sci 17:195–206

Giordani P (2007) Is the diversity of epiphytic lichens a reliable indicator of air pollution? A case study from Italy. Environ Pollut 146:317–323

Giordani P, Incerti G (2008) The influence of climate on the distribution of lichens: a case study in a borderline area (Liguria, NW Italy). Plant Ecol 195:257–272

Giordani P, Brunialti G, Modenesi P (2001) Applicability of the lichen biodiversity method (L.B.) in a mediterranean area (Liguria, NW Italy). Cryptogam Mycol. 22(3):193–208

Giordani P, Brunialti G, Alleteo D (2002) Effects of atmospheric pollution on lichen biodiversity (LB) in a Mediterranean region (Liguria, NW-Italy). Environ Pollut 118:53–64

Giordani P, Brunialti G, Benesperi R, Rizzi G, Frati L, Modenesi P (2009) Rapid biodiversity assessment in lichen biomonitoring surveys: implications for quality assurance. J Environ Monit 11:730–735

Giordani P, Incerti G, Rizzi G, Ginaldi F, Viglione S, Rellini I, Brunialti G, Malaspina P, Modenesi P (2010) Land use intensity drive the local variation of lichen diversity in Mediterranean ecosystems sensitive to desertification. In: Thomas H, Nash III et al (eds) Biology of lichens—symbiosis, ecology, environment monitoring, systematics, cyber applications. J. Cramer in der Gebrüder Borntraeger Verlagsbuchhandlung, Stuttgart. Biblio Lichenol 105:139–148

Giordani P, Brunialti G, Bacaro G, Nascimbene J (2012) Functional traits of epiphytic lichens as potential indicators of environmental conditions in forest ecosystems. Ecol Ind 18:413–420

Giordani P, Brunialti G, Frati L, Incerti G, Ianesch L, Vallone E, Bacaro G, Maccherini S (2013) Spatial scales of variation in lichens: implications for sampling design in biomonitoring surveys. Environ Monit Assess 185:1567–1576

Giordani P, Calatayud V, Stofer S, Seidling W, Granke O, Fischer R (2014a) Detecting the nitrogen critical loads on European forests by means of epiphytic lichens. A signal-to noise evaluation. Forest Ecol Manage 311:29–40

Giordani P, Matteucci E, Redana M, Ferrarese A, Isocrono D (2014b) Unsustainable cattle load in alpine pastures alters the diversity and the composition of lichen functional groups for nitrogen requirement. Fungal Ecol 9:69–72

Gottardini E, Cristofolini F, Cristofori A, Vannini A, Ferretti M (2009) Sampling bias and sampling errors in pollen counting in aerobiological monitoring in Italy. J Environ Monit 11:751–755

Goward T, Spribille T (2005) Lichenological evidence for the recognition of inland rain forests in western North America. J Biogeogr 32:1209–1219

Gunnar JB, Moen J (1998) Patterns in species associations in plant communities: the importance of scale. J Veg Sci 9:327–332

Gustafsson L, Eriksson I (1995) Factors of importance for the epiphytic vegetation of aspen *Populus tremula* with special emphasis on bark chemistry and soil chemistry. J Appl Ecol 32:412–424

Hall AR, Miller AD, Leggett HC, Roxburgh SH, Buckling A, Shea K (2012) Diversity-disturbance relationships: frequency and intensity interact. Biol Lett 8:768–771

Hauck M, Spribille T (2005) The significance of precipitation and substrate chemistry for epiphytic lichen diversity in spruce-fir forests of the Salish Mountains, Montana. Flora 200:547–562

Hawksworth DL, Rose L (1970) Qualitative scale for estimating sulphur dioxide air pollution in England and Wales using epiphytic lichens. Nature 227:145–148

Holt EA, Mccune B, Neitlich P (2008) Spatial scale of GIS-derived categorical variables affects their ability to separate sites by community composition. Appl Veg Sci 11:421–430

Humphrey JW, Newton AC, Peace AJ, Holden E (2000) The importance of conifer plantations in northern Britain as a habitat for native fungi. Biol Conserv 96:241–252

Humphrey JW, Davey S, Peace AJ, Ferris R, Harding K (2002) Lichens and bryophyte communities of planted and semi-natural forests in Britain: the influence of site type, stand structure and deadwood. Biol Conserv 107 (2002):165–180

Hyvärinen M, Halonen P, Kauppi M (1992) Influence of stand age and structure on the epiphytic lichen vegetation in the middle-boreal forests of Finland. The Lichenologist 24:165–180

Isocrono D, Matteucci E, Ferrarese A, Pensi E, Piervittori R (2007) Lichen colonization inthe city of Turin (N Italy) based on current and historical data. Environ Pollut 145:258–265

Johansson P, Rydin H, Thor G (2007) Tree age relationships with epiphytic lichen diversity and lichen life history traits on ash in southern Sweden. Écoscience 14:81–91

Jovan S, McCune B (2004) Regional variation in epiphytic macrolichen communities in northern and central California forests. The Bryologist 107:328–339

Jüriado I, Liira J, Paal J, Suija A (2009) Tree and stand level variables influencing diversity of lichens on temperate broad-leaved trees in boreo-nemoral flood-plain forests. Biodivers Conserv 18:105–125

Kermit T, Gauslaa Y (2001) The vertical gradient of bark pH of twigs and macrolichens in a Picea abies canopy not affected by acid rain. The Lichenologist 33:353–359

Köhl M, Innes JL, Kaufmann E (1994) Reliability of differing densities of sample grids used for the monitoring of forest condition in Europe. Environ Monit Assess 29:201–220

Köhl M, Traub B, Paivinen R (2000) Harmonisation and standardisation in multi- national environmental statistics—mission impossible? Environ Monit Assess 63:361

Kovacs M (1992) Biological indicators in environmental protection. Horwood, New York

Kricke R, Loppi S (2002) Bioindication: the I.A. P. approach. In: Nimis PL, Scheidegger C, Wolseley PA (eds) Monitoring with lichens—monitoring lichens. Kluwer Academic Publishers, Dordrecht, pp 21–37

Kuusinen M (1994) Epiphytic lichen diversity on Salix caprea in old-growth southern and middle boreal forests of Finland. Ann Botanici Fennici 31:77–92

Kuusinen M (1996) Epiphyte flora and diversity on basal trunks of six old-growth forest tree species in southern and middle boreal Finland. The Lichenologist 28:443–463

Laskowsky SL, Kutz FW (1998) Environmental data in decision making in EPA regional offices. Environ Monit Assess 51:15

Legg C, Nagy L (2006) Why most conservation monitoring is, but need not be, a waste of time. J Environ Manage 78:194–199

Leppik E, Jüriado I, Liira J (2011) Changes in stand structure due to the cessation of traditional land use in wooded meadows impoverish epiphytic lichen communities. The Lichenologist 43:257–274

Lévêque C, Mounolou J-C (2003) Biodiversity. Wiley, Chichester

Levin SA (1992) The problem of pattern and scale in ecology. Ecology 73:1943–1967

Lewis JEJ, Ellis CJ (2010) Taxon-compared with trait-based analysis of epiphytes, and the role of tree species and tree age in community composition. Plant Ecol Divers 3:203–210

Lisowska M (2011) Lichen recolonisation in an urban-industrial area of southern Poland as a result of air quality improvement. Environ Monit Assess 179:177–190

Loppi S De, Dominicis V (1996) Lichens as long-term biomonitors of air quality in central Italy. Acta Bot Neerl 45:563–570

Loppi S, Frati L (2004) Influence of tree substrates on the diversity of epiphytic lichens: comparison between Tilia platyphyllos and Quercus ilex (central Italy). Bryologist 107:340–344

Loppi S, Giordani P, Brunialti G, Isocrono D, Piervittori R (2002a) Identifying deviation from naturality of lichen diversity in heterogeneous situations. In: Nimis PL, Scheidegger C, Wolseley P (eds) Monitoring with lichens—monitoring lichens. Kluwer Academic Publishers, Dordrecht, pp 281–284

Loppi S, Giordani P, Brunialti G, Isocrono D, Piervittori R (2002b) A new scale of interpretation of lichen biodiversity in the Tyrrhenian side of Italy. Bibl Lichenol 82:235–243

Loppi S, Frati L, Paoli L, Bigagli V, Rossetti C, Bruscoli C, Corsini A (2004) Biodiversity of epiphytic lichens and heavy metal contents of Flavoparmelia caperata thalli as indicators of temporal variations of air pollution in the town of Montecatini Terme (central Italy). Sci Total Environ 326:113–122

Loppi S, Pirintsos SA, De Dominicis V (1997) Analysis of the distribution of epiphytic lichenson Quercus pubescens along an altitudinal gradient in a Mediterranean area (Tuscany, CentralItaly). Isr J Plant Sci 45:53–58

Loppi S, Bonini I, De Dominicis V (1999) Epiphytic lichens and bryophytes of forestecosystems in Tuscany (C. Italy). Cryptogam Mycol 20:127–135

Marchetto A, Mosello R, Tartari G, Tornimbeni O, Derome J, Derome K, Sorsa P, Konig N, Clarke N, Ulrich E, Kowalska A (2009) Influence of QA/QC procedures on non-sampling error in deposition monitoring in forests. J Environ Monit 11:745–750

McCune B, Lesica P (1992) The trade-off between species capture and quantitative accuracy in ecological inventory of lichens and bryophytes in forests in Montana. Bryologist 95:296

Mccune B, Mefford MJ (2004) Nonparametric multiplicative habitat modelling. MjM Software Design, Gleneden Beach

McCune B, Dey JP, Peck JE, Cassell D, Heiman K, Will-Wolf S, Neitlich PN (1997) Repeatability of community data: species richness versus gradient scores in large-scale lichen studies. Bryologist 100:40–46

Moe B, Botnen A (1997) A quantitative study of the epiphytic vegetation on pollarded trunks of Fraxinus excelsior at Havrå, Osterøy, western Norway. Plant Ecol 129:157–177

Moog O, Schmidt-Kloiber A, Ofenböck T, Gerritsen J (2004) Does the ecoregion approach support the typological demands of the EU 'water framework directive'? Hydrobiologia 516:21–33

Nascimbene J, Brunialti G, Ravera S, Frati L, Caniglia G (2010) Testing *Lobaria pulmonaria* (L.) Hoffm. as an indicator of lichen conservation importance of Italian forests. Ecol Ind 10:353–360

Nascimbene J, Benesperi R, Brunialti G, Catalano I, Dalle Vedove M, Grillo M, Isocrono D, Matteucci E, Potenza G, Puntillo D, Puntillo M, Ravera S, Rizzi G, Giordani P (2013a) Patterns and drivers of b-diversity and similarity of *Lobaria pulmonaria* communities in Italian forests. J Ecol 101:493–505

Nascimbene J, Thor G, Nimis PL (2013b) Effects of forest management on epiphytic lichens in temperate deciduous forests of Europe—a review. Forest Ecol Manage 298:27–38

Nilsson SG, Arup U, Baranowski R, Ekman S (1995) Tree-dependent lichens and beetles as indicators in conservation forests. Conserv Biol 9:1208–1215

Nimis PL (1999) Linee-guida per la bioindicazione degli effetti dell'inquinamento tramite la biodiversità dei licheni epifiti. In: Piccini C, Salvati S (eds) Proceedings workshop "Biomonitoraggio della qualità dell'aria sul territorio Nazionale". Roma, 26–27 November 1998. pp 267–277

Nimis PL, Martellos S (2001) Testing the predictivity of ecological indicator values. A comparison of real and "virtual" relevés of lichen vegetation. Plant Ecol 157:165–172

Nimis PL, Martellos S (2002) ITALIC—the information system on Italian lichens. In: Llimona X, Lumbsch HT, Ott S (eds) Progress and problems in lichenology at the turn of the millennium. Bibl Lichenologica 82:271–283

Nimis PL, Castello M, Perotti M (1990) Lichens as biomonitors of sulphur dioxide pollution in La Spezia (Northern Italy). The Lichenologist 22:333–344

Nimis PL, Lazzarin A, Lazzarin G, Gasparo D (1991) Lichens as bioindicators of air pollution by SO_2 in the Veneto region (NE Italy). Studia Geobotanica 11:3–76

Nimis PL, Scheidegger C, Wolseley PA (2002) Monitoring with lichens-monitoring lichens: an introduction. In: Nimis PL, Scheidegger C, Wolseley PA (eds) Monitoring with lichens-monitoring lichens. Nato science program-IV, vol VII. Kluwer Academic Publisher, The Netherlands, pp 408

Nimis PL, Riccamboni R, Martellos S (2012) Identification keys on mobile devices: The Dryades experience. Plant Biosystems 146(4):783–788

Oliver I, Beattie AJ, York A (1998) Spatial fidelity of plant, vertebrate, and invertebrate assemblages in multiple-use forest in Eastern Australia. Conserv Biol 12:822–835

Olsen AR et al (1999) Statistical issues for monitoring ecological and natural resources in the United States. Environ Monit Assess 54:1

Palmer MW (1990) Spatial scale and patterns of vegetation, flora and species richness in hardwood forests of the north Carolina Piedmont. Coenoses 5:89–96

Paoli L, Guttova A, Loppi S (2006) Assessment of environmental quality by the diversity of epiphytic lichens in a semi-arid Mediterranean area (Val Basento, South Italy). Biologia 61:353–359

Perry JN, Liebhold AM, Rosenberg MS, Dungan J, Miriti M, Jakomulska A, Citron-Pousty S (2002) Illustrations and guidelines for selecting statistical methods for quantifying spatial pattern in ecological data. Ecography 25:578–600

Pharo EJ, Beattie AJ, Pressey RL (2000) Effectiveness of using vascular plants to select reserves for bryophytes and lichens. Biol Conserv 96:371–378

Pinho P, Augusto S, Branquinho C, Bio A, Pereira MJ, Soares A, Catarino F (2004) Mapping lichen diversity as a first step for air quality assessment. J Atmos Chem 49:377–389

Pinho P, Augusto S, Máguas C, Pereira MJ, Soares A, Branquinho C (2008a) Impact of neighbourhood land-cover in epiphytic lichen diversity: analysis of multiple factors working at different spatial scales. Environ Pollut 151:414–422

Pinho P, Augusto S, Martins-Loucão M, Pereira MJ, Soares A, Máguas C, Branquinho C (2008b) Causes of change in nitrophytic and oligotrophic lichen species in a Mediterranean climate: impact of land cover and atmospheric pollutants. Environ Pollut 154:380–389

Pinho P, Dias T, Cruz C, Tang YS, Sutton MA, Martins-Loucão M-A, Máguas C, Branquinho C (2011) Using lichen functional diversity to assess the effects of atmospheric ammonia in Mediterranean woodlands. J Appl Ecol 48:1107–1116

Pinho P, Llop E, Ribeiro MC, Cruz C, Soares A, Pereira MJ, Branquinho C (2014) Tools for determining critical levels of atmospheric ammonia under the influence of multiple disturbances. Environ Pollut 188 (2014):88–93

Podani J, Csányi B (2010) Detecting indicator species: some extensions of the IndVal measure. Ecol Ind 10:1119–1124

Podani J, Schmera D (2011) A new conceptual and methodological frame-work for exploring and explaining pattern in presence–absence data. Oikos 120:1625–1638

Poikolainen J, Lippo H, Hongisto M, Kubin E, Mikkola K, Lindgren M (1998) On the abundance of epiphytic green algae in relation to the nitrogen concentration of biomonitors and nitrogen deposition in Finland. Environ Pollut 102:85–92

Posthumus AC (1988) Critical levels for effects of ammonia and ammonium. In: Proceedings of the bad Harzburg workshop. UBA, Berlin, pp 117–127

Purvis A, Hector A (2000) Getting the measure of biodiversity. Nature 405:212–219

Qian H, Klinka K, Song X (1999) Cryptogams on decaying wood in old-growth forests of southern coastal British Columbia. J Veg Sci 10:883–894

Ranius T, Johansson P, Berg N, Niklasson M (2008) The influence of tree age and microhabitat quality on the occurrence of crustose lichnes associated with old oaks. J Veg Sci 19:653–662

Ravera S, Brunialti G (2013) Epiphytic lichens of a poorly explored National Park: Is the probabilistic sampling effective to assess the occurrence of species of conservation concern? Plant Biosyst 147:115–124

Ribeiro MC, Pinho P, Llop E, Branquinho C, Sousa AJ, Pereira MJ (2013) Multivariate geostatistical methods for analysis of relationships between ecological indicators and environmental factors at multiple spatial scales. Ecol Ind 29:339–347

Richardson DHS (1993) Pollution monitoring with lichens. Richmond Publishers, Slough

Rose F (1976) Lichenological indicators of age and environmental continuity in woodlands. In: Brown DH, Hawksworth DL, Bailey RH (eds) Lichenology: progress and problems. Academic Press, London, pp 279–307

Rose F (1993) Ancient British woodlands and their epiphytes. Brit Wildl 5:83–93

Sætersdal M, Gjerde I, Blom HH (2005) Indicator species and the problem of spatial inconsistency in nestedness patterns. Biol Conserv 122:305–316

Santi E, Maccherini S, Rocchini D, Bonini I, Brunialti G, Favilli L, Perini C, Pezzo F, Piazzini S, Rota E, Salerni E, Chiarucci A (2010) Simple to sample: vascular plants as surrogate group in a nature reserve. J Nat Conserv 18:2–11

Sastre P, Lobo JM (2009) Taxonomist survey biases and the unveiling of biodiversity patterns. Biol Conserv 142:462–467

Seed L, Wolseley P, Gosling L, Davies L, Power SA (2013) Modelling relationships betweenlichen bioindicators, air quality and climate on a national scale: Results from the UK OPAL airsurvey. Environ Pollut 132:437–447

Stofer S, Catalayud V, Ferretti M, Fischer R, Giordani P, Keller C, Stapper N, Scheidegger C (2003) Epiphytic lichen monitoring within the EU/ICP forests biodiversity test-phase on level II plots. http://www.forest-biota.org

Stofer S, Calatayud V, Giordani P, Neville P (2012) Assessment of epiphytic lichen diversity. Manual Part VII.2, xx pp. In: ICP Forests (2010): Manual on methods and criteria for harmonized sampling, assessment, monitoring and analysis of the effects of air pollution on forests. UNECE ICP Forests Programme Co-ordinating Centre, Hamburg. ISBN: 978-3-926301-03-1. http://www.icp-forests.org/Manual.htm

Svoboda D, Peksa O, Vesela J (2010) Epiphytic lichen diversity in central European oakforests: assessment of the effects of natural environmental factors and human influences.Environ Pollut 158:812–819

Tarhanen S, Poikolainen J, Holopainen T, Oksanen J (2000) Severe photobiont injuries of lichens are strongly associated with air pollution. New Phytol 147:579–590

Tison J, Park Y-S, Coste M, Wasson JG, Rimet F, Ector L, Delmas F (2007) Predicting diatom reference communities at the French hydrosystem scale: a first step towards the definition of the good ecological status. Ecol Model 203:99–108

Uliczka H, Angelstam P (1999) Occurrence of epiphytic macrolichens in relation to tree species and age in managed Boreal forest. Ecography 22:396–405

Underwood AJ, Chapman MG, Connell SD (2000) Observations in ecology: you can't make progress on processes without understanding the patterns. J Exp Mar Biol Ecol 250:97–115

Van Dobben HF, DeBakker AY (1996) Re-mapping epiphytic lichen biodiversity in the Netherlands: effects of decreasing SO_2 and increasing NH3. Acta Bot Neerl 45:55–71

VDI 3957 Part 13 (2005) Biological measurement procedures for determining and evaluating the effects of ambient air pollutants by means of lichens (bioindication)—mapping the diversity of epiphytic lichens as an indicator of air quality. Beuth-Verlag, Berlin

Vitt DH, Yenhung L, Belland RJ (1995) Patterns of bryophyte diversity in peatlands of western Canada. Bryologist 98:218–227

Wasson JG, Chandesris A, Pella H, Blanc L (2002) Typology and reference conditions for surface water bodies in France: the hydro-ecoregion approach. In: Ruoppa M, Karttunen K (eds) Typology and classification of lakes and rivers, symposium on typology and classification of lakes and rivers. Helsinki, Finland, 24–26 Oct 2002. Tema Nord 566, pp 37–41

Wagner G (1995) Basic approaches and methods for quality assurance and quality control in samplecollection and storage for environmental monitoring, Sci Total Environ 264(3)

Werth S, Wagner HH, Gugerli F, Holderegger R, Csencsics D, Kalwij JM, Scheidegger C (2006) Quantifying dispersal and establishment limitation in a population of an epiphytic lichen. Ecology 87:2037–2046

Whittaker RH (1960) Vegetation of the Siskiyou mountains, Oregon and California. Ecol Monogr 30:279–338

Whittet R, Ellis CJ (2013) Critical tests for lichen indicators of woodland ecological continuity. Biol Conserv 168:19–23

Wilkie L, Cassis G, Gray M (2003) A quality control protocol for terrestrial invertebrate biodiversity assessment. Biodivers Conserv 12:121

Wirth V (2010) Ökologische Zeigerwerte von Flechten—erweiterte und aktualisierte Fassung. Herzogia 23:229–248

Wolseley PA, Stofer S, Mitchell A-M, Vanbergen A, Chimonides J, Scheidegger C (2006) Variation of lichen communities with land use in Aberdeenshire, UK. The Lichenologist 38:307–322

Yoccoz NG, Nichols JD, Boulinier T (2001) Monitoring of biological diversity in space and time. Trends Ecol Evol 16:446

Lichens and Particulate Matter: Inter-relations and Biomonitoring with Lichens

3

J. Garty and R.L. Garty-Spitz

Contents

J. Garty (✉)
Department of Molecular Biology and Ecology of Plants, Tel Aviv University, 69978 Tel Aviv, Israel
e-mail: garty@post.tau.ac.il

R.L. Garty-Spitz
Habanim Street 9, 45905 Ganei Am, Israel

Abstract

In the last three decades, considerable attention was paid to the inter-relations of lichens and particulate matter. This chapter highlights the determinant factors of the entrapment of airborne particles by the lichen, e.g., the growth form of the thallus. The soil which provides the greater part of the particles adhered-to and/or entrapped-in the thallus should be considered and characterized in biomonitoring studies which traditionally focus on airborne elements derived from anthropogenic activity, with an emphasis on heavy metals and radionuclides. Both visual data and analyses of coefficients of variation (CVs) contribute to knowledge of the nature of particulate matter entrapped by lichens. To date, a considerable number of laboratories studying the elemental content of lichens avoid the pre-treatment of thallial samples. A rough estimation based on a review of the relevant literature revealed that about 75 % of the laboratories used only dry cleaning and do not wash thalli prior to elemental determination. Other laboratories include in their protocol washing, sometimes several consecutive washings, combined with an ultrasonic treatment and/or shaking. This treatment has the potential to remove deposited material of natural and anthropogenic origin and induces the loss of elements. Lichenologists/bryologists need to face the dilemma of pre-treatment. In this context, scientists should determine whether the constituents of lichen are just and only algal and/or cyanobacterial cells and fungal hyphae, and particulates are "aliens." The location of biomonitoring studies is

D.K. Upreti et al. (eds.), *Recent Advances in Lichenology*,
DOI 10.1007/978-81-322-2181-4_3

of crucial importance: the increasing number of annual desert-dust storms especially in the Mediterranean basin and in southern Europe is the result of climate change. The temporal and spatial dimensions of bulk deposition are discussed, with relevance to change observed in biomonitoring studies.

Keywords
Lichens · Particulate matter · Entrapment · Soil · Dust · Biomonitoring

3.1 Introduction

Traditional definitions of lichens adopted for the greater part a biological ("organic") point of view focusing on the main components of the thallus. Writers on the subject defined lichens as stable, mutualistic, symbiotic, perennial, slow-growing associations of heterotrophic fungi (mycobionts), and photosynthetic green algae or cyanobacteria (photobionts). The fungal partner was described as belonging to a wide range of classes and orders, thus comprising an ecological non-systematic group. The greater part of this group consists of ascomycetes, whereas some are basidiomycetes (Gadd 2007). This is the accepted concept of a bipartite lichen. *Hypogymnia physodes*, for example, is a bipartite chlorolichen comprised of a fungus as mycobiont and a unicellular green alga as photobiont (Paul and Hauck 2006). *Nephroma helveticum* and *Leptogium saturnium* also contain a fungus as mycobiont and a *Nostoc* cyanobacterium as photobiont (Hauck et al. 2006) in a bipartite association.

The symbiotic association of one mycobiont and one or two photobionts forms a common thallus. The mycobiont receives fixed carbon produced by the population of photobiont cells. This thallus does not possess roots or a waxy cuticle. As a result of its high surface:volume ratio, its simple anatomy and the absence of a cuticle, lichens accumulate nanoparticulate matter derived from different natural sources, e.g., mineral dust (with a lithospheric signature), or organic dust from soil surface erosion and occasional wildfires, in addition to anthropogenic sources of pollution. Transport of pollutants is either long or short ranged. The factor of deposition, either dry or wet, is of relevance.

The reason for the adaptability of lichens to biomonitoring studies of air pollution is their extraordinary capability to accumulate mineral elements far above their needs, and their occurrence at a large geographical range. Lichens adjust to any terrestrial environment: an estimated percentage of 6 % of the Earth's land surface is covered by lichen-dominated vegetation (Haas and Purvis 2006; Gadd 2007).

Many publications dealt with the utilization of lichens as monitors of trace elements and accumulators of macro-nutrients in varying geographic areas, representing all continents. Lichens exhibit long-term deposition patterns and do not reflect necessarily the short-term deposition measured by most atmospheric scientists. The advantages of biomonitoring with lichens over instrumental monitoring are that lichens accumulate most of the elements of the periodic table, are usable at low expense, do not depend on electricity for their operation, do not need treatment, and are easy to hide, thus discouraging vandalism. Inexpensive biomonitoring systems also provide a great deal of information on the impact of heavy metals on physiological processes.

The greater share of the publications of the last five decades reports on the results of biomonitoring experiments of trace elements, radionuclides and macroelements, including nutrients. In the last three decades, however, considerable attention was paid to inter-relations between lichens and particulate matter. The present chapter deals specifically with the origin and mode of interception and incorporation of particulate matter with an emphasis on metal-containing particles. Of the large number of lichen species, only a few were used as biomonitors of trace-elemental deposition. However, the study of these few species yielded hundreds of publications.

3.2 Efficient Entrapment of Particulate Matter: Determinant Factors

Morphological and structural features of the lichen thallus determine the entrapment of airborne particles, in addition to the relevant features of surface and degree of attachment. The greater adaptability of foliose and fruticose lichens for biomonitoring over and above crustose lichens derives from ease of sampling. Since the interception of atmospheric particulates is dependent on a high ratio of surface to mass (Tretiach et al. 2007; Adamo et al. 2007), foliose lichens accumulated in certain cases more than twice the amount of Cu than fruticose lichens. St. Clair et al. (2002a) related the morphological features of the lichens to environmental conditions to explain this finding. Foliose thalli are continuous in space (within the same plane) and therefore more efficient at the interception of atmospheric deposition. In contrast, fruticose species with their three-dimensional open structure accommodate a greater flow-through of atmospheric deposition and their capability to entrap airborne elements is smaller.

A nationwide interspecies survey confirmed the higher affinity of the foliose *Flavoparmelia caperata* to Cd and Zn detected in a local study performed in northeast Italy by Nimis et al. (2001). The higher affinity of the foliose *Xanthoria parietina* to Al and Fe, on the other hand, did not coincide with nationwide data and was related to local factors and to the characteristics of the species. Previous studies suggest that Al and Fe are deposited for the greater part in the form of large particles. *Xanthoria* is assumed to endure a higher rate of eutrophication than *Flavoparmelia* and to be bound to communities with a strong accumulation of dust. Thus, the high Al and Fe content of *Xanthoria* is related to a study area with widespread deposits of Terra Rossa in northeast Italy, largely made up of Al- and Fe-hydroxides, and subject to strong, dry winds in winter. Physiological differences are thus assumed to account for the interspecies difference of affinity.

A different capability to entrap particles of lichens with a similar growth form, apparent in a study of two foliose lichens *F. caperata* and *Lobaria pulmonaria*, was attributed to ecological characteristics (Loppi and De Dominicis 1996). According to St. Clair et al. (2002b), elemental accumulative patterns are not the same for fruticose species either: for example, the epilithic *Usnea amblyoclada* and the epiphytic *Usnea hirta* present different accumulation patterns. Consequently, it is confirmed that analyses of elemental accumulation involve the consideration of variables other than growth form: the substrate and the inter-relation of growth form and substrate. This is true also for foliose species such as the epilithic *F. caperata* and the epiphytic *Flavopunctelia flaventior*: the growth form variable was found to be significant for K, Ca, Ti, B, Fe, Ni, Cu, Zn, Rb, and Sr, whereas the substrate variable was found to be significant for K, Ti, Mn, Fe, Ni, Rb, and Sr. The accumulation of S, on the other hand, appeared to be independent of both growth form and substrate (St. Clair et al. 2002b).

Foliose lichens such as *Parmelia sulcata* and *X. parietina*, both epiphytic, contained more ^{137}Cs than filamentose or fruticose epiphytic lichens in the same biotope in Macedonia, northern Greece (Sawidis et al. 1997). The authors suggested that the large area of attachment to the substrate of the foliose species may have helped to accumulate more ^{137}Cs from the tree canopy or from stemflow leachates, whereas filamentous or fruticose forms are attached to the substrate either at one or at a few points. For both filamentous and fruticose species, the accumulative capacity was found to accord with a high ratio of surface to biomass, e.g., *Usnea* sp. (Sawidis et al. 2010). Table 3.1 provides data of elemental content of lichens with different growth forms, while Table 3.2 presents the elemental content of lichens with similar growth forms.

Table 3.1 Elemental content of lichens with different growth forms

	Foliose lichens	Fruticose lichens	Elemental content	Remarks according to cited reference
Lawrey and Hale (1981)	*Flavoparmelia baltimorensis* *Xanthoparmelia conspersa*	*Cladonia subtenuis*	Higher in the foliose lichen	The prostrate growth form of the foliose lichens makes for a large, flat, particle-trapping surface. *C. subtenuis* is lesser structured to trap particles
Gough et al. (1988)	*Hypogymnia enteromorpha*	*Usnea* spp.	Higher in the foliose lichen	
Glenn et al. (1995)	*Parmelia subrudecta*	*Anaptychia cilliaris*	Higher in the foliose lichens	
	Parmelia sulcata			
Calvelo et al. (1997)	*Candelariella vitellina*	*Protousnea magellanica*	Higher in the foliose lichens	Findings are related to the greater capability of foliose lichens to entrap particulate matter
	Hypotrachyna brevirhiza			
	Parmelia cunninghamii	*Usnea fastigiata*		
	Physcia adscendens			
Aslan et al. (2004)	*Xanthoparmelia somloensis*	*Ramalina capitata*		Structural factors relevant to metal uptake are wider thallial surface, a thin or nonexistent upper cortex, a wide cellular space and a high cell permeability in the foliose lichens
	Dermatocarpon miniatum	*Pseudevernia furfuracea*		
Bennet and Wetmore (2003)	*Hypogymnia physodes*	*Evernia mesomorpha*	Higher in the foliose lichen	The *Hypogymnia* accumulated more Ca, Cd, Mg, and Mn of atmospheric origin, whereas the *Evernia* accumulated mostly soil elements
Garty et al.(1996a)	*Hypogymnia physodes*	*Usnea hirta*	Higher content of Fe, Zn, Ca, Mg, Cu, and Ni in the foliose lichen	
Käffer et al. (2012)	*Parmotrema tinctorum*	*Teloschistes exilis*	Higher content of Pb, Cr, Cd, and Zn in the foliose lichen	
Bergamaschi et al. (2007)	*Hypogymia physodes*	*Usnea hirta*	Similar accumulation capacity in the four lichens	
	Parmelia sulcata	*Pseudevernia furfuracea*		
Kahraman et al. (2009)	*Cladonia foliacea*	*Ramalina pollinaria*	Higher content of K, Zn, Rb, and Pb in *R. pollinaria*; higher Mn and Sr content in *C. foliacea*	
Bosserman and Hagner (1981)	*Parmelia* spp.	*Usnea* spp.	Higher content of dustfall elements such as P, Si, Al, Fe, and K in *Parmelia* spp. Higher content of Na, Ca, and Mg dominant in precipitation in *Usnea* spp.	
Singh et al. (2013)	Names of spp. not given	Names of spp. not given	The variation in the metal content was insignificant	
Bajpai et al. (2009)	*Foliose lichens*	*Leprose lichens*	Higher arsenic (As) in the foliose and the leprose lichen species than in squamulose and crustose lichen species	
	Phaeophyscia hispidula	*Lepraria lobificans*		
	Parmotrema praesorediosum	Squamulose lichens		
	Crustose lichens	*Peltula euploca*		
	Caloplaca subsoluta	*Phylliscum indicum*		
	Diploschistes candidissimus			

Table 3.2 Elemental content of lichens with similar growth forms

	Foliose lichens	Foliose lichens	Elemental content
Nimis et al. (2001)	*Flavoparmelia caperata*	*Xanthoria parietina*	Higher content of Fe and Al in *X. parietina*. Higher content of Cd and Zn in *F. caperata*
Bartók (1988)	*Xanthoparmelia conspersa*	*Peltigera canina*	Higher elemental content in *X. conspersa*
Upreti and Pandey (1994)	*Umbilicaria decussata*	*Umbilicaria aprina*	Higher content of Pb and Fe in *U. descussata*
Loppi and De Dominicis 1996	*Flavoparmelia caperata*	*Lobaria pulmonaria*	Higher ^{137}Cs in *L. pulmonaria*
Koz et al. (2010)	*Physcia adscendens*	*Flavoparmelia caperata*	Highest content of Cr, Mn, Ni, Zn, Sn, Ba, and Pb in *P. adscendens*; highest content of Cu and Co in *F. caperata*; highest content of Ti in *Xanthoparmelia conspersa*; highest content of Fe in *Parmotrema perlatum*
	Xanthoparmelia conspersa	*Parmotrema perlatum*	
Aslan et al. (2004)	Crustose lichens	Crustose lichens	*R. melanophtalma* accumulated more of K, Ca, Sr, and Ba than *R. crysoleuca*
	Rhizoplaca melanophtalma	*Rhizoplaca cryrsoleuca*	

3.3 Soil Particles Entrapped by Lichens: Apportionment and Characterization

Particulate matter of soil origin is very often adhered-to and/or entrapped-in the lichen thallus (e.g., Garty et al. 1979; Garty and Delarea 1987; Markert 1993; Dongarà et al. 1995; Dongarà and Varrica 1998; Bergamaschi et al. 2004; Giordano et al. 2013). Correlations of mineral elements such as Al and Fe accumulated in lichens are indicative of small entrapped aerosols (Nakajima et al. 2013) and of soil particulates (Doğrul et al. 2012). Particulate deposition can lead to damage in the lichen thallus. Some particulates are chemically inert, whereas others are highly reactive (Purvis and Pawlik-Skowrónska 2008). In the literature referring to lichens as biomonitors of air pollution, different definitions of soil particles are in use. Table 3.3 presents different definitions of particulate matter. An examination of the elemental composition of lichens should consider their dependence on the soil surface. A principal component analysis (PCA) of data obtained in Greenland of elemental concentrations in soil, humus, moss (*Rhacomitrium lanuginosum*), and lichens (*Flavocetraria nivalis*) indicated the soil–dust factor as representing one of three principal components explaining 87 % of the total variation. Dominant constituents of the soil-dust factor were Fe, Al, V, Ni, Cr, Cu, and Pb (Riget et al. 2000).

Data obtained in a monitoring survey on a national scale in The Netherlands were applied for the purpose of source apportionment (Sloof 1995b). A Monte Carlo-assisted factor analysis yielded 10 factors, interpreted in terms of 10 sources. One of these factors strongly depended on concentrations of Th, Sc, Al, Cr, Fe, and Cs, indicative of crustal material. Other significant constituents of this factor were La, Co, As, Se, and V, common trace elements in uncontaminated Dutch topsoils. Kuik and Wolterbeek (1995) indicated Sc, Sm, Ce, La, Th, Al, Fe, Cr, As, Co, and V as important constituents of crustal material accounting for 25 % of the total variance in the greater part of data obtained in 1992. Similarly, Reis et al. (1996) used the lichen *P. sulcata* to set up a database incorporating data for 43 elements, subjected to a Monte Carlo-aided target transform factor analysis (MCATTFA). Of nine factors obtained in this manner, four were interpreted as representing natural sources. Three were further interpreted by means of their Fe/Si concentration ratio, as limestone, basalt, and shale. Another factor, initially identified as anthropogenic, was found to

Table 3.3 Definitions of mineral elements originating in rock and soil and remarks given in biomonitoring studies using lichens

	Definition	Remarks
Loppi et al. (1999a)	Lithogene elements	Al, Fe
Loppi et al. (1999b)	Lithogene elements	Al, Fe, and Ti. In the Mediterranean environment the content of trace elements of unwashed samples is greatly affected by soil contamination
Nimis et al. (2001)	Terrigenous elements	Al, Fe
Loppi and Pirintsos (2003)	Terrigenous material	Al, Fe
Brunialti and Frati (2007)	Terrigenous material	The relationship of the terrigenous Al and Fe with Ni and Cr confirmed the presence of anthropogenic impact factors suggesting soil particulates trapped in the lichen
Ribeiro Guevara et al. (1995)	Lithophile elements	Al, Sc, Th, Hf, and Fe. Indicators of clastic inputs. Detritical particulate material could be trapped inside lichen tissues
Bargagli et al. (2002)	Lithophile elements	Al, Cr, Fe, Mn, Ni, and Ti
Bargagli (1995)	Lithophile elements	Al, Fe, Ba, Cr, and Mn. Lithophilic elements were higher at the end of the dry season, due to the quantities of soil particles trapped in the samples
Monna et al. (2006)	Lithophilic elements	Th, Sc, La, and Fe
	Chalcophilic elements	Pb, Zn, and Cu. Detected in mine dump and coals
Singh et al. (2013)	Lithophiles	Al, Cr, Cs, and V
	Siderophiles	Co, Fe, Mn, and Ni
	Chalcophiles	As, Cd, Cu, Pb, and Zn
Adamo et al. (2008a)	Lithophilic elements	Al, Fe, and K
Takala et al. (1998)	Geogenic element	Zn
Matschullat et al. (1999)	Geogenic elements	Ba, Mn, Rb, Sr
	"Intermediates"	Cr,Mo, Sn, V
	"Anthropogenically emitted"	Cd, Pb, Zn
Giordano et al. (2013)	Geogenic and anthropognic elements	Al, Ba, Ca, Cr, Cu, Fe, Ni, Pb, Ti, and Zn
		The high and significant correlations between geogenic and anthropogenic elements suggest that the polluted urban soil is a massive source of elements
Bermudez et al. (2009)		Mn is related to agricultural activity and Fe to industry and soil particles
Getty et al. (1999)		Fe is an important constituent of eolian dust
Pignata et al. (2007)		High levels of As were related either to As-rich loessic soil or to long-range transport of industrial contamination

resemble a soil factor. Its high P and K content in addition to the presence of S, Cu, and Zn referred to the resuspension of particles of viticultural origin. Freitas et al. (1999) using *P. sulcata* for a similar study in Portugal applied the elements Cr and Sb as markers of soil particles.

A comparison of the elemental content of the epiphytic foliose lichen *F. caperata* with that of the moss *Hypnum cupressiforme* found greater amounts of lithophile elements (Al, Cr, Fe, Mn, Ni, and Ti) in the moss, whereas the lichen contained a greater share of elements such as Hg, Cd, Pb, Cu, V, and Zn (Bargagli et al. 2002): the elemental composition of the moss was affected to a greater extent than that of the lichen by the geochemical features of the study area, in a former mining district in Italy. Another comparison of the metal bioaccumulation capacity of the epiphytic moss *Scorpiurum circinatum* and the epiphytic fruticose lichen *Pseudevernia furfuracea* exposed in bags in an urban area in S. Italy by Basile et al. (2008) revealed the following: (1) both species accumulated the entire range of the metals Al, As, Cd, Co, Cr, Cu, Fe, Mn, Mo, Ni, Pb, Ti, V, and Zn and (2) the moss, however, had the highest bioaccumulation capacity for all metals and showed a more constant, linear accumulation trend. These findings agree with a previous study located in Naples which used the lichen *P. furfuracea* (Giordano et al. 2005). The particulate matter contained Al, Si, P, S, Cl, Ca, Fe, and Cu deriving mostly from abrasion processes activated by vehicles: trams, railways, and cars. Additional studies performed by Sorbo et al. (2008) and by Aprile et al. (2010) confirmed that high values of Cu and Zn derive from abrasion as a result of vehicular activity. The varying results of similar studies using both an epiphytic lichen and an epigeic moss were ascribed by Basile et al. (2008) to the morphology and ecophysiology of the species: epigeic species were found to be susceptible to lithophile elements to a greater extent than epiphytic species, which are strictly dependent on atmospheric deposition. As contended previously by Bargagli et al. (2002), mosses and lichens cannot, apparently, be used interchangeably as biomonitors of trace metal accumulation. Moreover, in a comparative analysis of the elemental content of oak leaves (*Quercus pubescens*), epigeic mosses and epiphytic lichens in a former mining district in Italy, Bargagli et al. (2003) found that oak leaves were less affected by the elemental contribution of adsorbed soil particles, thus facilitating the evaluation of atmospheric inputs.

Two groups of elements were identified in lichens in the vicinity of Mt. Etna and on Vulcano island (Sicily, Italy) by Varrica et al. (2000). Of the lithophile elements derived from local crustal material, the average concentrations of Al, Ca, Mg, Fe, Na, K, P, and Ti were the most substantial. Typical crustal elements were found to account for over 95 % in weight of the total elements. On the other hand, enriched elements (Sb, Br, Zn, Au, and Cu) forming volatile compounds were found to originate mostly in anthropogenic sources and/or volcanic gases.

A clear distinction between "lithophilic" and "anthropogenic" elements was indicated by a cluster analysis following a study of trace elements in the lichen *P. furfuracea* and the moss *Sphagnum capillifolium* exposed in bags in the city of Naples (Italy). It was suggested that lithophilic elements were supplied by entrapped particulate matter circulated by erosive winds. The marked upward trend detected in *S. capillifolium* and *P. furfuracea* for elements such as Pb, Zn, Cr, Cu, As, Fe, Mo, and V was related to anthropogenic sources (Adamo et al. 2003). A comparative analysis of elemental levels in lichens *(H. physodes*) and mosses (*H. cupresiforme)* and/or *Pleurozium schreberi* collected in the same location yielded mean values of about 9 % soil-derived or lithophilic elements such as Cr, Fe, La, and Sc in *H. cupressiforme.* On the other hand, anthropogenic elements such as As, Br, Cd, and Zn attained significantly higher values in *H. physodes* (Jeran et al. 2003).

The contamination of lichen samples by soil particles is an apparent issue to be dealt with, the more so in studies focused on the use of cryptogams as biomonitors (Bargagli 1995; Bargagli et al. 1995; Loppi et al. 1999b; Bergamaschi et al. 2002, 2004; Loppi and Pirintsos 2003). To eliminate the contribution of local soil in calculations of the "net" elemental content of lichens used as biomonitors of anthropogenic pollutants, several studies analyzed both the elemental content of the lichen thallus and of the upper layer of soil in the adjacent collection/transplantation sites. Studies in this category should be classified as follows:

1. Descriptive studies of the elemental content of lichens, soil, and additional media performed to indicate the general status of the environment or to establish a set of markers on a local or regional scale: several studies in this category compared the elemental content of both soil and lichens (Garty 1985; Bargagli 1990a;

Garty et al. 1995, 1996b; Dillman 1996; Loppi 1997; Rizzio et al. 2001; Bačkor and Fahselt 2004; Bargagli et al. 2007; Leonardo et al. 2011; Grangeon et al. 2012); these studies showed mostly a high level of coincidence. Chemical analyses of both soil and the epiphytic foliose lichen *X. parietina* sampled in Luxemburg (Hissler et al. 2008) determined fractionated ratios of Sm/Nd as a function of distance from a steelworks. In remote areas, the lichens had Sm/Nd ratios which were comparable to those of corresponding soils. Lichens collected closer to steelwork sites showed decreasing Sm/Nd ratios.

2. Analytic studies determining significant direct correlations of elemental content of soil and lichens (Bargagli et al. 1987, 1993, 1999; Loppi et al. 1999a; Bari et al. 2001; Walker et al. 2006, 2009): in certain cases, significant correlations of metal content, such as Cu in soil and in lichens, are applicable as a tool for biogeochemical prospecting; Chettri et al. (1997) determined that lichens may be used to locate mining areas; other cases, however, of correlations of elemental content in both lichens and soil did not yield significant results (Rencz et al. 2003).

A comparison of the elemental content of soil, epiphytic lichens, tree bark, and leaves (or needles) has the potential to yield interesting results: according to Migaszewski et al. (2005), air and soil are the probable source for trace elements. The variability of S and trace metals in soil and bioindicators was found to be primarily governed by bedrock lithological factors in addition to anthropogenic factors. Another study (Purvis et al. 2008) showed highly significant correlations for Mn/Ca ratios in a lichen *(P. sulcata)*, versus oak-tree bark, for soil versus bark of the same tree, and for soil versus epiphytic lichens growing on the same tree. This study confirmed the existence of a cycling process via the soil–tree–lichen system.

Analyses of enrichment factors (EF) are rather helpful in an attempt to distinguish the contribution of air pollutants from that of crustal components associated with windblown soil particles (Nyarko et al. 2006). The calculation of EF was designed to discriminate between anthropogenic and natural sources of trace metals measured in lichens. Enrichment factors were calculated by a comparative analysis of the elemental composition of lichens and crustal rock or soil. Puckett and Finegan (1980) normalized these elemental levels to obtain a reference element considered an unambiguous indicator of the source of the material. The general formula presented by Puckett and Finegan to calculate the enrichment factor for any element (*X*) in the lichen relative to crustal rock or soil is as follows:

$$\mathrm{EF} = \frac{(X/\text{reference element})\ \text{in lichen}}{(X/\text{reference element})\ \text{in crustal rock or soil}}$$

Scandium (Sc) was commonly used as a crustal indicator by Puckett and Finegan and in following biomonitoring studies of air pollutants with lichens in different regions in the world (e.g., Olmez et al. 1985; Sloof 1995a; Ribeiro Guevara et al. 1995; Grasso et al. 1999; Bergamaschi et al. 2002, 2004; El-Khoukhi et al. 2004; Nyarko et al. 2006; Freitas et al. 2007). Puckett and Finegan determined that enrichment factors near unity imply non-enriched elements. Low EF values ranging from 1 to 5 obtained for Al, Cr, Fe, Na, Ti, and V negated the enrichment of the lichen.

Several authors applied Al as a crustal indicator (Bargagli 1990b; Wiersma et al. 1992; Calvelo et al. 1997; Varrica et al. 2000; Carignan et al. 2002; Cloquet et al. 2006; Rodushkin et al. 2007; Merešová et al. 2008; Boamponsen et al. 2010). In certain cases, Al is not available as a determinant factor of crustal identity and Mn is selected as a crustal reference (Machado et al. 2004). Manganese is unquestionably lithophilic, with negligible marine sources (Nriagu 1989).

Titanium was also used as a crustal indicator (Bargagli 1998; Bennett and Wetmore 1999; Purvis et al. 2006, 2007). Certain studies used titanium as an alternative to Al (Bettinelli et al. 2002; Minganti et al. 2003). Further applications of the EF factor, in addition to its use in lichen studies, were reported for mosses (Aničić et al. 2009) and trees: the reference element selected for sweet chestnut leaves was Mg, according to the following equation: (Martin et al. 2009)

$$\mathrm{EF} = \frac{(Y/\mathrm{Mg})\ \mathrm{plume}}{(Y/\mathrm{Mg})\mathrm{lava}}$$

Another reference element used for the calculation of elemental enrichment was zirconium (Zr): Dolgopolova et al. (2006) calculated the Pb and Zn enrichment relative to the local geological background using the following equation:

$$\mathrm{EF}_x = (X/\mathrm{Zr})\ \mathrm{sample}/(X/\mathrm{Zr})\ \mathrm{host}$$

As: X is Pb or Zn, (X/Zr) sample is the ratio obtained for the individual sample, and (X/Zr) is the ratio obtained for host rocks representative of the local background.

Rare earth elements (REE) are applicable reference elements for the calculation of elemental enrichment of lichens. According to Chiarenzelli et al. (2001), the advantageous properties of REE relevant to environmental studies are transport in the particulate phase, lack of significant anthropogenic sources, coherent group geochemistry, robust concentrations, and upper crustal signatures. The authors calculated the REE enrichment factor relative to the local geological background following the equation:

$$\mathrm{EF}_{\mathrm{REE}} = (\mathrm{Metal\ sample/REE\ sample/Metal\ UCC/REE\ UCC})$$

As UCC = upper continental crust.

REE analyses may be helpful for the identification of particulate deposition related to anthropogenic activities and enrichment by other elements originating in biogenic processes.

Although the calculation of EF is used widely to identify and quantify the influence of anthropogenic activity with respect to global elemental cycles, several authors (Reimann and De Caritat 2000; Simonetti et al. 2003) demonstrated the intrinsic flaws of the concept in studies of environmental geochemistry. The attempt to normalize elemental concentrations to an average value for the total crust was found to be rather theoretical, due to its neglect of the following: (1) The natural fractionation of elements upon transfer from the crust to the atmosphere, (2) the differential solubility of mineral elements upon alteration in the environment, and (3) the impact of biogeochemical processes.

3.3.1 Dependence of Soil-particle Contamination on Distance from Ground Level

Several factors should be considered when selecting the height above ground for sampling and exposure. If epigeic lichens are used, the material grows directly either on soil or on top of other lichens, mosses, or organic matter located on the forest floor. Lichen samples collected from a substrate like this are contaminated by the greatest amount of soil particles. Epilithic lichens located on substrates near the ground are also subject to soil particles to a great extent. This was the case for the epilithic desert lichen *Ramalina maciformis* growing on flintstones in the Negev Desert, Israel, and found to contain large amounts of mineral elements deriving partly from local soil (Garty 1985; Garty et al. 1995, 1996b, 2002b, c, 2007). Its transfer and relocation was done with its substrate and its new location in polluted sites was on the ground. In comparison, other studies using the epilithic lichen *Usnea amblycolada* growing near ground level detached it from its substrate (rock), put it in bags, and tied it to posts in biomonitoring sites at a height of 3 m, thus neglecting the relevance of the height factor (Carreras and Pignata 2002; Carreras et al. 2005, 2009a, b). Table 3.4 presents papers published in the years 2001–2014, referring to height above ground of collection and transplantation. In many cases, the data for height above ground for the transplants are not given (not only in Table 3.4) leaving the reader with incomplete data with reference to contamination by soil particles.

Table 3.4 Epiphytic lichens collected and transplanted: height above ground

	Collection (m)	Transplantation (m)
Adamo et al. (2003)	0.1–1.2	4–8
Yildiz et al. (2008)	Not given	above 3 m
Garty et al. (2001a)	3–4	2–3
Doğrul Demiray et al. (2012)	At least 1	–
Yenisoy-Karakaş and Tuncel (2004)	1–2	–
Affum et al. (2008)	2	approx. 1.5
Leonardo et al. (2011)	1.5	–
Farinha et al. (2009)	1.5	1.5
Sorbo et al. (2008)	Not given	5.0
Monna et al. (2006)	1–3	–
Boamponsen et al. (2010)	1.5–2	–
Basile et al. (2008)	Not given	4.0
Nyarko et al. (2006)	Approx. 1.5	–
Migaszewski et al. (2001)	0.25–2	–
Rosamilia et al. (2004)	Above 1	–
Sawidis et al. (2010)	1.5–2.5	–
Salo et al. (2012), Paoli et al. (2014)	0.5–2	–
Dzubaj et al. (2008)	1–1.5	–
Minganti et al. (2003)	1–1.7	–
Häffner et al. (2001)	Not given	1.5–2
Bari et al. (2001)	Not given	2
Nimis et al. (2001)	Above 1.2	–
Scerbo et al. (2002)	Approx. 1	–
Kularatne and de Freitas (2013)	Not given	approx. 2
Ayrault et al. (2007)	At least 1.5	1.5
Viera et al. (2004)	Approx. 1.5	–
Paoli and Loppi (2008)	1	2–2.5
Culicov and Yurukova (2006)	Not given	21
Otnyukova (2007)	2–5, 15–22	–

3.3.2 Particulate Matter Entrapped by Lichens: Visual Data and Coefficients of Variation (CV)

Mosses and lichens are preferable as biogenic dust collectors for detailed studies of airborne pollutants (Salo et al. 2012). Microscopic evidence of the entrapment of particulate matter by lichens is nowadays abundant. A large portion of the metal burden of lichens is found to occur as particulate matter trapped in the intercellular space of the medulla and/or on the cortex of the lichen thallus (Garty et al. 1979; Garty and Delarea 1987; Purvis et al. 2000; Clark et al. 2001; Bettinelli et al. 2002; Bačkor and Fahselt 2004; Vingniani et al. 2004; Williamson et al. 2004; Giordano et al. 2005, 2010; Ayrault et al. 2007; Adamo et al. 2008b; Bubach et al. 2012; Chaparro et al. 2013) and Fig. 3.1. The above-mentioned publications yield direct, visual evidence indicating that a certain part of the mineral elements, accumulated in the lichen thallus, displays indeed a particulate nature. The visual documentation of the presence of particulates, continuously provided by optical instrumentation, is complemented by an elaboration of quantitative data obtained by common chemical analyses of elemental content. Garty et al. (1977)

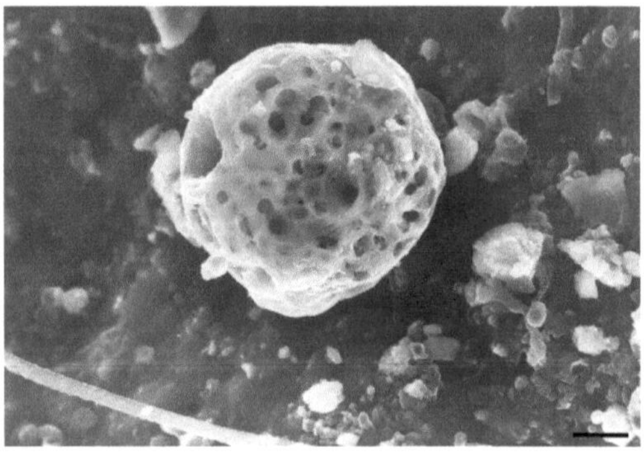

Fig. 3.1 A large particle, derived from heavy fuel combustion, detected on an unwashed thallus of the epiphytic lichen *Ramalina lacera* growing on *Faidherbia albida* trees close to the local power plant and the oil refineries, Ashdod, S.W. Israel. Scale bar = 1 μm (Garty et al., unpublished)

suggested that the occurrence of metal-containing particles in the lichen thallus is indicated by high coefficients of variation (CV) obtained by the ratio of values of standard deviation (SD) and mean metal content: high CVs indicate the probability of unevenly distributed particles in the thallus. Low CVs were assumed to indicate a low variation due to an even suspension of metal particles. CVs were further applied to assess the size of deposited/entrapped particles (Garty 1985, Garty and Ammann 1987; Loppi et al. 1994, 1997a; Garty et al. 1995, 1996b, 1997a, b, 1998a, b; Carreras and Pignata 2002; Purvis et al. 2004).

In a study, biomonitoring airborne heavy metals along a major road in Accra, Ghana, the CV values calculated for lichens (*Parmelia* sp.) were high for both sides of the road, indicating an irregular distribution. The very high CVs for Cr on both the eastern and western side (0.95 and 1.18, respectively) and for Ni on the eastern side (0.94) suggest an irregular pattern of pollution (Affum et al. 2008).

Lippo et al. (1995) applied CVs in a nationwide survey of atmospheric deposition in Finland. High CVs were found to reflect both regional differences in background areas and local emission sources. The CVs calculated for lichens in volcanic areas in Sicily, Italy, were high for several elements such as Au (1.48), Sb (1.30), Zn (1.07), Cu (1.00), and Pb (0.95) (Varrica et al. 2000). These ratios were not, however, related in this study to either particulate size or any other feature of particulate matter. In a survey of elemental deposition in the Aegean region of Turkey, Yenisoy-Karakaş and Tuncel (2004) applied CVs to determine local variation: morphology, size, and substrate of the lichen (*X. parietina*) and local pollution and negate its capability to interfere with regional elemental data.

To test the impact of pre-treatment, Adamo et al. (2008a) applied CVs in a study of the adaptability of the lichen *P. furfuracea*, growing on branches in NE Italy, for biomonitoring studies. The basal portion of washed thalli yielded lower CV ratios than unwashed basal parts for Al, Fe, Hg, K, Mn, Na, Pb, and V. The washing of this fruticose lichen not only made for a statistically significant decrease of Al, Fe, and Mn levels in the basal portion, which is assumed to contain high concentrations of lithophilic elements, relative to the lobes of the thallus, but also reduced the variability, as expressed by lower CVs following the washing treatment. CVs were proved to differentiate peripheral from central parts of foliose lichen thalli (Loppi et al. 1997b): the trace-elemental content of peripheral (younger) and central (older) parts of thalli of *F. caperata* in an unpolluted site in Central Italy was highly variable. Elements of limited metabolic significance such as Al, Cd, and Pb attained higher concentrations in the central part, thus suggesting their being trapped in the medulla; the CVs were 0.71 for Al, 0.54 for Cd, and 0.73 for Pb; the corresponding CVs for peripheral parts of the lichen were 0.48, 0.51, and 0.46, respectively. In a retrospective study performed by Loppi et al. (2004) in Toscana (Central Italy), the reduced use of leaded gasoline was found to produce in 1999 a significant decrease of Pb content of *F. caperata*, relative to 1993. The mean ± SD of Pb concentrations were 35.3 ± 24.9 $\mu g\ g^{-1}$ in 1999 versus 66.7 ± 51.3 $\mu g\ g^{-1}$ in 1993. The CVs obtained for 1999 (0.71) were a little lower than those obtained for 1993 (0.77). These CVs suggest a stability of environmental emissions. A similar stability was observed in an analysis of elemental content of the epiphytic *Bryoria fremontii* and *Letharia vulpina* investigated in Yellowstone National Park, Wyoming, USA, by Bennett and Wetmore (1999). The CVs for most of the elements were found to be equal or less than 0.50, indicating fairly constant values. According to Bergamaschi et al. (2004), the elemental content of different lichen species growing in three mountainous regions (Himalaya, Kenya and Italy) was quite similar, and the greater part of the CVs were around 0.50 or less, whereas some attained CVs of 0.70 or more. The average CVs obtained for 23 elements in the lichens in Kenya, Himalaya, and Italy were 0.65, 0.62, and 0.47, respectively. Since these lichen samples were collected in unpolluted areas, the different values obtained for elemental concentration should be attributed to the different composition of the soil, as the vegetative cover in these regions is sparse and the soil is exposed to wind erosion.

Windblown dust represents an important source of mineral elements in lichens (Bargagli 1995; Loppi et al. 1999b; Garty et al. 2002c, 2007, 2008; Bergamaschi et al. 2004; Jiries et al. 2008). Alterations in the annual number of dust storms are to be expected as a result of climate change, reflected by divergent parameters of which the increased number of dust storms originating in the great deserts of the world, like the Sahara and the Sahel in Africa is a prominent example. Data presented by Avila and Penũelas (1999) in northern Spain demonstrated the increased frequency of rainfall loaded with mineral dust (red rain) originating in North African deserts in the last decades: for the period 1983–1997, a clear upward trend of the annual median pH of rainwater was observed in Montseny, in addition to a double frequency of weeks with red rain, from 2.3 in 1944–1974 to 5.6 in 1983–1997, thus reflecting an increased transport of dust from Africa (Fig. 3.2) to southern Europe. Similarly, Ganor et al. (2007) and Garty et al. (2008) reported on an enlarged number of days with dust events in Israel, per year, from 5 days in 1958 to 30 days in 2006 (Figs. 3.3, 3.4, 3.5 and 3.6).

The average CVs calculated for the high mountains in Italy (0.47) based on the data obtained by Bergamaschi et al. (2004) were compared with the average CVs calculated by Pacheco et al. (2004) in 29 littoral-north sites and 29 littoral-center sites in mainland Portugal. The average CV of 0.58 determined for 17 elements detected in *Parmelia* spp. was a little higher than the average CV obtained for the elevated sites in Italy. The highest CV was determined for Al (0.87), a typical marker of soil contamination, whereas the lowest was obtained for K (0.30) which is reported consistently to be included in the thallus in a non-particulate form and attains very often low CVs. An additional case of low CVs for K was observed for *Ramalina lacera* in the North-Eastern part of Israel which has a Mediterranean climate. The CVs obtained for K in this lichen growing in an unpolluted site (HaZorea) were 0.10 (Garty et al. 1997a), 0.07 (Garty et al. 1998b), 0.08 (Garty et al. 2001a), and 0.14 (Garty et al. 2002a). Another study (Garty et al. 1998a) yielded a CV of 0.09 for K, whereas the CVs for lithophilic elements such as Al, Si, Ti, and Fe were 0.27, 0.27, 0.25, and 0.21, respectively. Thus, the status of entrapped elements such as Al, Si, Ti, and Fe as markers of soil-derived particulate matter was further established.

Fig. 3.2 A dust storm in the Sahara Desert, Erfoud, an oasis town, Morocco, April 2005

Fig. 3.3 Saharan dust over Hod Hasharon, W. Israel, March 1998

Fig. 3.4 Saharan dust over Petach Tiqwa, W. Israel, April 2010

The CV obtained for a different *Ramalina* species, *R. maciformis*, collected in an unpolluted site in the Negev Desert, was 0.10 for K (Garty et al. 1996b), whereas Fe and Mn attained a CV of 0.17 and 0.24, respectively (Garty et al. 1995). A low CV for K (0.08) was obtained also in the epiphytic fructicose lichen *Protousnea magellanica* in a pristine, unpolluted area in Northern Patagonia, in the Argentine Andean Mountains, whereas the CVs obtained for Al, Mn, Ti, and Fe were 0.33, 0.79, 026, and 0.35, respectively. An additional subject investigated in this study was the distribution of the elements within the thallus. A comparison of the CVs for K in primary branches of the thallus versus secondary branches yielded ratios of 0.07 versus 0.03, respectively, whereas Al attained a CV of 0.20 for primary branches versus 0.27 for secondary branches. Based on additional data obtained in this study, the elemental distribution was found to be mostly homogenous. Na, however, attained a CV of 0.43 for primary branches and 0.73 for secondary branches (Bubach et al. 1998). Low CVs for K and higher CVs for lithophilic elements obtained

Fig. 3.5 Saharan dust over Wadi Ara, C. Israel, April 2011

Fig. 3.6 Saharan dust over Tel Aviv University, Tel Aviv, W. Israel, March 2012

for *R. lacera* in Israel and for *P. magellanica* in Patagonia are attributed to the difference between soluble and insoluble particulates.

In transplant studies, it is apparent that high CVs refer to short exposures. Based on the assumption that lichens capture a great deal of their metal content by particulate impaction and retain most of their metals by physical entrapment, Gailey and Lloyd (1986) contended that short exposures yield low values for metal content (near the detection limit) and high CVs. It was assumed that a lichen needs time to adapt to a new environment and reach an equilibrium at which it accumulates metals efficiently and evenly.

Another study calculated CVs for unwashed lichen samples (*F. caperata*) to provide evidence for the presence of soil particles. Soil contamination, associated with an enlarged Al or Fe content, was indeed indicated by variable concentrations in lichens, tree bark, and leaves. The CVs ranged from 0.43 to 1.81. The CVs obtained for the lichens were 0.50 (Al), 0.43 (Fe), 0.77 (Mn), and

0.60 (Mo). Thus, crustal material was found to interfere with baseline concentrations and with inter-relations of trace elements (Loppi et al. 1997a).

3.4 Pre-treatment of Lichen Thalli Prior to Elemental Determination

Although lichens were used for biomonitoring for five decades, a standardized protocol for the preparation of thalli for elemental determination is still missing. There is, however, some consistency in the literature with regard to the cleaning of lichen samples. It is generally accepted that detritus should be removed. Many laboratories report on a procedure of dry cleaning which leaves a cover of dust, yet separates remains of the substrate, mineral particles, foreign lichens, mosses, insects, and litter.

Markert (1993) defines "cleaning" as performing mechanical steps, such as using a dry/damp cloth, shaking, blowing, bursting, or washing, either in solvents (common for vascular plants and less for lichens and mosses) or in distilled water. This pre-treatment has, however, the potential to remove deposited material of natural and anthropogenic origin (Wadleigh and Blake 1999). This factor is particularly relevant in the case of stable isotope studies, as removed material might be of a different isotopic composition and could bias the interpretation of sources. The washing procedure was found to induce the loss of elements in *X. parietina* (Bargagli et al. 1985) but not in *Cladonia* (Nieboer et al. 1972). Washing was proved to remove nitrogen from *Physcia adsendens,* which according to Gombert et al. (2003), is a part of the nitrogen deposited on the surface of thalli, e.g., as fine particles adsorbed by the lichen surface or as gaseous nitrogen compounds, such as HNO_3 which is also adsorbed but not assimilated.

The washing of exposed lichen material led to a substantial loss of elements such as Al, Cr, Mn, and Ni in washed *P. furfuracea* following a transplantation experiment (Adamo et a. 2007). Water-washing made for a statistically significant decrease of Al, Fe, and Mn in the basal portion of the lichen that contains normally higher concentrations of lithophile elements (Adamo et al. 2008a). Water-washing was found also to lead to a statistically significant decrease of C and N in the lobes of the lichen, explained by the washing away of adsorbed organic or carbonate particles and/or the extraction of water soluble compounds (Adamo et al. 2008a). It was found also that concentrations of F decreased significantly in washed lichen thalli (Palomäki et al. 1992). The washing procedure induced also a loss of elements in *Evernia prunastri* in urban and highly industrialized areas in and around Padova, Italy. In the vicinity of a steel works, however, the loss was lower than in a site located near a crossroads (Caniglia et al. 1994).

Certain elements such as Pb and S which are tightly bound to cell components are little changed by washing (Prussia and Killingbeck 1991; Richardson et al. 1995). Indeed, Pb (together with Cu, Fe, and Zn) was retained in thalli of *P. furfuracea,* washed upon exposure, in considerable quantities (ca. 60–80 %) in all lichen material (Adamo et al. 2007). Unwashed samples of *Canoparmelia texana* presented elemental concentrations at the same magnitude or slightly higher than the washed samples (Saiki et al. 1997). However, the results were statistically insignificant, with the exception of Al, K, Na, Se, and Ti. These results indicate that the washing procedure can be used to eliminate adhering material.

In the absence of any form of washing, great amounts of airborne dust and soil particles will be left on the thallial surface. As much as the collection site is exposed to dust storms and/or emissions of industrial or vehicular dust, the determination of elemental content is influenced by external contamination. If a period of dry windy weather precedes collection, significant surface dust should be expected, whereas rain preceding collection removes most of the dust. According to Richardson et al. (1995), differences of weather conditions preceding collection provide an unwanted source of variability of elemental levels in mosses, lichens, and hairy leaf samples.

A survey of literature including hundreds of publications dealing with lichens as biomonitors of air quality revealed a very wide range of pretreatments prior to determination of elemental content, ranging from no cleaning to wet cleaning. The use of dry cleaning is established in laboratories applying lichens washed in the field by rain and snow which is fitting in Nordic countries. In quite dry and relatively unpolluted terrestrial ecosystems of the Mediterranean, the elemental composition of unwashed epigeic lichens was affected to a greater extent by soil and rock dust than that of unwashed epiphytic lichen samples (Bargagli et al. 2002). Lithophile elements dominate in mosses, whereas atmophile elements dominate in epiphytic lichens in central Italy (Bargagli 1995; Loppi and Bonini 2000; Bargagli et al. 2002). In Mediterranean countries, levels of airborne soil dust rise in summer and precautions should be taken in order to gather comparable data at different times (Bargagli 1998; Bettinelli et al. 2002).

In the last two decades (1994–2014), the greater part of studies in this field applied a pretreatment of dry cleaning. The reasons cited for the omission of a washing procedure were to avoid the leaching of soluble substrates (Carreras and Pignata 2002), to avoid losing particles trapped on the lichen surface (Bergamaschi et al. 2007), and to avoid unpredictable alterations of the chemical composition of the thalli (Bettinelli et al. 1996; Paoli et al. 2014). Jia et al. (2006) further explained that the lichen samples were not washed in order to measure the elements that were physically trapped on the surface of the thallus, as well as those chemically bound to cell walls.

The objective of washing is to produce initial, homogeneous results (Tretiach et al. 2007; Adamo et al. 2008a; Ares et al. 2012). Ares et al. (2012) evaluating the degree of standardization for mosses, suggested that the outcome of the washing procedure is determined by the following parameters: (1) number of washes; (2) duration of washing; (3) use (or not) of shaking; (4) type of water; and (5) relation between the weight of the moss and the volume of water. Data referring to these parameters were presented in one of the papers reviewed, whereas in 42 % of the papers, none of these were specified.

In lichens, it seems that the number of protocols is close to the number of laboratories that use lichens as biomonitors. The variability is such that it is difficult to compare results of elemental determination even if they are collected in the same location. Adamo et al. (2003) submitted lichen and moss samples to seven consecutive washings with distilled water prior to exposure in bags and performed an additional washing after exposure. The washing treatment prior to transplantations of cryptogams is not very common in biomonitoring studies (e.g., Gonzáles et al. 1996, 1998, 2003; Godinho et al. 2009a; Spagnuolo et al. 2011; Giordano et al. 2013). In certain cases, lichen and moss materials, collected in the field, were rehydrated overnight in a humid chamber and washed four times. Following this procedure, bags were prepared, exposed in transplantation sites, and collected for vitality assessment and for a determination of total content of C, N, and S (Tretiach et al. 2007). Adamo et al. (2007) applied washing for the determination of trace-elemental content of lichens and mosses in addition to a washing in distilled water post exposure, in order to evaluate the contribution of surface particulate matter. Water-washing of exposed material leached similar amounts of most elements from mosses and lichens, thus indicating a similar mechanism. Water-washings of *P. furfuracea* prior and post transplantation were applied also in Ankara, Turkey, by Yildiz et al. (2008). Upon collection, the transplants were washed in tap water and then in distilled water. This study, however, did not compare the impact of washing prior and post transplantation.

Table 3.5 presents selected studies describing the wet cleaning of lichen samples prior to elemental determination. It is apparent that almost every laboratory treats the lichens arriving from the field in a different manner. Standardization of procedures (Seaward 1995) is called for, in order to attain comparable results. Wolterbeek and Bode (1995) contended that problems of comparability are introduced as soon as more than a single sampling-method and/or sample pre-treatment/

Table 3.5 Washing procedures applied in lichen biomonitoring studies (2001–2012)

	Washing procedure
Leonardo et al. (2011)	Cleaning in distilled water to remove dust, bark, and insects
Backor et al. (2009)	Rinsing in deionized water
Dowdal et al. (2005)	Rinsing in distilled water to remove detritus and remaining soil
Fenn et al. (2007)	Sonicating five times in one liter of distilled water, 1 min, then air drying to remove mineral debris trapped externally between overlapping lobes
Farinha et al. (2009)	Washing, 30 s, in deionized water
Freitas et al. (2007)	Two washing cycles in deionized water with agitation
Nyarko et al. (2006)	Washing approx. one min. in deionized water
Sorbo et al. (2008)	Five consecutive washings in distilled water
Garty et al. (2001a, b, 2002a, b)	Three consecutive rinsings, 5 s each
Godinho et al. 2004, 2008, 2009a, b)	As above
Paoli and Loppi (2008)	As above
Pignata et al. (2007)	Washing in double distilled water and leaving to dry at room temperature in a clean area
Golubev et al. (2005)	After collection, washing in bi-distilled water
Affum et al. (2008)	Washing in water, 30 s
Giordano et al. (2005)	Removing soil particles and water washing
Monna et al. (2006)	Rinsing in Milli-Q water in an ultrasonic bath, 1 min
Doğrul Demiray et al. (2012)	Stirring clean samples in deionized water, 1 min to remove dust
Boamponsen et al. (2010)	Washing approx. 60 s in deionized water to remove dust and sand
Yenisoy-Karakaş and Tuncel (2004)	Floating and stirring lichens and substrate in a glass beaker approx. 30 s to remove soil particles and debris
Basile et al. (2008)	Seven consecutive washings in distilled water (50 mL), 5 min, to remove soil particles

analysis is applied. The authors suggested a single laboratory should take care of all pre-treatments. It seems that a unification of pre-treatment is problematic to a greater extent than the standardization of a qualitative determination of accumulated mineral elements.

The International Atomic Energy Agency (IAEA) supports the biomonitoring of air pollution since 1997 (Smodiš 2003; Smodiš and Bleise 2002). Members of the working group, representing 14 countries and different biomonitors, suggested a procedure for sample preparation that includes the cleaning and separation of samples from the substrate. Smodiš and Bleise (2002) suggested the following steps: (1) dry cleaning (2) rinsing in distilled water for 5–10 s (3) drying at 40 °C for 48 h (or freeze-drying), and (4) grinding using liquid nitrogen. We recommended that the rinsing (washing) procedure will be done by handshaking the samples with double distilled water at a temperature of 20 °C for 15 s. The rinsing procedure has to be repeated three times of 5 s each in order to minimize the loss of water soluble elements. We suggest to apply additional washes in the same manner if the water used for the third rinsing is still unclear. The above-mentioned procedure used in Israel yielded low coefficients of variation (CV) for Ca (an important constituent of desert dust and soils) calculated for the lichen *Ramalina lacera*: 0.10–0.26 (Garty et al. 1997a, b, 1998a, 2000, 2001a, b, 2003, 2007, 2008).

3.5 Elemental Content of Lichens and Bulk Deposition: Temporal and Spatial Dimensions

The validity of lichens as environmental biomonitors depends on a positive correlation of the elemental content of the thallus and the average concentration of atmospheric deposits. Jenkins and Davies (1966) detected a linear positive correlation for elemental content of lichens and average concentration of ashed atmospheric deposits, thus paving the way for subsequent studies. Today, the notion that the chemical composition of lichens reflects the availability of trace elements in the environment is established (Bačkor and Loppi 2009; Paoli et al. 2012). In the present section, we deal with lichens applied as biological receptors in addition to man-made receptors, collectors, or filters designed or applied to accommodate bulk deposition. Bulk deposition is defined in the relevant literature either as "airborne particulates" (Bari et al. 2001), "air particulate matter" (Costa et al. 2002), "airborne particulate matter" (Carreras et al. 2009a), aerosol samples (Farinha et al. 2009), dust samples (Rossbach et al. 1999), particulate matter (Farinha et al. 2009), and dry dust deposition (Olszowski et al. 2012).

The ratio of concentrations of certain elements such as Cr, Zn, and Fe in dust collected on air conditioners of large hotel buildings and in samples of *Usnea* spp. was found to be comparatively high relative to the ratio of concentrations of Ca, Rb, and Sr in the same media (Rossbach et al. 1999): data of this kind imply the necessity to complete the available information relating to the span of concentrations and address modifying factors. A comparison of the uranium content of *H. physodes* and air sampled in filters, on the other hand, yielded good results for the lichen-indication method (Golubev et al. 2005), enabling the diagnosis of a long-term atmospheric uranium contamination and an identification of its source.

Another study demonstrated the applicability of continuously operating particulate samplers set up near transplanted lichens from an uncontaminated area. The use of this instrumental apparatus in combination with the fruticose epiphytic lichen *P. furfuracea* exposed in bags yielded significant correlations for atmospheric concentrations and for the greater part of the metals (Cd, Pb, Zn, Fe, Cu, and Ni) accumulated in the thalli. This study thus corroborated the assertion that correlations of this kind are indicative of the efficacy of the lichen and its ability to represent atmospheric contamination (Bari et al. 2001). However, several groups of lichenologists operating in different study areas encountered the issue of modified availability of airborne particulate matter. For example, the observed $^{206}Pb/^{207}Pb$ range for urban and industrial aerosols in Sicily, closely matched the anthropogenic local signature, whereas the range obtained for the lichens was closer to the compositional field of ^{206}Pb rich geogenic sources (Monna et al. 1999). Following analyses of elemental content of lichens (*P. sulcata*), filters, and collectors, it was possible to state that lichens do not function as a measuring instrument for many elements available in the atmospheric environment. The information presented by lichens with reference to availability is definitely biased by biological effects (Reis et al. 2002). This modified availability termed equivalent constant availability (ECA) represents a cluster of properties similar to the equivalent aerodynamic diameter for airborne particles' size. It is apparent that attempts to establish correlations for lichen content and bulk deposition involve the consideration of two dimensions: (a) the temporal dimension and (b) the spatial dimension.

(a) The temporal dimension

The following study, referring to the temporal dimension, presented a compilation of data obtained in a rural area in Brasil (Dias da Cunha et al. 2004). Analyses of elemental content of the lichen *Ramalina gracilis* in addition to dust collected in diverse man-made receptors yielded information referring to human health issues in a span of 15 years. The analysis of lichen content indicated that in this period, prior to the investigation, the health of the inhabitants of a certain village in the region was endangered by monazite particles. Another study applying lichens to testify of the atmospheric chemistry on a temporal axis was that performed by Cloquet et al. (2006) in Metz, France. Lichens sampled in the city close to traffic roads displayed $^{206}Pb/^{207}Pb$ ratios of ca. 1.13, whereas urban aerosols displayed homogenous $^{206}Pb/^{207}Pb$ ratios of 1.153 ± 0.003. The ratios obtained for lichens thus suggested that the Pb was reflective of "old Pb pollution."

The capability of lichens to tell of episodic contamination periods in the past is further illustrated by the model of Reis et al. (1999): the performance of the lichen is viewed as a dynamic process of uptake and release which operate simultaneously and continuously. Reis et al. (2002) introduced the concept of "remembrance time" which describes the time span in which lichens reflect the environmental availability of heavy metals at a certain point in time, based on the rate of uptake and release. It was shown (Wolterbeek et al. 2003) that lichens need time to "follow" a change of availability: the rate of uptake and release is conclusive for the time span needed to "forget" previous conditions. Reis et al. (1999) suggested a mathematical approach to calculate the "memory-loss" function. The model suggests that every alteration of rate-constant will lead to a change of both lichen equilibrium levels and remembrance times.

Transplanted lichens may bear witness to extended periods of elemental deposition, relative to the limited period of transplantation, apparently preserving background information. On the other hand, lichens may reflect shorter periods than the total transplantation time or subperiods of exposure (Godinho et al. 2008). The foliose lichen *F. caperata* is rather much in use in biomonitoring studies (Loppi et al. 2004; Aslan et al. 2006, Pacheco et al. 2007, 2008; Baptista et al. 2008; Godinho et al. 2009a, b). Godinho et al. (2008) used it applying the cross-transplant technique to determine "lichen remembrance" along different exposure periods. The lichens were sampled every two months. The ultimate period of exposure was 12 months. The atmospheric bulk deposition was measured concurrently. The data thus obtained, differentiated between short periods of exposure and cumulative exposure-periods suggesting the presence of acclimatization patterns. The study concluded that the content of the lichen transplants does not unequivocally represent the average or cumulative environmental availability along the exposure period. The above-mentioned study included in addition to chemical analyses of elemental content of both lichens and bulk deposition, physiological parameters of viability such as membrane integrity (see reviews Garty 2000, 2001, 2002; Conti and Cecchetti 2001; Wolterbeek et al. 2003; Bačkor and Loppi 2009). Other applications of physiological parameters of vitality were described in additional studies. Godinho et al. (2008) concluded that the elemental content of lichens is dependent on both elemental availability and the physiological condition of the lichen. Recent studies comparing the accumulation of trace elements in the peripheral and central parts of epiphytic foliose lichens transplanted in a polluted site in Portugal (Godinho et al. 2009a, b) suggested that the distribution data of the elements demonstrate the biological regulation of internal concentrations. For thallus parts, this internal translocation should be taken into account as an additional factor affecting "memory length."

How long is the time span needed to "forget" contaminating events under semiarid conditions? A retrospective study performed in Israel (Garty et al. 2009) compared the Pb content of *Ramalina lacera* in an unpolluted site in the NE of the country along the years 1974–2002. Relative high Pb values were detected in the lichen in this site in July 1982, subsequent to an extraordinary large range of vehicular activity, mostly using leaded gasoline, in June of the same year. The Pb values remained relatively high until 1993. Only in the years 1994–2002, the Pb values decreased to a level resembling the first series of data obtained in the years 1974–1981. It is apparent that a decade and more were needed to "forget" the polluting event and that the data obtained in this period (1982–1993) were irrelevant to the availability of Pb pollutants.

The lichen *X. parietina* was taken from a herbarium to compare early twentieth century and present-day atmospheric pollution in SW France. The atmospheric environment characterized by a high dust-content was impacted in the beginning of the twentieth century mainly by coal combustion, indicated by findings of As, Pb, and Cd of both local and long-range origins. Recently, however, a different mixture of pollutants, such as Sb, Sn, Pb and Cu originating in local factories and in car traffic has emerged (Agnan et al. 2013).

(b) The spatial dimension

The pattern of increase near the source of metal/ash content and of decrease away from it refers to the particulate nature of the metals accumulated in the lichen. A significant decrease correlating with distance from point emission sources refers to coarse particles that carry to a limited extent. In many cases, the decrease occurs at a distance of less than 100 m from the source of emission (Nieboer et al. 1982). In other cases, coarse particles were found to carry to a greater distance (Garty, 2001; Sensen and Richardson 2002; Branquinho et al. 2008). In certain cases, the bulk deposition of trace metals such as Cu or Ni increases exponentially toward a large source of pollution, i.e., a smelter. Biomonitors, however, accumulate these metals at a different rate. Salemaa et al. (2004) detected the highest trace metal concentrations in the vicinity of a large Cu–Ni smelter in Finland in bryophytes, whereas lichens and vascular plants accumulated lower quantities.

The relevance of temporal and spatial factors is accentuated by environmental factors such as the interference of humidity with atmospheric bulk deposition. Walker et al. (2003) investigated the chemical composition of snow and terricolous lichens in the USA river basin in the northeastern part of European Russia. Both mediae provided evidence for pollution gradients on a spatial scale, produced by local deposition and by the long-range transport of pollutants. However, data obtained in analyses of wet media such as rainwater and snow do not always coincide with data of elemental content of lichen thalli in the same study area. A comparison of plant and precipitation chemistry in catchments with different levels of pollution on the Kola Peninsula, Russia (Reimann et al. 1999) detected the highest levels of Cr, Fe, Mg, S, in lichens, mosses and vascular plants near a nickel roaster. This study detected also the highest concentrations within the impact zone for Ba, Bi, Cd, and Na in mosses and lichens. The importance of studies of precipitation chemistry derives from their capability to point to the well-represented elements in each plant group. It appears that mosses react most strongly to precipitation, whereas lichens are probably the most efficient "integrators" and the vascular plant *Emperetum nigrum* reflects best the particulate input. Of the total atmospheric input by precipitation, nickel appeared to be the element reflected directly by the three plant groups. The relevance of these findings appears to be restricted to the above-mentioned study area, as it involves three representatives of the local biota. It is to be expected that the use of other media will yield different results. Aubert et al. (2006), for example, found a similar distribution of rare earth elements (REE) in snow, lichens, and peat, in the Black Forest, Germany, suggesting that the complete range of atmospheric signals is present in the peat profile. A comparative analysis of total deposition samples (wet plus dry) and metal content of lichens may yield interesting results: Carreras et al. (2009b) found that the lichen *Ramalina celastri* attained the highest correlation rates (in comparison with other epiphytes) with samples of total deposition in Córdoba City, Argentina. It appears that measurements of metal content of bulk deposition and/or biological receptors are also valuable for assessments of human risk in cases of exposure. The significant correlations observed in this case for the metal content of both *Ramalina celastri* and the flowering epiphyte *Tillandsia capillaris* and respiratory diseases in children accentuated the applicability of biomonitoring systems, in spite of their characteristic shortcomings.

Søndergaard et al. (2010) performed a study of Pb isotopes as tracers of mining-related Pb in lichens, seaweed, and mussels along a temporal-spatial axis. The lichen used, *F. nivalis*, growing on dead organic matter near an abandoned mine in West Greenland was contaminated by atmospheric deposition, spreading Pb as dust. Following the determination of stable Pb isotope ratios and total Pb concentrations in lichens, the natural background Pb ($^{207}Pb/^{206}Pb$: 0.704–0.767) and the Pb originating from the mine ore ($^{207}Pb/^{206}Pb$: 0.955) were found to possess a distinct isotopic fingerprint, applicable as an identification tool. The Pb in *F. nivalis* was found to amount to 633 mg kg^{-1} dry weight and was shown to be comprised of natural background Pb and of mining-related ore-Pb. A significant linear correlation between metal concentrations in transplanted thalli of *F. nivalis* and atmospheric metal deposition in Greenland was found by Søndergaard et al. (2013), thus obtaining data to estimate the dust deposition/dispersion period in the snow-free period of the year. The quantification of mining-related Pb provides the temporal dimension, valuable in the absence of baseline information. The applicability of Pb isotope-data is made evident by their capability to determine the spatial variation in Pb mine-related contamination in accordance with distance from the mine. For additional information on Pb isotopes as tracers of anthropogenic activities, see the review by Komárek et al. (2008). Fine particulates containing potentially harmful metals such as Pb detected in street sediments are possibly re-suspended in the air, as suggested by their presence in lichens. Previously, particulates containing Pb in street sediments were attributed to vehicular exhaust and break wear. However, yellow road paint containing $PbCrO_4$ as a pigment was mostly ignored (LeGalley et al. 2013).

Lichens offer an additional advantage for the detection of both spatial and temporal change produced by airborne Cr, Cu, Pb, and Zn. The lichen *Parmotrema reticulatum* accumulated these metals even when they were present in the air in very low concentrations which is the case in different study areas in New Zealand (Kularatne and de Freitas 2013).

3.6 Radionuclides in Lichens and in Soil: Sources of Pollution

Studies performed in the last decades yielded several publications focused on both radioactivity in lichen biomonitors and soil contaminated by radionuclides. In cases of radioactive fallout as a result of nuclear tests or nuclear accidents such as Chernobyl, emissions of several short-living radionuclides are assumed to make for a readily available distinction between the overall natural nanoparticular atmospheric load and pollution by anthropogenic sources in addition to enrichment factors. The sources of radionuclides as reported for both lichens and soil may be categorized as follows.

3.6.1 Global Background Activity

The testing of nuclear weapons led to the introduction of large amounts of radionuclides into the environment. Samples of lichens, mosses, soil, and air collected since 1961 in Greenland, Svalbard, Iceland, the Faroe Islands, Sweden, and Denmark testified to the presence of ^{207}Bi, a nuclide not reported previously as part of the worldwide fallout (Aarkrog et al. 1984). It was assumed to be produced by thermonuclear test explosions and in particular by the 55 Mt detonation of October 30, 1961, at Novaya Zemlya. With regard to the same part of the world, Nifontova (1995) presented data referring to the ^{90}Sr and ^{137}Cs content of moss–lichen vegetation and other components of the soil–vegetation cover of the Yamal Peninsula, Russia. The last fifteen years yielded decreased concentrations of both irradiants in mosses and lichens, as well as a lowered degree of radionuclide pollution of the superficial soil layer. The present data were found to be applicable as a base for further assessments of the presence of radionuclides in the Far North.

Following the intensive testing of nuclear weapons which started in the early years of 1950, efforts were made to detect the deposition of radionuclides in other parts of the world, i.e., in the Antarctic Peninsula area. Analyses focusing on ^{210}Pb, ^{137}Cs, $^{239+240}Pu$, ^{238}Pu, and ^{241}Am in lichens, mosses, grass, and soil samples in the South Shetland Islands in the Antarctic revealed a significant difference of $^{137}Cs/^{239+240}Pu$ ratios between lichens and moss/grass/soil (Roos et al. 1994), related to the effect of melt water and submerging. In other parts of the Antarctic region, King George Island and other South Shetland Islands, mosses accumulated ^{137}Cs to a greater extent than lichens, thus enlarging their relative applicability as bio-indicators (Godoy et al. 1998).

3.6.2 Coal-fired Power Plants

Several studies dealt with coal-fired power plants as sources of pollution (Daillant et al. 1996; Uğur et al. 2003, 2004). The amount of natural radionuclides discharged into the atmosphere by power plants depends on factors such as concentration in coal, ash content, temperature of combustion, partitioning of bottom and fly ash, and efficiency of the emission-control device. Uğur et al. (2003) observed a maximum of ^{210}Po and ^{210}Pb activity in lichens in the vicinity of ash stacks of power stations. The concentrations of ^{210}Pb in the soil decreased in accordance with increasing depth, as a result of the excessive deposition of atmospheric ^{210}Pb on the surface soil. The ratio $^{210}Pb/^{226}Ra$ referring to activity in soil indicated that the main source of ^{210}Pb, in addition to the radioactive decay of ^{226}Ra, is the atmospheric deposition of ^{210}Pb produced by coal combustion. Uğur et al. (2004) observed a correlation of ^{210}Po activity and Pb and Fe content in lichens around coal-fired power plants, attributed to the emissions produced in the process of combustion of fossil fuels. Sheppard et al. (2008) determined the activity ratio $^{210}Po/^{210}Pb$ for Canadian background sites indicating unity (1.0) in soils and lichens, whereas annual plant tissue diverged to an extent of 0.6. Lichens and mosses attained as expected, exceptionally high plant/soil concentration ratios (CRs) for ^{210}Pb, ^{210}Po, and ^{226}Ra, as cryptogams retain atmospheric particles rather effectively. Coal used for the generation of electric power apparently affects the distribution of different radioactive elements. As the combustion of coal is supposed to increase in the next century, it is rather important to observe the enlarged amounts of natural radionuclides, e.g., ^{210}Pb around coal-fired power plants near populated areas (Sert et al. 2011).

3.6.3 Redistribution of Radioactive Fallout by Fire

Dry summers with numerous lightening storms inevitably turn forested areas into zones of high fire-frequency. Shcherbov et al. (2008) investigated the radioactive fallout of the Chernobyl accident, transferred by smoke plumes, in a study of the ecogeochemical role of forest fires in the Baikal region. Paliouris et al. (1995) investigated the ecological implication of ^{137}Cs cycling in ecosystems in the Canadian boreal forest. The effect of fire was found to be decisive for the redistribution of fallout ^{137}Cs. Shcherbov et al. (2008) contended that forest fires accompanied by the airborne migration of several heavy metals (Cd, Hg, Pb) and artificial nuclides produce pollution in adjacent areas: up to 40 % of ^{137}Cs and ^{90}Sr and to a lesser extent $^{239,240}Pu$, originally contained in the soil-vegetation cover in the burned area, were deposited in forest litter, mosses, lichens, and tree needles in its vicinity. Paliouris et al. (1995) observed relatively high concentrations of ^{137}Cs in the organic surface soil of burned stands. In the unburnt stands on the other hand, the highest ^{137}Cs concentrations were identified in epiphytic lichens and in mosses. The fire was assumed to mobilize part of the ^{137}Cs bound to the above-ground matter and concentrate it in the ash layer of burned surface soil. Forest fires presumably make for the loss of part of the ^{137}Cs, through volatilization and fly ash or surface runoff, thus contaminating other ecosystems, initially unaffected, as well as previously contaminated ecosystems.

3.6.4 Mining and Metal Processing

A study performed in Brasil indicated significantly high mean values for ^{210}Pb concentrations in lichens and bryophytes in the vicinity of uranium mining and milling plants relative to the control region (Santos et al. 1993): soil and vegetables from kitchen gardens located near the mills contained much higher concentrations of ^{210}Pb than those observed for other regions. A similar study performed in Brazil (Dias da Cunha et al. 2004) detected ^{210}Pb and ^{226}Ra in aerosols and in samples of the lichen *Ramalina gracilis* in the exclusion area of a mineral sands processing plant and in the adjacent village.

To evaluate the applicability of lichens as bioindicators of air pollution by analyses of radionuclides and REE in the vicinity of a tin and lead industry in Brazil, Leonardo et al. (2011) used the large foliose epiphytic (or epilithic) lichen *Canoparmelia texana*. The authors examined the composition of the piles disposed in open air to obtain the fingerprint of the pollutant and the composition of the raw material used in the industry. As the lichen is assumed to derive nutrients mainly from wet- and dry precipitates, it was postulated that the re-suspension of soil-dust particles and the emissions from the stacks will accumulate in the lichen. Lichen, soil, raw material, and residue samples were analyzed for uranium, thorium, ^{226}Ra, ^{228}Ra, ^{210}Pb, and REEs. Lichen samples were found to concentrate radionuclides in amounts that were 15-fold higher than the background on average, whereas REEs attained amounts that were 10-fold higher on the average.

An example of a radioecological investigation under different climatic conditions is the study performed in the vicinity of prospected uranium mining sites in a subarctic environment (Northern Sweden). Radionuclides in the uranium series were detected in both soil, epigeic lichens (*Cladonia* sp.) and epiphytic lichens (*Alectoria* sp.). Aerial dispersion led to ten times higher concentrations of uranium, thorium, and radium in lichens in the vicinity of the mineralizations. At different distances from the mineralizations, the relevance of ground surface contamination was apparent. Plant-to-soil-activity ratios for *Alectoria*, *Cladonia,* and the mosses *Polytrichum* and *Sphagnum* varied to a great extent. The quotients of plant-to-soil ratios for ^{226}Ra and plant-to-soil ratios for ^{238}U averaged 1.7, 11.4, and 2.6 for *Alectoria*, *Cladonia,* and mosses, respectively. The higher quotient for *Cladonia* was produced by a higher transfer of radium (relative to uranium) from the soil. In the case of the mosses and the epiphytic *Alectoria*, the transfer of soil particles was insignificant (Pettersson et al. 1988).

Nineteen years after the closure of a lead–zinc mine in West Greenland, elevated concentrations of Pb in transplants of *F. nivalis* after one year of exposure indicated that the dispersal of contaminated dust still occurred within a distance of 20 km from the mine: dust is considered to be redeposited continuously. Additional dust is generated by the weathering of the waste rock left on the steep mountain sides (Søndergaard et al. 2011).

3.6.5 Underground Nuclear Explosion Sites

Plutonium (Pu) is a toxic radioactive element dispersed widely in the environment since 1945, mostly through testing of nuclear weapons and nuclear reactor accidents (Bu et al. 2013). Nuclear underground explosions were conducted in Sakha (Yakutia), North East Siberia, in the years 1974–1987. Consequently, radioactive products were discharged into the environment leading to residual radioactive contamination. Geodonov et al. (2002) investigated nuclear explosion sites and determined the concentrations and activity ratios for ^{137}Cs, $^{239+240}$Pu, and ^{238}Pu in lichens and soil. In a contaminated area, the $^{239+240}$Pu in soil in two sites exceeded the limit of 0.37 Bq/g determined for material, thus defined as radioactive waste. Clean-up efforts of these sites were found to be inadequate.

3.6.6 The Chernobyl Accident

The fallout of the Chernobyl accident on April 26, 1986, was subject to many studies investigating the distribution of radionuclides. Studies focused on the presence of radionuclides in both lichens and soil in the wake of the Chernobyl accident reported mostly on the detection of ^{137}Cs. The presence of radionuclides such as

^{137}Cs and others (for plutonium see review of Mietelski 2001) in soil and in epiphytic lichens was reported by van den Berg et al. (1992); Barci-Funel et al. (1995); Heinrich et al. (1999); Yoshida et al. (2004); Celik et al. (2009). Additional studies focused on radionuclides such as ^{137}Cs in soil and in both epigeic and epiphytic lichens (Steinnes and Njåstad 1993; Strandberg 1994). The sequence of accumulation of radionuclides in epiphytic lichens and its relevance to the radioactivity of soil, reported in the literature, was studied by Adamo et al. (2004). Following the Chernobyl accident, lichens sampled in Italy in October 1986 were found to contain lower concentrations of radionuclides than soil. Of the total ^{137}Cs measured in the lichens, only 14 % was attributed to the fallout of the 1950s and 1960s nuclear tests. A report presented by Mitchell et al. (1998) of different radionuclides such as curium (Cm) detected in soil and in lichens excluded, however, the relevant data of lichen identity and substrate. Several studies dealt with radionuclides in epigeic lichens providing forage, and in soil, in Nordic countries (Varskog et al. 1994; Pálsson et al. 1994; Outola et al. 2003; Skuterud et al. 2005; Dowdall et al. 2005; Korobova et al. 2007; Lehto et al. 2008).

The behavior of radioactive Cs in soil–plant systems was extensively studied in the literature, motivated by ^{137}Cs being one of the major long-lived constituents of the fallout from the atmospheric nuclear weapons testing in the 1950s and 1960s (Varskog et al.1994). Although correlating results referring to lichens and soil appear to corroborate the assertion that the contribution of soil particles is decisive, several studies pointed to the atmospheric origin of metal particles common to both media. In a study conducted in Norway in the years 1987–1989, Varskog et al. (1994) analyzed soil and lichen (*F. nivalis*) samples with respect to the total amount of radiocesium (^{137}Cs and ^{134}Cs), Rb, stable Cs and exchangeable ^{137}Cs (in soil only), K, Ca, and Mg. The radiocesium activity in lichens decreased by at least 40 % in the sampling period. This decrease was explained by the growth of new low-activity material at the top of the lichen in 1988 and 1989, thus diluting the radioactivity in the living part of the lichen. The fast decrease of activity in the lichen indicates that some of the radiocesium, accumulated on the outer surface, was washed out by rain. The results of the average vertical distribution of ^{137}Cs in the soil profile sampled in the above-mentioned study did not show a downward movement of ^{137}Cs. These findings are in accordance with the slow movement of ^{137}Cs in soil observed, for example, by Squire and Middleton (1966).

The activity of radionuclides in soil depends on its composition. Strandberg (1994) investigated a podsolic soil which consists mostly of sand and is covered by a ~5-cm-thick layer of decomposed organic matter with a low clay content of 0–2 %. This study, performed in Denmark, found that the Chernobyl cesium is distributed as follows: 20 % in the litter layer and 80 % in the upper 5 cm of the soil profile. The insignificant penetration observed for lower soil layers should be attributed to a certain form of fixation or binding to soil compounds. Since clay is almost completely absent, the author suggested that humus compounds are responsible for this binding of cesium. Similarly, Yoshida et al. (2004) found that the ^{137}Cs in soil was the highest in near-surface organic layers sampled in Belarus, decreasing with depth in the mineral layers, whereas the concentrations of stable Cs were almost constant in the entire soil profile.

Certain factors referring to the ^{137}Cs budget in the environment, such as industrial activity in regions subjected to the fallout of radionuclides, make for rather complex analyses of results. Such is the case in a study investigating the impact of a major Cu–Ni smelter in Harjavalta, SW Finland (Outola et al. 2003): the authors contended that the greater part of the ^{137}Cs bound to different environmental media, derived from the Chernobyl accident. The atmospheric pollution produced by the smelter damages the microbial population in its vicinity, thus inhibiting the decomposition of litter into organic material. This inhibition of decomposition produced a decrease of ^{137}Cs activity in the organic soil layer in the immediate vicinity of the smelter, whereas away of it the ^{137}Cs activity increased. The ^{137}Cs concentrations in lichens were supposed to increase in accordance with their proximity to the smelter in the same way they did in the litter layer. As the opposite situation was apparent, it is suggested that large amounts of trace metals accumulating in the lichens result in the removal of ^{137}Cs. Several studies demonstrated an enlarged loss of potassium following an exposure of lichens to trace elements (Puckett 1976; Nieboer et al. 1979; Tarhanen et al. 1999; Hyvärinen et al. 2000; Cabral 2003; Mrak et al. 2007) or to SO_2 (Puckett et al. 1977; Tomassini et al. 1977). Low K contents were observed also for lichens in environments polluted by trace metals and other contaminants (Garty et al. 1985, 1997c, 1998a, 2002c; Adamo et al. 2003; Pacheco et al. 2004; Williamson et al. 2004; Tretiach et al. 2007). Studies comparing the K content of lichen thalli and total elemental deposition, measured by air samplers, reveal sometimes a strong inverse correlation (Marques et al. 2009).

Circumpolar environmental conditions were illuminated by comparative analyses of the uptake of anthropogenic radionuclides such as ^{137}Cs and natural radionuclides such as ^{40}K, ^{238}U, ^{226}Ra, and ^{232}Th. Lichens, mosses, vascular plants, and soil in a High Arctic location (Svalbard) were analyzed for their comparative content of the above-mentioned elements. According to Dowdall et al. (2005), the levels of ^{137}Cs in vegetation generally followed the order: mosses > lichens > vascular plants. This sequence was explained by the different mechanism of uptake of this radionuclide. The concentration in vegetation showed little correlation with associated soil concentrations, whereas the uptake of ^{238}U, ^{226}Ra, and ^{232}Th by vascular and non-vascular plants was generally low. Plant species forming important components of Arctic food chains in the Yenisey Estuary (Russia) were ranked according to their ^{137}Cs activity in the following sequence: grasses < alder, willow (leaves), lichens < mosses (upper part) < mosses (lower part), litter (Korobova et al. 2007): the ^{137}Cs contamination of soil indicated both regional and global sources of anthropogenic pollution. The association of radionuclide activity in soil and epigeic lichens was not considered. Studies performed in Denmark (Strandberg 1994), Iceland (Pálsson et al. 1994), and Norway (Skuterud et al. 2005), however, demonstrated the implausibility of the transfer of ^{137}Cs from soil to lichens. The concentrations of ^{137}Cs activity in lichens declined due to washout, dilution by growth, and removal of contaminated parts by grazing reindeer (Skuterud et al. 2005). Lichens were the primary source for ^{137}Cs in reindeer in winter until 1998–1999. Since then, vascular plants became a major source for ^{137}Cs in winter (Skuterud et al. 2005).

A comparison of the ^{137}Cs concentrations in two lichens grazed by reindeer in Iceland found higher concentrations in *Cetraria delisei* than in *Flavocetraria islandica* (Pálsson et al. 1994). Although similar in structure and appearance, *C. delisei* was observed to be smaller and to have consequently a larger surface area per unit of mass. A relatively efficient collector of deposition, *C. delisei* apparently accumulated higher concentrations of ^{137}Cs.

Lehto et al. (2008) used epigeic lichens to analyze the activity of radionuclides in lichen-soil columns in order to determine the transfer of radionuclides from lichen to soil and from the organic soil layer to the rock surface in order to predict the performance of radionuclides in lichen carpets. This information was needed to determine, for example, the period of time at which the lichen is representative of total deposition values. Four different procedures were determined for radionuclides to be removed or transferred downward in lichen-soil columns: (1) Particles attached to the lichen surface removed mechanically by rain and wind, (2) radionuclides taken up into the lichen along with nutrients and transported within the lichen, (3) radionuclides, present in surface-attached particles, dissolved and transported toward the underlying soil, and (4) radionuclides transferred in a process of decomposition and growth of lichens. Twenty years after the Chernobyl accident, epilithic species in W. Macedonia, N. Greece, were contaminated to a greater extent by ^{137}Cs than epiphytic species (Sawidis et al. 2010). Epiphytic lichens depend on airborne nutrients, whereas lichens inhabiting rock or soil depend on the proximity of their substrate. Thus, the epilithic lichens presumably reflect conditions in the soil.

A study performed in France was designed to follow the transfer of radionuclides from soil to trees with lichens on them. The radionuclide activity in the tree rings and in the deposition of fallout did not correlate and no direct track soil–tree–lichen could be detected (Barci-Funel et al. 1995). Another study found that the location of epiphytic lichens on tree trunks is a determinant of the degree of activity of radionuclides in the thallus. According to Heinrich et al. (1999), the epiphytic fruticose lichen *P. furfuracea* growing on dead trunks in the Styria Province, Austria, was contaminated about three times as much at the top-end relative to the lower section of the tree. This was explained by the vertical rainfall making for a greater rate of absorption at the upper end. In certain parts of Styria, a high degree of ^{137}Cs contamination was detected in soil relative to the ^{90}Sr soil contamination, following the Chernobyl accident. The assumption that the greater part of the radionuclide contamination derives from atmospheric deposition was corroborated by others (Strandberg 1994), whereas the possible contribution of soil contamination was largely dismissed.

In Turkey, Celik et al. (2009) determined the sequence of concentrations of ^{137}Cs activity in soil, moss, and epiphytic lichens in the following descending order: ^{137}Cs moss > ^{137}Cs lichen > ^{137}Cs soil explained by factors such as removal from soil by washing-off versus a continued fixation of radionuclides in cryptogams. Indeed, a strong correlation was obtained for concentrations of ^{137}Cs activity in lichens and moss samples. According to van den Berg et al. (1992), the averaged levels of ^{137}Cs of atmospheric origin in lichens (*P. sulcata*) around Chernobyl were generally in proportion to soil surface deposition unlike higher soil ^{137}Cs activity classes (>500 kBq m^{-2}).

The radionuclide activity of epiphytic lichens depends also on the characteristics of the sampled tree: coniferous trees and birches yield different results. According to Steinnes and Njåstad (1993), the epiphytic foliose lichen *H. physodes* growing on birch in Norway and collected in July 1986, shortly after the Chernobyl accident, was exposed to a greater extent to the precipitation that carried the radioactive fallout than the deciduous leafless birch. On coniferous trees, on the other hand, a significant fraction of the radionuclides was fixed to the tree needles. The use of *H. physodes* to biomonitor Cs was therefore discouraged.

3.6.7 Depleted Uranium in the Balkan War Zone

Lichens are known to accumulate uranium emitted by various sources (Gough and Erdman 1977; Beckett et al. 1982; Boileau et al. 1982; Looney et al. 1985; Sheard 1986a, b; Pettersson et al. 1988; Singh and Wrenn 1988; Fahselt et al. 1995; Jeran et al. 1995, 1996, 2002, 2003, 2007; Daillant et al. 1996; Trembley et al. 1997; Gueidan et al. 1997; Horvat et al. 2000; Golubev et al. 2005; Purvis et al. 2006). In the war fought in the Balkan area in the years 1995–1999, depleted uranium (DU) was emitted by anti-tank ammunition fired by NATO aircrafts. DU is the by-product of the industrial process applied to enrich natural uranium for use in nuclear reactors and nuclear weapons (Jia et al. 2006). DU is an extremely dense (19.05 g cm^{-3}), hard, autopyrophoric, relatively cheap metal used for military applications such as piercing of munitions and enhancement of armor protection (Jia et al. 2005). Soil analyses performed in the Kosovo area indicated that U concentrations exceeding normal environmental values (~2–3 mg/kg) were due to DU contamination (Danesi et al. 2003a). An additional study of the same area yielded results which did not permit a legitimate extrapolation for all war sites hit by DU in the conflict, but indicated the probability of "spots" where hundreds of thousands of particles are present in a few milligrams of DU-contaminated soil (Danesi et al. 2003b), About 50 % of the DU particles collected in the Kosovo war zone were characterized as UO_2, whereas the remaining DU particles were either U_3O_8 or a mixture of oxidized forms (Salbu et al. 2003).

Following the Kosovo conflict, Loppi et al. (2003) observed that the average U concentrations found in lichens collected in the Balkan were in line with values reported for lichens in other countries. The measurement of uranium concentrations in lichen thalli, in addition to a calculation of $^{235}U/^{238}U$ isotopic ratios, denied the presence of depleted uranium in the Balkan area. These findings were supported by Di Lella et al. (2003) who detected species-specific differences of total uranium concentrations in the lichens *X. parietina, Phaeophyscia orbicularis, Physcia adscendens,* and *Physcia biziana* contaminated by soil particles. The results of another study (Di Lella et al. 2004) suggested that the use of DU ammunitions in Kosovo did not cause a diffuse environmental contamination capable of producing a detectable U enrichment in lichens. Isotopic $^{235}U/^{238}U$ did not indicate the presence of DU particles in lichens either. Sansone et al. (2001), however, reported on a large degree of variability in the concentrations of uranium activity in the surface soil. As DU was detected in lichens, their role as sensitive bioindicators of DU dust or aerosols, formed at targets hit by DU ammunition, was accentuated. Uranium isotopes detected in lichens and tree bark in Bosnia-Herzegovina (Rosamilia et al. 2004) still bore evidence, eight years after the conflict, of past airborne contamination by depleted uranium in target sites. Similarly, Žunić et al. (2008) detected traces of DU in lichens and soil in target sites in the Balkan and recommended the use of lichens as bioindicators for the presence of DU in the environment.

3.7 Conclusions and Open Questions

The traditional definition of lichens describes different organisms living in a stable symbiotic association. Recent knowledge indicates, however, that an important share of additional, non-cellular constituents is integrated in the lichen thallus. Mineral particulate matter originating in the environment apparently derives from both crustal and anthropogenic sources. The significance of trace metals produced by anthropogenic activity cannot be overstated: many trace metals are toxic for living organisms. The historical reason to apply lichens as biomonitors was their capability to absorb and accumulate airborne anthropogenic particulate matter, whereas the inclusion of soil particles entrapped-in the thallus was considered to be of lesser significance.

Lichenologists focused in the 1960s and the 1970s on the capacity of lichens as active or passive biomonitors to yield data referring to air quality based on analyses of elemental content. The 1980s and 1990s of the last century introduced, however, innovative studies including, in addition to chemical analyses of lichen samples, data referring to the physiological status of the lichen. Lichen physiology and cell biology of the thallus were extensively studied also in the first decade of the new millennium. In the last 20 years, combined chemical and physiological studies of lichens used as biomonitors, with an extensive use of collectors, were designed and operated to detect particulate matter derived from both crustal sources and anthropogenic activity. Combined studies of this kind increase the validity of lichen biomonitoring, as chemical analyses of particulate matter collected by samplers yield corroborating data. The significant progress made in recent years makes for a better understanding of the capability of lichens to demonstrate alterations and enhance the applicability of bioaccumulators and bioindicators.

Our understanding of the strategy applied in the domain of chemical analysis and biomonitoring is at a challenging junction. It appears that the second decade of the new millennium will yield several comparative studies investigating chemical content and physiological status in addition to a detailed description of the composition of the particulate matter trapped by collectors. Future studies should be dedicated also to an effort to quantify the share of soil in the composition of epigeic lichens, in order to accomplish an evaluation of the upward flow of soil particles into lichen carpets. The immediacy of this subject is accentuated by herbivores (reindeer and others) grazing in areas infested by radionuclides.

Lichenologists/bryologists need to face the dilemma of pre-treatment: to wash or not to wash? Particulate matter adhered and/or included in the plant material provides data referring to the environmental quality and pre-treatment by washing eliminates an important constituent of the lichen or the moss. Furthermore, is it legitimate to ask: what is the organism we want to analyze? Are the constituents of a lichen just and only algal and/or cyanobacterial cells and fungal hyphae? Is there an organism which does not contain any mineral component? And what about calcium oxalate crystals (Fig. 3.7), sometimes produced in great amounts in and on the thallial surface (Garty et al. 2002a)? What about calcium sulfate crystals (Fig. 3.8) that emerge from the thallial surface of lichens exposed to SO_2 in polluted industrial sites (Garty et al. 2002a)? Do we and can we eliminate such structures by 5–8 washings, sonication, and shaking in an aquatic medium and other drastic treatments and then determine chemical elements located in a "pure" organic association?

It appears improbable to expect a definite conclusion: is the lichen indeed the sum of its "pure," "traditional" components and is there room for particulate matter as a third component? Should scientists refer to particulates as "aliens" or are these mineral structures legitimate members of the symbiotic association? Finally, Table 3.4 and other references indicate that epiphytic lichens are collected in many cases close to ground level. The issue of pre-treatment by washing (Table 3.5) needs to be addressed especially in these cases. Of crucial importance in this context is also the increasing number of annual desert-dust storms, especially in the Mediterranean basin, in southern Europe and probably in other parts of the world.

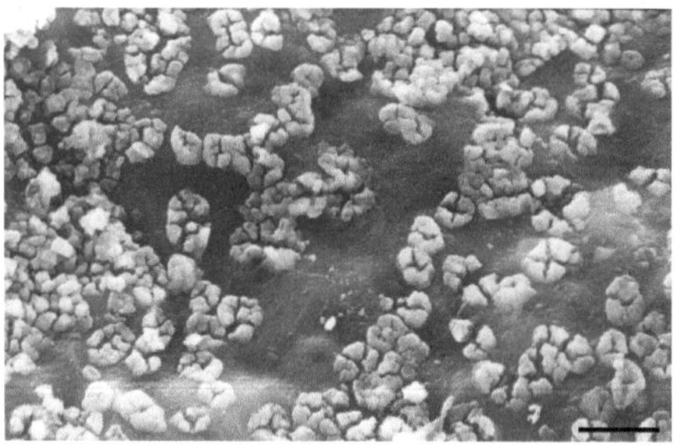

Fig. 3.7 Crater-like structures on thallial surface of in situ samples of *Ramalina lacera* collected in HaZorea Forest, N.E. Israel. Significant amounts of calcium were demonstrated by EDX analyses. Infrared spectroscopy of powders obtained by scraping the thallial surface found crystals of calcium oxalate monohydrate (whewellite: $CaC_2O_4 \cdot H_2O$) and calcium oxalate dihydrate [weddellite: $CaC_2O_4 \cdot (X + 2)H_2O$]. Scale bar = 10 μm (Garty et al., unpublished)

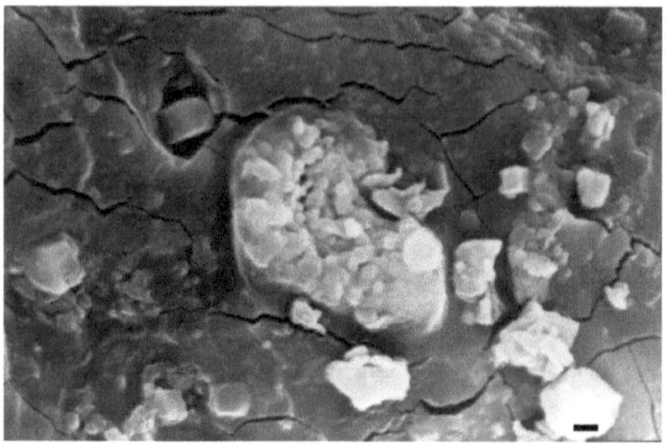

Fig. 3.8 Edge of *Ramalina lacera* thallus collected in HaZorea Forest, transplanted and retrieved after 10 months of exposure near the oil refineries in Ashdod, SW Israel, covered by crater-like structures. EDX spectra showed these structures to consist mainly of Ca and S. Powder samples of scraped thalli revealed crystals of whewellite, weddellite, and gypsum (calcium sulfate dihydrate—$CaSO_4 \cdot 2H_2O$). Scale bar = 1 μm (Garty, Kunin, Delarea, Weiner, unpublished)

References

Aarkrog A, Dahlgaard H, Holm E, Hallstadius L (1984) Evidence for bismuth-207 in global fallout. J Environ Radioact 1:107–117

Adamo P, Giordano S, Vingiani S, Castaldo Cobianchi R, Violante P (2003) Trace element accumulation by moss and lichen exposed in bags in the city of Naples (Italy). Environ Pollut 122:91–103

Adamo P, Arienzo M, Pugliese M, Roca V, Violante P (2004) Accumulation history of radionuclides in the lichen *Stereocaulon vesuvianum* from Mt. Vesuvius (south Italy). Environ Pollut 127:455–461

Adamo P, Crisafulli P, Giordano S, Minganti V, Modenesi P, Monaci F et al (2007) Lichen and moss bags as monitoring devices in urban areas. Part II. Trace element content in living and dead biomonitors and comparison with synthetic materials. Environ Pollut 146:392–399

Adamo P, Bargagli R, Giordano S, Modenesi P, Monaci F, Pittao E et al (2008a) Natural and pre-treatments induced variability in the chemical composition and morphology of lichens and mosses selected for active monitoring of airborne elements. Environ Pollut 152:11–19

Adamo P, Giordano S, Naimo D, Bargagli R (2008b) Geochemical properties of airborne particulate matter (PM10) collected by automatic device and biomonitors in a Mediterranean urban environment. Atmos Environ 42:346–357

Affum HA, Oduro-Afriyie K, Nartey VK, Adomako D, Nyarko BJB (2008) Biomonitoring of airborne heavy metals along a major road in Accra, Ghana. Environ Monit Assess 137:15–24

Agnan Y, Séjalon-Delmas N, Probst A (2013) Comparing early twentieth century and present-day atmospheric pollution in SW France: a story of lichens. Environ Pollut 172:139–148

Aničić M, Tasić M, Frontasyeva MV, Tomaševic M, Rajšić S, Mijić Z et al (2009) Active moss biomonitoring of trace elements with *Sphagnum girgensohnii* moss bags in relation to atmospheric bulk deposition in Belgrade, Serbia. Environ Pollut 157:673–679

Aprile GG, Di Salvatore M, Carratù G, Mingo A, Carafa AM (2010) Comparison of the suitability of two lichen species and one higher plant for monitoring airborne heavy metals. Environ Monit Assess 162:291–299

Ares A, Aboal JR, Carballeira A, Giordano S, Adamo P, Fernández JA (2012) Moss bag biomonitoring: a methodological review. Sci Total Environ 432:143–158

Aslan A, Budak G, Karabulut A (2004) The amounts of Fe, Ba, Sr, K, Ca and Ti in some lichens growing in Erzurum province (Turkey). J Quant Spectrosc Radiat Transfer 88:423–431

Aslan A, Budak G, Tiraşoğlu E, Karabulut A (2006) Determination of elements in some lichens growing in Giresum and Ordu province (Turkey) using energy dispersive X-ray fluorescence spectrometry. J Quant Spectrosc Radiat Transfer 97:10–19

Aubert D, Le Roux G, Krachler M, Cherbukin A, Kober B, Shotyk W et al (2006) Origin and fluxes of atmospheric REE entering an ombrotrophic peat bog in Black Forest (SW Germany): Evidence from snow, lichens and mosses. Geochim Cosmochim Acta 70:2815–2826

Avila A, Penũelas J (1999) Increasing frequency of Saharan rains over northeastern Spain and its ecological consequences. Sci Total Environ 228:153–156

Ayrault S, Clochiatti R, Carrot F, Daudin L, Bennett JP (2007) Factors to consider for trace element deposition biomonitoring surveys with lichen transplants. Sci Total Environ 372:717–727

Bačkor M, Fahselt D (2004) Physiological attributes of the lichen *Cladonia pleurota* in heavy metal-rich and control sites near Sudbury (Ont. Canada). Environ Exp Bot 52:149–159

Bačkor M, Loppi S (2009) Interaction of lichens with heavy metals. Biol Plant 53:214–222

Bačkor M, Klejdus B, Vantová I, Kováčik J (2009) Physiological adaptations in the lichens *Peltigera rufescens* and *Cladina arbuscula* var. *mitis*, and the moss *Racomitrium lanuginosum* to copper-rich substrate. Chemosphere 76:1340–1343

Bajpai R, Upreti DK, Dwiveki SK (2009) Arsenic accumulation in lichens of Mandav monuments, Dhar district, Madhya Pradesh, India. Environ Monit Assess 159:437–442

Baptista MS, Vasconcelos MTSD, Cabral JP, Freitas MC, Pacheco AMG (2008) Copper, nickel and lead in lichen and tree bark transplants over different periods of time. Environ Pollut 151:408–413

Barci-Funel G, Dalmasso J, Barci VL, Ardisson G (1995) Study of the transfer of radionuclides in trees at a forest site. Sci Total Environ 173/174:369–373

Bargagli R (1990a) Mercury emission in an abandoned mining area: assessment by epiphytic lichens. In: Cheremisinoff PN (ed) Encyclopedia of environmental control technology, vol 4. Hazardous waste containment and treatment. Gulf Publishing Company, Houston, pp 613–640

Bargagli R (1990b) Assessment of metal air pollution by epiphytic lichens: The incidence of crustal materials and of possible uptake from substrate barks. Stud Geobotanica 10:97–103

Bargagli R (1995) The elemental composition of vegetation and the possible incidence of soil contamination of samples. Sci Total Environ 176:121–128

Bargagli R (1998) Trace elements in terrestrial plants. An ecophysiological approach to biomonitoring and biorecovery, Springer, Berlin

Bargagli R, Iosco FP, Leonzo C (1985) Monitarggio di elementi in trace mediante licheni epifiti. Inquinamento 2:33–37

Bargagli R, Iosco FP, Barghigiani C (1987) Assessment of mercury dispersal in an abandoned mining area by soil and lichen analysis. Water Air Soil Pollut 36:219–225

Bargagli R, Battisti E, Focardi S, Formichi P (1993) Preliminary data on environmental distribution of mercury in northern Victoria Land, Antarctica. Antarct Sci 5:3–8

Bargagli R, Brown DH, Nelli L (1995) Metal biomonitoring with mosses: procedures for correcting for soil contamination. Environ Pollut 89:169–175

Bargagli R, Sanchez-Hernandez JC, Monaci F (1999) Baseline concentrations of elements in the Antarctic macrolichen *Umbilicaria decussata*. Chemosphere 38:475–487

Bargagli R, Monaci F, Borghini R, Bravi F, Agnorelli C (2002) Mosses and lichens as biomonitors of trace metals. A comparison study on *Hypnum cupressiforme* and *Parmelia caperata* in a former mining district in Italy. Environ Pollut 116:279–287

Bargagli R, Monaci F, Agnorelli C (2003) Oak leaves as accumulators of airborne elements in an area with geochemical an geothermal anomalies. Environ Pollut 124:321–329

Bargagli R, Monaci F, Bucci C (2007) Environmental biogeochemistry of mercury in Antarctic ecosystems. Soil Biol Biochem 39:352–360

Bari A, Rosso A, Minciardi MR, Troiani F, Piervittori R (2001) Analysis of heavy metals in atmospheric particulates in relation to their bioaccumulation in explanted *Pseudevernia furfuracea* thalli. Environ Monit Assess 69:205–220

Bartók K (1988) Heavy metal distribution in several lichen species in a polluted area. Rev Roum Biol—Biol Végét 33:127–134

Basile A, Sorbo S, Aprile G, Conte B, Castaldo Cobianchi R (2008) Comparison of the heavy metal bioaccumulation capacity of an epiphytic moss and an epiphytic lichen. Environ Pollut 151:401–407

Beckett PJ, Boileau LJR, Padovan D, Richardson DHS, Nieboer E (1982) Lichens and mosses as monitors of industrial activity associated with uranium mining in northern Ontario, Canada—part 2: distance dependent uranium and lead accumulation patterns. Environ Pollut B 4:91–107

Bennett JP, Wetmore CM (1999) Geothermal elements in lichens of Yellowstone National Park, USA. Environ Exp Bot 42:191–200

Bennett JP, Wetmore CM (2003) Elemental chemistry of four lichen species from the Apostle Islands, Wisconsin, 1987, 1995 and 2001. Sci Total Environ 305:77–86

Bergamaschi L, Rizzio E, Valcuvia MG, Verza G, Profumo A, Gallorini M (2002) Determination of trace elements and evaluation of their enrichment factors in Himalayan lichens. Environ Pollut 120:137–144

Bergamaschi L, Rizzio E, Giavery G, Profumo A, Loppi S, Gallorini M (2004) Determination of baseline element composition of lichens using samples from high elevations. Chemosphere 55:933–939

Bergamaschi L, Rizzio E, Giaveri G, Loppi S, Gallorini M (2007) Comparison between the accumulation capacity of four lichen species transplanted to an urban site. Environ Pollut 148:468–476

Bermudez GMA, Rodriguez JH, Pignata ML (2009) Comparison of the air pollution biomonitoring ability of three *Tillandsia* species and the lichen *Ramalina celastri* in Argentina. Environ Res 109:6–14

Bettinelli M, Spezia S, Bizzarri G (1996) Trace elements determination in lichens by ICP-MS. Atom Spectrosc 17:133–141

Bettinelli M, Perotti M, Spezia S, Baffi C, Beone GM, Alberici F et al (2002) The role of analytical methods for the determination of trace elements in environmental biomonitors. Microchem J 73:131–152

Boamponsem LK, Adam JI, Dampare SB, Nyarko BJB, Essumang DK (2010) Assessment of atmospheric heavy metal deposition in the Tarkwa gold mining area of Ghana using epiphytic lichens. Nucl Instrum Methods Phys Res B 268:1492–1501

Boileau LJR, Beckett PJ, Lavoie P, Richardson DHS, Nieboer E (1982) Lichens and mosses as monitors of industrial activity associated with uranium mining in northern Ontario, Canada—part 1: field procedures, chemical analysis and interspecies comparisons. Environ Pollut B 4:69–84

Bosserman RW, Hagner JE (1981) Elemental composition of epiphytic lichens from Okefenokee Swamp. Bryologist 84:48–58

Branquinho C, Gaio-Oliveira G, Augusto S, Pinho P, Máguas C, Correia O (2008) Biomonitoring spatial and temporal impact of atmospheric dust from a cement industry. Environ Pollut 151:292–299

Brunialti G, Frati L (2007) Biomonitoring of nine elements by the lichen *Xanthoria parietina* in Adriatic Italy: a retrospective study over a 7-year time span. Sci Total Environ 387:289–300

Bu K, Cizdziel JV, Dasher D (2013) Plutonium concentration and $^{240}Pu/^{239}Pu$ atom ratio in biota collected from Amchitka Island, Alaska: recent measurements using ICP-SFMS. J Environ Radioact 124:29–36

Bubach D, Arribére MA, Calvelo S, Ribeiro Guevara S, Román Ross G (1998) Characterization of minor and trace element contents of *Protousnea magellanica* from pristine areas of Northern Patagonia. Lichens 2:1–7

Bubach D, Pérez Catán S, Arribére M, Ribeiro Guevara S (2012) Bioindication of volatile elements emission by the Puyehue-Cordón Caulle (North Patagonia) volcanic event in 2011. Chemosphere 88:584–590

Cabral JP (2003) Copper toxicity to five Parmelia lichens in vitro. Environ Exp Bot 49:237–250

Calvelo S, Baccala N, Arribere MA, Ribeiro Guevara S, Bubach D (1997) Analytical and statistical analysis of elemental composition of lichens. J Radioanal Nucl Ch 222:99–104

Caniglia G, Calliari I, Celin L, Tollardo AM (1994) Metal determination by EDXBF in lichens. A contribution to pollutants monitoring. Biol Trace Elem Res 43–45:213–221

Carignan J, Simonetti A, Gariépy C (2002) Dispersal of atmospheric lead in northeastern North America as recorded by epiphytic lichens. Atmos Environ 36:3759–3766

Carreras HA, Pignata ML (2002) Biomonitoring of heavy metals and air quality in Córdoba City, Argentina, using transplanted lichens. Environ Pollut 117:77–87

Carreras HA, Rodriguez JH, González CM, Wannaz ED, Garcia Ferreyra F, Perez CA et al (2009a) Assessment of the relationship between total suspended particles and the response of two biological indicators transplanted to an urban area in central Argentina. Atmos Environ 43:2944–2949

Carreras HA, Wannaz ED, Pignata ML (2009b) Assessment of human health risk related to metals by the use of biomonitors in the province of Córdoba, Argentina. Environ Pollut 157:117–122

Carreras HA, Wannaz ED, Perez CA, Pignata ML(2005) The role of urban air pollutants on the performance of heavy metal accumulation in *Usnea amblyoclada*. Environ Res 97:50–57

Celik N, Cevik U, Celik A, Koz B (2009) Natural and artificial radioactivity measurements in Eastern Black Sea region in Turkey. J Hazard Mater 162:146–153

Chaparro MAE, Lavornia JM, Chaparro MAE, Sinito AM (2013) Biomonitors of urban air pollution: Magnetic studies and SEM observations of corticolous foliose and microfoliose lichens and their suitability for magnetic monitoring. Environ Pollut 172:61–69

Chettri MK, Sawidis T, Karataglis S (1997) Lichens as tool for biogeochemical prospecting. Ecotoxicol Environ Saf 38:322–335

Chiarenzelli J, Aspler L, Dunn C, Cousens B, Ozarko D, Powis K (2001) Multi-element and rare earth element composition of lichens, mosses and vascular plants from the Central Barrenlands, Nunavut, Canada. Appl Geochem 16:245–270

Clark BM, St Clair LL, Mangelson NF, Rees LB, Grant PG, Bench GS (2001) Characterization of mycobiont adaptations in the foliose lichen *Xanthoparmelia chlorochroa* (Parmeliaceae). Am J Bot 88:1742–1749

Cloquet C, Carignan J, Libourel G (2006) Atmospheric pollutant dispersion around an urban area using trace metal concentrations and Pb isotopic composition in epiphytic lichens. Atmos Environ 10:574–587

Conti ME, Cecchetti G (2001) Biological monitoring: lichens as bioindicators of air pollution assessment—a review. Environ Pollut 114:471–492

Costa CJ, Marques AP, Freitas MC, Reis MA, Oliveira OR (2002) A comparative study for results obtained using biomonitors and PM10 collectors in Sado Estuary. Environ Pollut 120:97–106

Culicov OA, Yurukova L (2006) Comparison of element accumulation of different moss—and lichen bags, exposed in the city of Sofia (Bulgaria). J Atmos Chem 55:1–12

Daillant O, Kirchner G, Jacquiot L, Lecointe A, Tillier C, Van Haluwyn C (1996) Recherche d'élément radioactifs natureles dans quelques lichens foliaces. Pollut Atmos 65–79

Danesi PR, Bleise A, Burkart W, Cabianca T, Campbell MJ, Makarewicz M et al (2003a) Isotopic composition and origin of uranium and plutonium in selected soil samples collected in Kosovo. J Environ Radioactiv 64:121–131

Danesi PR, Markowicz A, Chinea-Cano E, Burkart W, Salbu B, Donohue D et al (2003b) Depleted uranium particles in selected Kosovo samples. J Environ Radioactiv 64:143–154

Di Lella LA, Frati L, Loppi S, Protano G, Riccobono F (2003) Lichens as biomonitors of uranium and other trace elements in an area of Kosovo heavily shelled with depleted uranium rounds. Atmos Environ 37:5445–5449

Di Lella LA, Frati L, Loppi S, Protano G, Riccobono F (2004) Environmental distribution of uranium and other trace elements at selected Kosovo sites. Chemosphere 56:861–865

Dias da Cunha K, Barros Leite CV, Zays Z (2004) Exposure to mineral sands dust particles. Nucl Instrum Methods Phys Res B 217:649–656

Dillman KL (1996) Use of the lichen *Rhizoplaca melanophthalma* as a biomonitor in relation to phosphate refineries near Pocatello, Idaho. Environ Pollut 92:91–96

Doğrul Demiray A, Yolcubal I, Akyol NH, Çobanoğlu G (2012) Biomonitoring of airborne metals using the lichen *Xanthoria parietina* in Kocaeli Province, Turkey. Ecol Ind 18:632–643

Dolgopolova A, Weiss DJ, Seltmann R, Kober B, Mason TFD, Coles B et al (2006) Use of isotope ratios to assess sources of Pb and Zn dispersed in the environment during mining and ore processing within the Orlovka-Spokoinoe mining site (Russia). Appl Geochem 21:563–579

Dongarrà G, Varrica D (1998) The presence of heavy metals in air particulate at Vulcano island (Italy). Sci Total Environ 212:1–9

Dongarrà G, Ottonello D, Sabatino G, Triscari M (1995) Use of lichens in detecting environmental risk and in geochemical prospecting. Environ Geol 26:139–146

Dowdall M, Gwynn JP, Moran C, O'Dea J, Davids C, Lind B (2005) Uptake of radionuclides by vegetation at a High Arctic location. Environ Pollut 133:327–332

Dzubaj A, Bačkor M, Tomko J, Peli E, Tuba Z (2008) Tolerance of the lichen *Xanthoria parietina* (L.) Fr. to metal stress. Ecotoxicol Environ Saf 70:319–326

El Khoukhi T, Cherkaoui RM, Gaudry A, Ayrault S, Senhou A, Chouak A et al (2004) Air pollution biomonitoring survey in Morocco using ko-INAA. Nucl Instrum Methods Phys Res B 213:770–774

Fahselt D, Wu T-W, Mott B (1995) Trace element patterns in lichens following uranium mine closures. Bryologist 98:228–234

Farinha MM, Verburg TG, Freitas MC, Wolterbeek HTh (2009) Local and regional sources as given by aerosol measurements and biomonitoring at Sado estuary area, Portugal. J Radioanal Nucl Ch 281:299–305

Fenn ME, Geiser L, Bachman R, Blubaugh TJ, Bytnerowicz A (2007) Atmospheric deposition inputs and effects on lichen chemistry and indicator species in the Columbia River Gorge, USA. Environ Pollut 146:77–91

Freitas MC, Reis MA, Alves LC, Wolterbeek HTh (1999) Distribution in Portugal of some pollutants in the lichen *Parmelia sulcata*. Environ Pollut 106:229–235

Freitas MC, Pacheco AMG, Baptista MS, Dionisio I, Vasconcelos MTSD, Cabral JP (2007) Response of exposed detached lichens to atmospheric elemental deposition. Ecol Chem Eng 14:631–644

Gadd GM (2007) Geomycology: biogeochemical transformations of rocks, minerals, metals and radionuclides by fungi bioweathering and bioremediation. Mycol Res 111:3–49

Gailey FAY, Lloyd OLI (1986) Methodological investigations into low technology monitoring of atmospheric metal pollution: part 2—the effect of length of exposure on metal concentrations. Environ Pollut B 12:61–74

Ganor E, Stupp A, Alpert P, Osetinsky I (2007) Increasing trends in dust cloud intrusion from the Sahara over Israel. Abstracts of the IUGG 27, Perugia, Italy, 2–13 July 2007, p 210

Garty J (1985) The amounts of heavy metals in some lichens in the Negev Desert. Environ Pollut B 10:287–300

Garty J (2000) Trace metals, other chemical elements and lichen physiology: research in the nineties. In: Markert B, Friese K (eds) Trace elements—their distribution and effects in the environment. Elsevier, Amsterdam, pp 277–322

Garty J (2001) Biomonitoring atmospheric heavy metal with lichens: theory and application. Crit Rev Plant Sci 20:309–371

Garty J (2002) Biomonitoring heavy metal pollution with lichens. In: Kranner I, Beckett RP, Varma AK (eds) Protocols in lichenology. Springer, Culturing, biochemistry, ecophysiology and use in biomonitoring. Berlin, pp 458–482

Garty J, Amman K (1987) The amount of Ni, Cr, Zn, Pb, Cu, Fe, and Mn in some lichens growing in Switzerland. Environ Exp Bot 27:127–138

Garty J, Delarea J (1987) Some initial stages in the formation of epilithic crustose lichens in nature: a SEM study. Symbiosis 3:49–56

Garty J, Fuchs C, Zisapel N, Galun M (1977) Heavy metals in the lichen *Caloplaca aurantia* from urban, suburban and rural regions in Israel (a comparative study). Water Air Soil Pollut 8:171–188

Garty J, Galun M, Kessel M (1979) Localization of heavy metals and other elements accumulated in the lichen thallus. New Phytol 82:159–168

Garty J, Ronen R, Galun M (1985) Correlation between chlorophyll degradation and the amount of some elements in the lichen *Ramalina duriaei* (De Not.) Jatta. Environ Exp Bot 25:67–74

Garty J, Harel Y, Steinberger Y (1995) The role of lichens in the cycling of metals in the Negev Desert. Arch Environ Contam Toxicol 29:247–253

Garty J, Kauppi M, Kauppi A (1996a) Accumulation of airborne elements from vehicles in transplanted lichens in urban sites. J Environ Qual 25:265–272

Garty J, Steinberger Y, Harel Y (1996b) Spatial and temporal changes in the concentration of K, Na, Mg and Ca in epilithic and in decomposing detached thalli of the lichen *Ramalina maciformis* and its potential role in the cycling of these elements in the Negev Desert. Environ Exp Bot 36:83–97

Garty J, Cohen Y, Kloog N, Karnieli A (1997a) Effects of air pollution on cell membrane integrity, spectral reflectance and metal and sulfur concentrations in lichens. Environ Toxicol Chem 16:1396–1402

Garty J, Karnieli A, Wolfson R, Kunin P, Garty-Spitz R (1997b) Spectral reflectance and integrity of cell membranes and chlorophyll relative to the concentration of airborne mineral elements in a lichen. Physiol Plant 101:257–264

Garty J, Kloog N, Wolfson R, Cohen Y, Karnieli A, Avni A (1997c) The influence of air pollution on the concentration of mineral elements, on the spectral reflectance response and on the production of stress-ethylene in the lichen *Ramalina duriaei*. New Phytol 137:587–597

Garty J, Cohen Y, Kloog N (1998a) Airborne elements, cell membranes, and chlorophyll in transplanted lichens. J Environ Qual 27:973–979

Garty J, Kloog N, Cohen Y (1998b) Integrity of lichen cell membranes in relation to concentration of airborne elements. Arch Environ Contam Toxicol 34:136–144

Garty J, Weissman L, Tamir O, Beer S, Cohen Y, Karnieli A et al (2000) Comparison of five physiological parameters to assess the vitality of the lichen *Ramalina lacera* exposed to air pollution. Phys Plant 109:410–418

Garty J, Tamir O, Hassid I, Eshel A, Cohen Y, Karnieli A et al (2001a) Photosynthesis, chlorophyll integrity, and spectral reflectance in lichens exposed to air pollution. J Environ Qual 30:884–893

Garty J, Weissman L, Cohen Y, Karnieli A, Orlovsky L (2001b) Transplanted lichens in and around the Mount Carmel National Park and the Haifa Bay industrial region in Israel: physiological and chemical responses. Environ Res A85:159–176

Garty J, Kunin P, Delarea J, Weiner S (2002a) Calcium oxalate and sulphate-containing structures on the thallial surface of the lichen *Ramalina lacera*: response to polluted air and simulated acid rain. Plant, Cell Environ 25:1591–1604

Garty J, Levin T, Cohen Y, Lehr H (2002b) Biomonitoring air pollution with the desert lichen *Ramalina maciformis*. Physiol Plant 115:265–275

Garty J, Tamir O, Cohen Y, Lehr H, Goren AI (2002c) Changes in the potential quantum yield of photosystem II and the integrity of cell membranes relative to the elemental content of the epilithic desert lichen *Ramalina maciformis*. Environ Toxicol Chem 21:848–858

Garty J, Tomer S, Levin T, Lehr H (2003) Lichens as biomonitors around a coal-fired power station in Israel. Environ Res 91:186–198

Garty J, Tamir O, Levin T, Lehr H (2007) The impact of UV-B and sulphur—or copper—containing solutions in acidic conditions on chlorophyll fluorescence in selected *Ramalina* species. Environ Pollut 145:266–273

Garty J, Lehr H, Garty-Spitz RL, Ganor E, Stupp A, Alpert P, Osetinsky I (2008) Temporal fluctuations of mineral concentrations as related to the physiological status of the lichen *Ramalina lacera* (With) J.R. Laund. Isr J Plant Sci 56:361–369

Garty J, Lehr H, Garty-Spitz RL (2009) Three decades of biomonitoring Pb in a rural area with the epiphytic lichen *Ramalina lacera*: a retrospective study. Isr J Plant Sci 56:25–34

Geodonov AD, Petrov ER, Alexeev VG, Kuleshova IN, Savopulo ML, Burtsev IS et al (2002) Residual radioactive contamination of the peaceful underground nuclear explosion sites "Craton-3" and "Crystal" in the republic of Sakha (Yakutia). J Environ Radioactiv 60:221–234

Getty SR, Gutzler DS, Asmeron Y, Shearer CK, Free SJ (1999) Chemical signals of epiphytic lichens in southwestern North America; natural versus man-made sources for airborne particulates. Atmos Environ 33:5095–5104

Giordano S, Adamo P, Sorbo S, Vingiani S (2005) Atmospheric trace metal pollution in the Naples urban area based on results from moss and lichen bags. Environ Pollut 136:431–442

Giordano S, Adamo P, Spagnuolo V, Vaglieco BM (2010) Instrumental and bio-monitoring of heavy metal and nanoparticle emissions from diesel engine exhaust in controlled environment. J Environ Sci 22:1357–1363

Giordano S, Adamo P, Spagnuolo V, Tretiach M, Bargagli R (2013) Accumulation of airborne trace elements in mosses, lichens and synthetic materials exposed at urban monitoring stations: towards a harmonization of the moss-bag technique. Chemosphere 90:292–299

Glenn MG, Gomez-Bolea A, Lobello R (1995) Metal content and community structure of cryptogam bioindicators in relation to vehicular traffic in Montseny Biospheric Reserve (Catalonia, Spain). Lichenol 27:291–304

Godinho RM, Freitas MC, Wolterbeek HT (2004) Assessment of lichen vitality during a transplantation experiment to a polluted site. J Atmos Chem 49:355–361

Godinho RM, Wolterbeek HT, Verburg T, Freitas MC (2008) Bioaccumulation behaviour of transplants of the lichen *Flavoparmelia caperata* in relation to total deposition at a polluted location in Portugal. Environ Pollut 151:318–325

Godinho RM, Verburg TG, Freitas MC, Wolterbeek HT (2009a) Accumulation of trace elements in the peripheral and central parts of two species of epiphytic lichens transplanted to a polluted site in Portugal. Environ Pollut 157:102–109

Godinho RM, Wolterbeek HT, Pinheiro MT, Alves LC, Verburg TG, Freitas MC (2009b) Micro-scale elemental distribution in the thallus of *Flavoparmelia caperata* transplanted to polluted site. J Radioanal Nucl Chem 281:205–210

Godoy JM, Schuch LA, Nordemann DJR, Reis VRG, Ramalho M, Recio JC et al (1998) ^{137}Cs, ^{226,228}Ra, ^{210}Pb and ^{40}K concentrations in Antarctic soil, sediment and selected moss and lichen samples. J Environ Radioactiv 41:33–45

Golubev AV, Golubeva VN, Krylov NG, Kuznetsova VF, Mavrin S, Aleinikov YuA et al (2005) On monitoring anthropogenic airborne uranium concentrations and ^{235}U/^{238}U isotopic ration by Lichen-bio-indicator technique. J Environ Radioactiv 84:333–342

Gombert S, Asta J, Seaward MRD (2003) Correlation between the nitrogen concentration of two epiphytic lichens and the traffic density in an urban area. Environ Pollut 123:281–290

González CM, Casanovas S, Pignata ML (1996) Biomonitoring air pollutants from traffic and industries employing *Ramalina ecklonii* (Spreng.) Mey. and Flot. in Córdoba. Argentina. Environ Pollut 91:269–277

González CM, Orellana LC, Casanovas SS, Pignata ML (1998) Environmental conditions and chemical response of a transplanted lichen to an urban area. J Environ Manage 53:73–81

González CM, Pignata ML, Orellana L (2003) Applications of redundancy analysis for the detection of chemical response patterns to air pollution in lichen. Sci Total Environ 312:245–253

Gough LP, Jackson LL, Sacklin JA (1988) Determining baseline element composition of lichens II. *Hypogymnia enteromorpha* and *Usnea* spp. at Redwood National Park, California. Water Air Soil Pollut 38:169–180

Gough LP, Erdman JA (1977) Influence of a coal-fired powerplant on the element contentof Parmelia chlorochroa. Bryologist 80:492–501

Grangeon S, Guédron S, Asta J, Sarret G, Charlet L (2012) Lichen and soil as indicators of an atmospheric mercury contamination in the vicinity of a chlor-alkali plant (Grenoble, France). Ecol Ind 13:178–183

Grasso MF, Clocchiatti R, Carrot F, Deschamps C, Vurro F (1999) Lichens as bioindicators in volcanic areas: Mt. Etna and Vulcano Island (Italy). Environ Geol 37:207–217

Gueidan C, Daillant O, Tillier C (1997) Observation de la qualité de l'air par les lichen à Autun. Bull Soc Nat Hist Autun 164:15–32

Haas JR, Purvis OW (2006) Lichen biogeochemistry. In: Gadd GM (ed) Fungi in biochemical cycles. Cambridge University Press, Cambridge, pp 344–376

Häffner E, Lomský B, Hynek V, Hällgren JE, Batič F, Pfanz H (2001) Air pollution and lichen physiology. Water Air Soil Pollut 131:185–201

Hauck H, Paul A, Spribille T (2006) Uptake and toxicity of manganese in epiphytic cyanolichens. Environ Exp Bot 56:216–224

Heinrich G, Oswald K, Müller HJ (1999) Lichens as monitors of radiocesium and radiostrontium in Austria. J Environ Radioactiv 45:13–27

Hissler C, Stille P, Krein A, Lahd Geagea M, Perrone T, Probst J-L et al (2008) Identifying the origins of local atmospheric deposition in the steel industry basin of Luxembourg using the chemical and isotopic composition of the lichen *Xanthoria parietina*. Sci Total Environ 405:338–344

Horvat M, Jeran Z, Špirič Z, Jaćimović R, Miklavčič V (2000) Mercury and other elements in lichens near the INA Naftaplin gas treatment plant, Molve, Croatia. J Environ Monitor 2:139–144

Hyvärinen M, Roitto M, Ohtonen R, Markkola A (2000) Impact of wet deposited nickel on the cation content of a mat-forming lichen *Cladina stellaris*. Environ Exp Bot 43:211–218

Jenkins DA, Davis RI (1966) Trace element content of organic accumulation. Nature 210:1296–1297

Jeran Z, Byrne AR, Batič F (1995) Transplanted epiphytic lichens as biomonitors of air-contamination by natural radionuclides around the Žirovski VRH uranium mine, Slovenia. Lichenol 27:375–385

Jeran Z, Jaćimović R, Batič F, Smodiš B, Wolterbeek HT (1996) Atmospheric heavy metal pollution in Slovenia derived from results for epiphytic lichens. Fresen J Anal Chem 354:681–687

Jeran Z, Jaćimović R, Batič F, Mavsar R (2002) Lichens as integrating air pollution monitors. Environ Pollut 120:107–113

Jeran Z, Jaćimović R, Pavsic Mikuz P (2003) Lichens and mosses as biomonitors. J Phys IV Fr 107:675–678

Jeran Z, Mrak T, Jaćimović R, Batič F, Kastelec D, Mavsar R, Simončič P (2007) Epiphytic lichens as biomonitors of atmospheric pollution in Slovenian forests. Environ Pollut 146:324–331

Jia G, Belli M, Sansone U, Rosamilia S, Gaudino S (2005) Concentration and characteristics of depleted uranium in water, air and biological samples collected in Serbia and Montenegro. Appl Radiat Isotopes 63:381–399

Jia G, Belli M, Sansone U, Rosamilia S, Gaudino S (2006) Concentration and characteristics of depleted uranium in biological and water samples collected in Bosnia and Herzegovina. J Environ Radioactiv 89:172–187

Jiries AG, Batarseh MI, El-Hassen T, Ziadat AH, Al-Nasir F, Berdanier BW (2008) Lichens (*Rhizocarpon geographicum*) (L.) DC) as a biomonitor for atmospheric pollution in Amman City. Jordan. Environ Bioindic 3:106–113

Käffer MI, Lemos AT, Anders Apel M, Vaz Rocha J, de Azevedo Martins SM, Ferrão Vargas VM (2012) Use of bioindicators to evaluate air quality and genotoxic compounds in an urban environment in Southern Brazil. Environ Pollut 163:24–31

Kahraman A, Kaynak G, Gurler O, Yalcin S, Ozturk S, Gundogdu O (2009) Investigation of environmental contamination of lichens of Gökçeada (Imbroz) Island in Turkey. Rad Meas 44:199–202

Komárek M, Ettler V, Chrastný V, Mihaljevič M (2008) Lead isotopes in environmental sciences: a review. Environ Int 34:562–577

Korobova EM, Brown JB, Ukraintseva NG, Surkow VV (2007) ^{137}Cs and ^{40}K in the terrestrial vegetation of the Yenisey Estuary: landscape, soil and plant relationships. J Environ Radioactiv 96:144–156

Koz B, Celik N, Cevik U (2010) Biomonitoring of heavy metals by epiphytic lichen species in Black Sea region of Turkey. Ecol Indic 10:762–765

Kuik P, Wolterbeek HTh (1995) Factor analysis of atmospheric trace-element deposition data in the Netherlands obtained by moss monitoring. Water Air Soil Pollut 84:323–346

Kularatne KIA, de Freitas CR (2013) Epiphytic lichens as biomonitors of airborne heavy metal pollution. Environ Exp Bot 88:24–32

Lawrey JD, Hale ME Jr (1981) Retrospective study of lichen lead accumulation in the northeastern United States. Bryologist 84:449–456

LeGalley E, Widom E, Krekeler MPS, Kuentz DC (2013) Chemical and lead isotope constraints on sources of metal pollution in street sediment and lichens in Southwest Ohio. Appl Geochem 32:195–203

Lehto J, Paatero J, Pehrman R, Kulmala S, Suski J, Koivula T et al (2008) Deposition of gamma emitters from Chernobyl accident and their transfer in lichen-soil columns. J Environ Radioactiv 99:1656–1664

Leonardo L, Paci Mazzilli B, Damatto SR, Saiki M, Barros de Oliveira SM (2011) Assessment of atmospheric pollution in the vicinity of a tin and lead industry using lichen species *Canoparmelia texana*. J Environ Radioactiv 102:906–910

Lippo H, Poikolainen J, Kubin E (1995) The use of moss, lichen and pine bark in the nationwide monitoring of atmospheric heavy metal deposition in Finland. Water Air Soil Pollut 85:2241–2246

Looney JHH, Kershaw KA, Nieboer E, Webber C, Stetsko PI (1985) The distribution of uranium and companion elements in lichen heath associated with undisturbed uranium deposits in the Canadian Arctic. In: Brown DH (ed) Lichen physiology and cell biology. Plenum Press, New York, pp 193–209

Loppi S (1997) Monitoring of arsenic, boron and mercury by lichen and soil analysis in the Mt. Amiata geothermal area (Central Italy). Trans Geotherm Resour Counc 21:137–140

Loppi S, Bonini I (2000) Lichens and mosses as biomonitors of trace elements in areas with thermal springs and fumarole activity (Mt. Amiata, central Italy). Chemosphere 41:1333–1336

Loppi S, De Dominicis V (1996) Lichens as long-term biomonitors of air quality in central Italy. Acta Bot Neer 45:563–570

Loppi S, Pirintsos SA (2003) Epihytic lichens as sentinels for heavy metal pollution at forest ecosystems (central Italy). Environ Pollut 121:327–332

Loppi S, Chiti C, Corsini A, Bernardi L (1994) Lichen biomonitoring of trace metals in the Pistoia area (central northern Italy). Environ Monit Assess 29:17–27

Loppi S, Nelli L, Ancora S, Bargagli R (1997a) Passive monitoring of trace elements by means of tree leaves, epiphytic lichens and bark substrate. Environ Monit Assess 45:81–88

Loppi S, Nelli L, Ancora S, Bargagli R (1997b) Accumulation of trace elements in the peripheral and central parts of a foliose lichen thallus. Bryologist 100:251–253

Loppi S, Giomarelli B, Bargagli R (1999a) Lichens and mosses as biomonitors of trace elements in a geothermal area (Mt. Amiata, central Italy). Cryptogam Mycol 20:119–126

Loppi S, Pirintsos SA, De Dominicis V (1999b) Soil contribution to the elemental composition of epiphytic lichens (Tuscany, Central Italy). Environ Monit Assess 58:121–131

Loppi S, Riccobono F, Zhang ZH, Savic S, Ivanov D, Pirintsos SA (2003) Lichens as biomonitors of uranium in the Balkan area. Environ Pollut 125:277–280

Loppi S, Frati L, Paoli L, Bigagli V, Rossetti C, Bruscoli C et al (2004) Biodiversity of epiphytic lichens and heavy metal contents of *Flavoparmelia caperata* thalli as indicators of temporal variations of air pollution in the town of Montecatini Terme (central Italy). Sci Total Environ 326:113–122

Machado A, Freitas MC, Pacheco AMG (2004) Relative response of biological and non-biological monitors in a coastal environment. J Atmos Chem 49:161–173

Markert B (1993) Instrumental analysis of plants. In: Markert B (ed) Plants as biomonitors, indicators for heavy metals in the terrestrial environment. VCH, Weinheim, NewYork, pp 65–103

Marques AP, Freitas MC, Wolterbeek HTh, Verburg T (2009) *Parmelia sulcata* lichen transplants positioning towards wind direction (part I): precipitation volumes, total element deposition and lichen element content. J Radioanal Nucl Chem 281:225–230

Martin RS, Mather TA, Pyle DM, Watt SFL, Day JA, Collins SJ et al (2009) Sweet chestnut (*Castanea sativa*) leaves as a bio-indicator of volcanic gas, aerosol and ash deposition onto the flanks of Mt. Etna in 2005–2007. J Volcanol Geotherm Res 179:107–119

Matschullat J, Scharnweber T, Garbe-Schönberg D, Walther A, Wirth V (1999) Epilithic lichen-atmospheric deposition monitors of trace elements and organohalogens? J Air Waste Manage Assoc 49:1201–1211

Merešová J, Florek M, Holý K, Ješkovský M, Sýkora I, Frontasyeva MV et al (2008) Evaluation of elemental content in airborne particulate matter in low-level atmosphere of Bratislava. Atmos Environ 42:8079–8085

Mietelski JW (2001) Plutonium in the environment of Poland (a review). In: Kudo A (ed) Plutonium in the environment. Elsevier, Amsterdam, pp 401–412

Migaszewski ZM, Galuszka A, Świercz A, Kucharzyk J (2001) Element concentrations in soils and plant bioindicators in selected habitats of the Holy Cross Mountains, Poland. Water Air Soil Pollut 129:369–386

Migaszewski ZM, Galuszka A, Paslawski P (2005) The use of the barbell cluster ANOVA design for the assessment of environmental pollution: a case study, Wigierski National Park, NE Poland. Environ Pollut 133:213–223

Minganti V, Capelli R, Drava G, De Pellegrini R, Brunialti B, Giordani P et al (2003) Biomonitoring of trace metals by different species of lichen (*Parmelia*) in North-West Italy. J Atmos Chem 45:219–229

Mitchell PI, Holm E, León Vintró L, Condren OM, Roos P (1998) Determination of the ^{243}Cm/^{244}Cm ratio alpha spectrometry and spectral deconvolution in environmental samples exposed to discharges from the nuclear fuel cycle. Appl Radiat Isotopes 49:1283–1288

Monna F, Aiuppa A, Varrica D, Dongarà G (1999) Pb isotope composition in lichens and aerosols from Eastern Sicily: insights into the regional impact of volcanoes on the environment. Environ Sci Technol 33:2517–2523

Monna F, Poujol M, Losno R, Dominik J, Annegarn H, Coetzee H (2006) Origin of atmospheric lead in Johannesburg, South Africa. Atmos Environ 40:6554–6566

Mrak T, Simčič J, Pelicon P, Jeran Z, Reis MA, Pinheiro T (2007) Use of micro-PIXE in the study of arsenate uptake in lichens and its influence on element distribution and concentrations. Nucl Instrum Methods Phys Res B 260:245–253

Nakajima H, Yamamoto Y, Yoshitani A, Itoh K (2013) Effect of metal stress on photosynthetic pigments in the Cu-hyperaccumulating lichens *Cladonia humilis* and *Stereocaulon japonicum* growing in Cu-polluted sites in Japan. Ecotoxicol Environ Saf 97:154–159

Nieboer E, Ahmed HM, Puckett KJ, Richardson DHS (1972) Heavy metal content of lichens in relation to distance from a nickel smelter in Sudbury, Ontario. Lichenologist 5:292–304

Nieboer E, Richardson DHS, Lavoi P, Padovan D (1979) The role of metal ion binding in modifying the toxic effects of sulphur dioxide on the lichen *Umbilicaria muhlenbergii* I. Potassium efflux studies. New Pythol 82:621–632

Nieboer E, Richardson DHS, Boileau LJR, Beckett PJ, Lavoi P, Padovan D (1982) Lichens and mosses as monitors of industrial activity associated with uranium mining in northern Ontario, Canada—Part 3: accumulations of iron and titanium and their mutual dependence. Environ Pollut B 4:181–192

Nifontova M (1995) Radionuclides in the moss-lichen cover of tundra communities in the Yamal Peninsula. Sci Total Environ 160(161):749–752

Nimis PL, Andreussi S, Pittao E (2001) The performance of two lichen species as bioaccumulators of trace metals. Sci Total Environ 275:43–51

Nriagu JO (1989) A global assessment of natural sources of atmospheric metals. Nature 338:47–49

Nyarko BJB, Adomako D, Serfor-Armah Y, Dampare SB, Adotey D, Akaho EHK (2006) Biomonitoring of atmospheric trace element deposition around an industrial town in Ghana. Radiat Phys Chem 75:954–958

Olmez I, Cetin Gulovali M, Gordon GE (1985) Trace element concentrations in lichens near a coal-fired power plant. Atmos Environ 19:1663–1669

Olszowski T, Tomaszewska B, Góralna-Wlodarscyk K (2012) Air quality in non-industrialized area in the typical Polish countryside based on measurements of selected pollutants in immission and deposition phase. Atmos Environ 50:139–147

Otnyukova T (2007) Epiphytic lichen growth abnormalities and element concentrations as early indicators of forest decline. Environ Pollut 146:359–365

Outola I, Pehrman R, Jaakkola T (2003) Effect of industrial pollution on the distribution of ^{137}Cs in soil and the soil-to-plant transfer in a pine forest in SW Finland. Sci Total Environ 303:221–230

Pacheco AMG, Freitas MC (2004) Are lower epiphytes really that better than higher plants for indicating airborne contaminants? An insight into the elemental content of lichen thalli and tree bark by INAA. J Radioanal Nucl Chem 259:27–33

Pacheco AMG, Freitas MC, Ventura MG (2004) A nonparametric assessment of the relative output of INAA and PIXE on joint determinants in environmental samples (atmospheric biomonitors). Nucl Instrum Methods Phys Res B 222:587–592

Pacheco AMG, Freitas MC, Sarmento S (2007) Nuclear and non-nuclear techniques for assessing the differential uptake of anthropogenic elements by atmospheric biomonitors. Nucl Instrum Methods Phys Res A 579:499–502

Pacheco AMG, Freitas MC, Baptista MS, Vasconcelos MTSD, Cabral JP (2008) Elemental levels in tree-bark and epiphytic-lichen transplants at a mixed environment in mainland Portugal, and comparisons with an in situ lichen. Environ Pollut 151:326–333

Paliouris G, Taylor HW, Wein RW, Svoboda J, Mierzynski B (1995) Fire as an agent in redistributing fallout ^{137}Cs in the Canadian boreal forest. Sci Total Environ 160(161):153–166

Palomäki V, Tynnyrinen S, Holopainen T (1992) Lichen transplantation in monitoring fluoride and sulphur deposition in the surroundings of a fertilizer plant and a strip mine in Siilinjärvi. Ann Bot Fennici 29:25–34

Pálsson SE, Egilsson K, Pórisson S, Magnússon SM, Ólafsdóttir ED, Indridason K (1994) Transfer of radiocaesium from soil and plants to reindeer in Iceland. J Environ Radioactiv 24:107–125

Paoli L, Loppi S (2008) A biological method to monitor early effects of the air pollution caused by the industrial exploitation of geothermal energy. Environ Pollut 155:383–388

Paoli L, Corsini A, Bigagli V, Vannini J, Bruscoli C, Loppi S (2012) Long-term biological monitoring of environment quality around a solid waste landfill assessed with lichens. Environ Pollut 161:70–75

Paoli L, Guttova A, Grassi A, Lackovičová A, Senko D, Loppi S (2014) Biological effects of airborne pollutants released during cement production assessed with lichens (SW Slovakia) Ecol Indic 40:127–135

Paul A, Hauck M (2006) Effects of manganese on chlorophyll fluorescence in epiphyticcyano- and chlorolychens. Flora 201:451–460

Pettersson HBL, Hallstadius L, Hedvall R, Holm E (1988) Radioecology in the vicinity of prospected uranium mining sites in a subarctic environment. J Environ Radioactiv 6:25–40

Pignata ML, Plá RR, Jasan RC, Martinez MS, Rodriguez JH, Wannaz ED et al (2007) Distribution of atmospheric trace elements and assessment of air quality in Argentina employing the lichen, *Ramalina celastri*, as a passive biomonitor: detection of air pollution emission sources. Int J Environ Heal 1:29–46

Prussia CM, Killingbeck KT (1991) Concentrations of ten elements in two common foliose lichens: leachability, seasonality, and the influence of rock and tree bark substrates. Bryologist 94:135–142

Puckett KJ (1976) The effect of heavy metals on some aspects of lichen physiology. Can J Bot 54:2695–2703

Puckett KJ, Finegan EJ (1980) An analysis of the element content of lichens from the Northwest Territories, Canada. Can J Bot 58:2073–2089

Puckett KJ, Tomassini FD, Nieboer E, Richardson DHS (1977) Potassium efflux by lichen thalli following exposure to aqueous sulphur dioxide. New Phytol 9:135–145

Purvis OW, Pawlik-Skowrónska B (2008) Lichens and metals. In: van West P, Statford M, Avery S (eds) Stress in fungi. Elsevier, Mycological Society Symposia Series, Amsterdam, pp 175–200

Purvis OW, Williamson BJ, Bartok K, Zoltani N (2000) Bioaccumulation of lead by the lichen *Acarospora smagardula* from smelter emissions. New Phytol 147:591–599

Purvis OW, Chimonides PJ, Jones GC, Mikhailova IN, Spiro B, Weiss DJ et al (2004) Lichen biomonitoring near Karabash Smelter Town, Ural Mountains, Russia, one of the most polluted areas in the world. Proc R Soc Lond B 271:221–226

Purvis OW, Longden J, Shaw G, Chimonides PDJ, Jeffries TE, Jones GC et al (2006) Biogeochemical signatures in he lichen *Hypogymnia physodes* in the mid Urals. J Environ Radioactiv 90:151–162

Purvis OW, Chimonides PDJ, Jeffries T, Jones GC, Rusu A-M et al (2007) Multi-element composition of historical lichen collections and bark samples, indicators of changing atmospheric conditions. Atmos Environ 41:72–80

Purvis OW, Dubbin W, Chimonides PDJ, Jones GC, Read H (2008) The multi-element content of the lichen *Parmelia sulcata*, soil, and oak bark in relation to acidification and climate. Sci Total Environ 390:558–568

Reimann C, De Caritat P (2000) Intrinsic flaws of element factors (EFs) in environmental geochemistry. Environ Sci Technol 34:5084–5091

Reimann C, Halleraker JH, Kashulina G, Bogatyrev I (1999) Comparison of plant and precipitation chemistry in catchments with different levels of pollution on the Kola Peninsula. Russia. Sci Total Environ 243 (244):169–191

Reis MA, Alves LC, Wolterbeek HTh, Verburg T, Freitas MC, Gouveia A (1996) Main atmospheric heavy metal sources in Portugal by biomonitor analysis. Nucl Instrum Methods Phys Res B 109(110):493–497

Reis MA, Alves LC, Freitas MC, van Os B, Wolterbeek HTh (1999) Lichens (*Parmelia sulcata*) time response model to environmental elemental availability. Sci Total Environ 232:105–115

Reis MA, Alves LC, Freitas MC, Van Os B, de Goeij J, Wolterbeek HT (2002) Calibration of lichen transplants considering faint memory effects. Environ Pollut 120:87–95

Rencz AN, O'Driscoll NJ, Hall GEM, Peron T, Telmer K, Burgess NM (2003) Spatial variation and correlations of mercury levels in the terrestrial and aquatic components of a wetland dominated ecosystem: Kejimkujik Park, Nova Scotia, Canada. Water Air Soil Pollut 143:271–288

Ribeiro Guevara S, Arribére MA, Calvelo S, Román Ross G (1995) Elemental composition of lichens at Nahuel Huapi National Park, Patagonia, Argentina. J Radioanal Nucl Ch 198:437–448

Richardson DHS, Shore M, Hartree R, Richardson RM (1995) The use of X-ray spectrometry for the analysis of plants, especially lichens, employed in biological monitoring. Sci Total Environ 176:97–105

Riget F, Asmund G, Aastrup P (2000) The use of lichen (*Flavocetraria nivalis*) and moss (*Rhacomitrium lanuginosum*) as monitors for atmospheric deposition in Greenland. Sci Total Environ 245:137–148

Rizzio E, Bergamaschi L, Valcuvia MG, Profumo A, Gallorini M (2001) Trace elements determination in lichens and in the airborne particulate matter for the evaluation of the atmospheric pollution in a region of northern Italy. Environ Int 26:543–549

Rodushkin I, Engström E, Sörlin D, Pontèr C, Baxter DC (2007) Osmium in environmental samples from Northeast Sweden. Part I. Evaluation of background status. Sci Total Environ 386:145–158

Roos P, Holm E, Persson RBR, Aarkrog A, Nielsen SP (1994) Deposition of ^{210}Pb, ^{137}Cs, $^{239+240}$Pu, ^{238}Pu and ^{241}Am in the Antarctic Peninsula area. J Environ Radioact 24:235–251

Rosamilia S, Gaudino S, Sansone U, Belli M, Jeran Z, Ruisi S et al (2004) Uranium isotopes, metals and other elements in lichens and tree barks collected in Bosnia-Herzegovina. J Atmos Chem 49:447–460

Rossbach M, Jayasekera R, Kniewald G, Thang NH (1999) Large scale air monitoring: lichen vs. air particulate matter analysis. Sci Total Environ 232:59–66

Saiki M, Chaparro CG, Vasconcellos MBA, Marcelli MP (1997) Determination of trace elements in lichens by instrumental neutron activation analysis. J Radioana Nucl Chem 217:111–115

Salbu B, Janssens K, Lind OC, Proost K, Danesi PR (2003) Oxidation states of uranium in DU particles from Kosovo. J Environ Radioactiv 64:167–173

Salemaa M, Derome J, Helmisaari H-S, Nieminen T, Vanha-Majamaa I (2004) Element accumulation in boreal bryophytes, lichens and vascular plants exposed to heavy metal and sulfur deposition in Finland. Sci Total Environ 324:141–160

Salo H, Bućko MS, Vaahtovuo E, Limo J, Mäkinen J, Pesonen LJ (2012) Biomonitoring of air pollution in SW Finland by magnetic and chemical measurements of moss bags and lichens. J Geochem Explor 115:69–81

Sansone U, Danesi PR, Barbizzi S, Belli M, Campbel M, Gaudino S et al (2001) Radioecological survey at selected sites hit by depleted uranium ammunitions during the 1999 Kosovo conflict. Sci Total Environ 281:23–25

Santos PL, Gouvea RC, Dutra IR (1993) Lead-210 in vegetables and soils from an area of high natural radioactivity in Brazil. Sci Total Environ 138:37–46

Sawidis T, Heinrich G, Chettri MK (1997) Cesium-137 monitoring using lichens from Macedonia, northern Greece. Can J Bot 75:2216–2223

Sawidis T, Tsigaridas K, Tsikritzis L (2010) Cesium-137 monitoring using lichens from W. Macedonia. N. Greece. Ecotoxicol Environ Saf 73:1789–1796

Scerbo R, Ristori T, Possenti L, Lampugnani L, Barale R, Barghigiani C (2002) Lichen (*Xanthoria parietina*) biomonitoring of trace element contamination and air quality assessment in Pisa Province (Tuscany, Italy). Sci Total Environ 286:27–40

Seaward MRD (1995) Use and abuse of heavy metal bioassays in environmental monitoring. Sci Total Environ 176:129–134

Sensen M, Richardson DHS (2002) Mercury levels in lichens from different host trees around a chlor-alkali plant in New Brunswick, Canada. Sci Total Environ 293:31–45

Sert E, Uğur A, Őzden B, Saç MM, Camgöz B (2011) Biomonitoring of ^{210} Po and ^{210}Pb using lichens and mosses around coal-fired power plants in Western Turkey. J Environ Radioact 102:535–542

Shcherbov BL, Strakhovenko VD, Sukhorukov FV (2008) The ecogeochemical role of forest fires in the Baikal region. Geogr Nat Resour 29:150–155

Sheard JW (1986a) Distribution of uranium series radionuclides in upland vegetation of northern Saskatchewan. I. Plant and soil concentrations. Can J Bot 64:2446–2452

Sheard JW (1986b) Distribution of uranium series radionuclides in upland vegetation of northern Saskatchewan. II. Pattern of accumulation among species and localities. Can J Bot 64:2453–2463

Sheppard SC, Sheppard MI, Ilin M, Tait J, Sanipelli B (2008) Primordial radionuclides in Canadian background sites: secular equilibrium and isotopic differences. J Environ Radioactiv 99:933–946

Simonetti A, Gariépy C, Carignan J (2003) Tracing sources of atmospheric pollution in Western Canada using the Pb isotopic composition and heavy metal abundances of epiphytic lichens. Atmos Environ 37:2853–2865

Singh NP, Wrenn ME (1988) Determination of actinides in biological and environmental samples. Sci Total Environ 70:187–203

Singh SM, Sharma J, Gawas-Sakhalkar P, Upadhyay AK, Naik S, Pendnecker SM, Ravindra R (2013) Atmospheric deposition studies of heavy metals in Arctic by comparative analysis of lichens and cryoconite. Environ Monit Assess 185:1367–1376

Skuterud L, Gaare E, Eikelmann IM, Hove K, Steinnes E (2005) Chernobyl radioactivity persists in reindeer. J Environ Radioactiv 83:231–252

Sloof JE (1995a) Lichens as quantitative biomonitors for atmospheric trace-element deposition, using transplants. Atmos Environ 29:11–20

Sloof JE (1995b) Pattern recognition in lichens for source apportionment. Atmos Environ 29:333–343

Smodiš B (2003) IAEA Approaches to assessment of chemical elements in atmosphere. In: Markert BA, Breure AM, Zechmeister HG (eds) Bioindicators and biomonitors, principles, concepts and applications. Elsevier, Amsterdam, pp 875–902

Smodiš B, Bleise A (2002) Internationally harmonized approach to biomonitoring trace element atmospheric deposition. Environ Pollut 120:3–10

Søndergaard J, Asmund G, Johansen P, Elberling B (2010) Pb isotopes as tracers of mining-related Pb in lichens, seaweed and mussels near a former Pb–Zn mine in West Greenland. Environ Pollut 158:1319–1326

Søndergaard J, Johansen P, Asmund G, Riget F (2011) Trends of lead and zinc in resident and transplanted *Flavocetraria nivalis* lichens near a former lead-zinc mine in West Greenland. Sci Total Environ 409:4063–4071

Søndergaard J, Bach L, Asmund G (2013) Modelling atmospheric bulk deposition of Pb, Zn and Cd near a former Pb–Zn mine in West Greenland using transplanted *Flavocetraria nivalis* lichens. Chemosphere 90:2549–2556

Sorbo S, Aprile G, Strumia S, Castaldo Cobianchi R, Leone A, Basile A (2008) Trace element accumulation in *Pseudevernia furfuracea* (L.) Zopf exposed in Italy's so called Triangle of Death. Sci Total Environ 407:647–654

Spagnuolo V, Zampella M, Giordano S, Adamo P (2011) Cytological stress and element uptake in moss and lichen exposed in bags in urban area. Ecotoxicol Environ Saf 74:1434–1443

Squire HM, Middleton LJ (1966) Behavior of Cs^{137} in soils and pastures a long term experiment. Radiat Bot 6:413–423

St.Clair SB, St.Clair LL, Mangelson NF, Weber DJ (2002a) Influence of growth form on the accumulation of airborne copper by lichens. Atmos Environ 26:5637–5644

St.Clair SB, St.Clair LL, Weber DJ, Mangelson NF, Eggett DL (2002b) Element accumulation patterns in foliose and fruticose lichens from rock and bark substrates in Arizona. Bryologist 105:415–421

Steinnes E, Njåstad O (1993) Use of mosses and lichens for regional mapping of ^{137}Cs fallout from Chernobyl accident. J Environ Radioactiv 21:65–73

Strandberg M (1994) Radiocesium in a Danish pine forest ecosystem. Sci Total Environ 157:125–132

Takala K, Salminen R, Olkkonen H (1998) Geogenic and anthropogenic zinc in epiphytic and terricolous lichens in Finland. J Geochem Explor 63:57–66

Tarhanen S, Metsärinne S, Holopainen T, Oksanen J (1999) Membrane permeability response of lichen *Bryoria fuscescens* to wet deposited heavy metals and acid rain. Environ Pollut 104:121–129

Tomassini FD, Lavoie P, Puckett KJ, Nieboer E, Richardson DHS (1977) The effect of time of exposure to sulphur dioxide on potassium loss from and photosynthesis in the lichen *Cladina rangiferina* (L.) Harm. New Phytol 79:147–155

Trembley ML, Fahselt D, Madzia S (1997) Localization of uranium in *Cladina rangiferina* and *Cladina mitis* and removal by aqueous washing. Bryologist 100:368–376

Tretiach M, Adamo P, Bargagli R, Baruffo L, Carletti L, Crisafulli P et al (2007) Lichen and moss bags as monitoring devices in urban areas. Part I: influence of exposure on sample vitality. Environ Pollut 146:380–391

Uğür A, Özden B, Saç MM, Yener G (2003) Biomonitoring of ^{210}Po and ^{210}Pb using lichens and mosses around a uraniferous coal-fired power plant in western Turkey. Atmos Environ 37:2237–2245

Uğür A, Özden B, Saç MM, Yener G, Altinbaş Ü, Kurucu Y et al (2004) Lichens and mosses for correlation between trace elements and ^{210}Po in the areas near coal-fired power plant at Yatağan, Turkey. J Radioanal Nucl Ch 259:87–92

Upreti DK, Pandey V (1994) Heavy metals of Antarctic lichens 1. Umbilicaria. Feddes Repert 105:197–199

Van den Berg GJ, Tyssen TPM, Ammerlaan MJJ, Vokers KJ, Woroniecka UD, de Bruin M et al (1992) Radiocesium and lead in the lichen species *Parmelia sulcata* sampled in three regions around Chernobyl: assessment of concentrations in 1990. J Environ Radioactiv 17:115–127

Varrica D, Aiuppa A, Dongarrà G (2000) Volcanic and anthropogenic contribution to heavy metal content in lichens from Mt. Etna and Volcano island (Sicily). Environ Pollut 108:153–162

Varskog P, Naeumann R, Steinnes E (1994) Mobility and plant availability of radioactive Cs in natural soil in relation to stable Cs, other alkali elements and soil fertility. J Environ Radioactiv 12:43–53

Vieira BJ, Freitas MC, Rodrigues AF, Pacheco AMG, Soares PM, Correia N (2004) Element-enrichment factors in lichens from Terceira, Santa Maria and Madeira Island (Azores and Madeira Archipelagoes). J Atmos Chem 49:231–249

Vingiani S, Adamo P, Giordano S (2004) Sulphur, nitrogen and carbon content of *Sphagum capillifolium* and *Pseudevernia furfuracea* exposed in bags in Naples urban area. Environ Pollut 129:145–158

Wadleigh MA, Blake DM (1999) Tracing sources of atmospheric sulphur using epiphyticlichens. Environ Pollut 106:265–271

Walker TR, Crittenden PD, Young SD (2003) Regional variation in the chemical composition of winter snow pack and terricolous lichens in relation to sources of acid emissions in the Usa river basin, northeast European Russia. Environ Pollut 125:401–412

Walker TR, Crittenden PD, Young SD, Prystina T (2006) An assessment of pollution impacts due to the oil and gas industries in the Pechora basin, north-eastern European Russia. Ecol Ind 6:369–387

Walker TR, Crittenden PD, Dauvalter VA, Jones V, Kuhry P, Loskutova O et al (2009) Multiple indicators of human impacts on the environment in the Pechora Basin, north-eastern European Russia. Ecol Indic 9:765–779

Wiersma GB, Finley K, Whitworth C, Bruns DA, McAnulty L, Boelcke C (1992) Elemental composition of lichens from a remote *Nothofagus* forest site in Southern Chile. Chemosphere 24:155–167

Williamson BJ, Mikhailova I, Purvis OW, Udachin V (2004) SEM-EDX analysis in the source apportionment of particulate matter on *Hypogymnia physodes* lichen transplants around the Cu smelter and former mining town of Karabash, South Urals, Russia. Sci Total Environ 322:139–154

Wolterbeek HTh, Bode P (1995) Strategies in sampling and sample handling in the context of large-scale plant biomonitoring surveys of trace element air pollution. Sci Total Environ 176:33–43

Wolterbeek HTh, Garty J, Reis MA, Freitas MC (2003) Biomonitors in use: lichens and metal air pollution. In: Markert BA, Breure AM, Zechmeister HG (eds) Bioindicators and biomonitors, principles, concepts and applications. Elsevier, Amsterdam, pp 377–419

Yenisoy-Karakaş S, Tuncel SG (2004) Geographic patterns of elemental deposition in the Aegean region of Turkey indicated by the lichen, *Xanthoria parietina* (L.) Th. Fr. Sci Total Environ 329:43–60

Yildiz A, Aksoy A, Tug GN, Islek C, Demirezen D (2008) Biomonitoring of heavy metals by *Pseudevernia furfuracea* (L.) Zopf in Ankara (Turkey). J Atmos Chem 60:71–81

Yoshida S, Muramatsu Y, Dvornik AM, Zhuchenko TA, Linkov I (2004) Equilibrium of radiocesium with stable cesium within the biological cycle of contaminated forest ecosystems. J Environ Radioact 75:301–313

Žunic ZS, Mietelski JW, Blazej S, Gaca P, Tomankiewicz E, Ujić P et al (2008) Traces of DU in samples of environmental bio-monitors (non flowering plants, fungi) and soil from target sites of the Western Balkan region. J Environ Radioact 99:1324–1328

Monitoring Lichen as Indicators of Atmospheric Quality

4

Gintarė Sujetovienė

Contents

Abstract

Bioindicators are living organisms that react to environmental pollution with their life functions. Lichens are reviewed for their potential to reflect air pollution. This chapter highlights the usefulness of lichens as bioindicators of air pollution and recent advances in usage of lichens in the biomonitoring studies. The attention is focused on lichens biological performance at physiological and biochemical levels, the accumulation of substances and on community level. Laboratory approaches related assessment of lichen reactions along with results from field studies point out the main potential characteristics of lichens as indications of air pollution. Discussed data on the lichen studies indicated that analysis of some physiological parameters has given clear picture about air quality of site. Recently, more studies analyse changes at biochemical level that lichens gives more reasonable clarification of induced changes in lichens by pollutants. Review on lichen monitoring contributes to our understanding of change in environmental conditions caused by air pollution.

Keywords

Bioindicator · Biomonitoring · Heavy metals · Lichens · Nitrogen · Persistent organic pollutants · Pollution · Sulphur

G. Sujetovienė (✉)
Department of Environmental Sciences, Vytautas Magnus University, Vileikos 8, 44404 Kaunas, Lithuania
e-mail: g.sujetoviene@gmf.vdu.lt

D.K. Upreti et al. (eds.), *Recent Advances in Lichenology*,
DOI 10.1007/978-81-322-2181-4_4

4.1 Introduction

Air pollution is an essential problem in many urban, industrialized and densely populated areas. Pollution is increasing as a consequence of combustion for transport power generation and other human activities. Combustion process produce a mixture of pollutants including gaseous sulphur and nitrogen compounds in association commonly have high concentrations of heavy metals, organic compounds and others. Air pollution is a major environmental-related health treat. In order to protect living organisms from exposure to environmental hazards requires a better understanding of the relationship between environmental conditions and response of organisms.

As environmental contamination becomes actual problem, scientists have to find methods of detecting and quantifying atmospheric pollution. An automated monitoring system is a useful tool for the determination of levels of atmospheric contamination. One of the limitations of such system is large number of sampling sites that leads to high cost. Alternatively, many recent studies on atmospheric pollution have used biological samples such as lichens or mosses (Conti and Cecchetti 2001; Wolterbeek et al. 2003; Seaward 2004). Biological monitoring is based on the assumption that changes in environment affect the biota and these changes can be effectively used as an early warning signal in order to detect environmental changes (Garty 2001). The advantage of the biomonitors usage is the monitoring of several sites at low cost. In addition, their reaction reflects the effects of integrated pollutants impact over long period.

Lichens are one of the most sensitive and valuable biomonitors of atmospheric pollution and can be used as indicators of air pollution at community-level mapping lichen species present in an area, transplanting healthy lichens and measuring deterioration in thallus, assessing the contaminants accumulated in lichen thallus. Lichens have been effective biomonitoring tools for decades because they absorb substances directly form the air and are sensitive to them.

Many lichen communities are affected by various pollutants deposition which resulted alteration in the community structure. The most sensitive species tend to disappear from the urban and industrial areas, whereas tolerant species exist in moderately polluted areas (Ranta 2001; Davies et al. 2007; Larsen et al. 2007). At individual-level biomonitoring studies on the physiological effects of pollution has often emphasized the inhibition of photosynthesis, respiration, fluorescence processes, chlorophylls content, membrane damage and oxidative stress (Garty et al. 2001; Weissman et at. 2006; Bermudez et al. 2009; Bajpai et al. 2010; Oztetik and Cicek 2011; Paoli et al. 2011; Majumder et al. 2013; Karakoti et al. 2014). As lichens absorb elements through their entire surface, even in quantities above their metabolic requirements (Gonzalez et al. 1998; Adamo et al. 2003; Bergamaschi et al. 2007; Shukla and Upreti 2009; Bajpai and Upreti 2012; Kularatne and de Freitas 2013; Sondergaard 2013; Stamenkovic et al. 2013). The absorbed substances are stored for a long time and can be used for bioaccumulation studies. This allows to assess the different level of accumulated certain elements in an altered environment. Due to their accumulation capacity and their sensitivity to various pollutants, lichens can be used either as accumulation indicators or as sensitive indicators of environmental quality.

The review presented in the chapter deals with response of lichens to pollution as indicators of air quality.

4.2 Lichens as Bioindicators of Air Pollution

Lichens are widely used in the biomonitoring studies as an effective tool for mapping spatial and temporal changes of atmospheric contamination (Nimis et al. 2002). Due to the unique biology and their sensitivity to pollutants, lichens respond to variations in atmospheric composition and may be a useful measure for monitoring environmental quality. Lichens might be as a warning indicator of the changes in the environmental parameters and the exceedance of critical levels of pollutants in the atmosphere before the damage become evident. Sensitive lichen species respond rapidly to pollution and helps in detection of unrecognized pollution sources.

Most of the studies have been found that epiphytic lichens are the first to be affected by environmental contamination and the high sensitivity of lichens are related to its biology (Conti and Cecchetti 2001; Giordani et al. 2002; Nash 2010). Lichens have no deciduous parts of their structure, thus, can not avoid damaged or accumulated parts by shading them. Besides, they are slow growers and lacking any root, cuticle (lipid surface layer) and stomata. This means that the unique way for their nutrition is through atmospheric deposition as lichens absorb water and nutrients including pollutants through entire thallus surface.

Lichens do not react to a particular pollutant but the reaction includes the general effect of complex of pollutants. Lichens also have different sensitivity to specific pollutants. Tolerant lichen species may persist in polluted environment, while sensitive species would become extinct. Such differential sensitivity of lichens is very useful in biomonitoring studies (Seaward 2004; Hauck 2008; Mayer et al. 2009). However, the sensitivity can be influenced by different morphological and biochemical properties. Blasco et al. (2011) found that foliose lichen *Lobaria pulmonaria* was more enriched in PAHs compared with *Parmelia sulcata* due to the fact that the former was more aerohydrophytic—absorbing more water and air moisture and consequently more airborne PAHs. The sensitivity to different pollutants varies among different species, and the response to different pollutants and substrates may vary even within a species (Sigal and Nash 1983; Nash 2010).

Other features that contribute to the success of lichens as biomonitors are related to their morphology which is constant around the year, which allows collecting them at any season and are also widely distributed over the Earth. Lichens are long-lived perennial organisms that accumulate pollutants and are subject of cumulative effects of pollutants over decades thus being indicators of long-term and complex atmospheric pollution (Nash 2010).

4.3 Biomonitoring Methods with Lichens

Lichens may be used for biological monitoring studies in two ways: (a) by mapping species in an area, this technique include distribution maps of individual species or the whole lichen flora; (b) lichen species sampling growing naturally or by transplanting lichens from relatively clean environment to a contaminated one, evaluating the accumulation of pollutants, morphological and/or physiological parameters changes (Conti and Cecchetti 2001).

The composition of species in an area reflects the quality of atmosphere. Therefore, changes in lichen communities are correlated with changes in levels of atmospheric pollution. The mapping of lichen diversity is quick and inexpensive method, provides an estimate of the biological impact of air pollution and gives an integrative picture of air quality. Quantitative information could be obtained by calculating different indices to estimate air quality. The most frequently used are the index of atmospheric purity (IAP) and index of lichen diversity (ILD). The elemental composition of lichen thalli both naturally growing and/or transplanted could indicate the accumulation of certain elements providing means of the sources of pollution. More recently and widely physiological parameters have been tested in order to assess the vitality of lichens.

Index of Atmospheric Purity. In order to assess the quality of air pollution, one of approaches could be the use of the IAP: $\text{IAP} = \frac{1}{10} \times \left(\sum_1^n \frac{Q \times f}{n}\right)$ (where n—the number of species, Q—degree of toxiphoby and f—the frequency or abundance of each species) (LeBlanc and De Sloover 1970). IAP is a quantitative approach combining the number, frequency and tolerance of the lichens species present in a study area with their sensitivity to environmental stress. Recently, the IAP was reviewed by Kricke and Loppi (2002).

The index values obtained may be plotted to obtain air quality map. Several diverse modifications from the original formula of the IAP were proposed in many studies (Yule and Lloyd 1984; Batic and Mayrhofer 1996; Asta and Rolley 1999; Fernandez-Salegui et al. 2006b). For example, the index of poleotolerance was developed in Estonia (Trass 1973)—a similar approach to the IAP but with a value of poleotolerance: $IP = \sum_1^n \left(a_i \times \frac{c_i}{C_i}\right)$ (where n—the number of species, a_i—species-specific factor expressing tolerance to pollutants (scale from 1 to 10), c_i—degree of cover for each species (scale from 1 to 10) and C_i—degree of cover for all species.

A number of IAP studies have been applied in different regions of the world, around a paper mill (Das et al. 2013), thermal plants (Romeralo et al. 2012), geothermal power plants (Loppi 1996; Loppi and DeDominicis 1996; Loppi and Nascimbene 1998), municipal solid waste incinerator (Loppi et al. 1995), industrial area (Sim-Sim et al. 2000) and in the National Park (Gibson et al. 2013). A large number of IAP studies were performed in urban areas (Siena, Italy: Monaci et al. 1997; Thessaloniki, Greece: Vokou et al. 1999; Grenoble, France: Gombert et al. 2004; San Carlos de Bariloche and Villa Regina, Argentina: Calvelo and Liberatore 2004; Ohio, USA: Washburn and Culley 2006; Linz, Austria: Zechmeister and Hohenwallner 2006; Aragon, Spain: Blasco et al. 2008; Bariloche, Argentina: Calvelo et al. 2009; Dimitrovgrad, Serbia: Stamenkovic et al. 2010; Knjazevac, Serbia: Stamenkovic and Cvijan 2010; Viamao, Brazil: Kaeffer et al. 2011; Blace, Serbia: Stamenkovic et al. 2013). Some IAP studies have involved a larger geographical areas (Finland and Russia: Mayer et al. 2009; Mediterranean area: Giordano et al. 2004; Slovenian forests: Jeran et al. 2002). The results of the studies suggest that the IAP may be a useful air pollution indicator tool in combination with modelled data, even in relatively clean regions and it was strongly correlated with sulphur and nitrogen concentrations (Zechmeister and Hohenwallner 2006; Mayer et al. 2009). IAP values were compared with data of multi-elemental composition of lichen thalli and the resulted in negative correlation indicated that the IAP is an accurate, reliable and relatively inexpensive method to detect and evaluate atmospheric pollution (Calvelo and Liberatore 2004; Blasco et al. 2008).

The IAP value also showed no clear relation with sources pollution, such as roads or industry, and was nor correlated with element concentrations in the atmosphere (Gombert et al. 2004). It initially created to characterize atmospheric pollution or atmospheric purity by means of the lichen flora and was shown to be influenced by other parameters such as environmental and landscape factors. At higher IAP values, lichen frequencies did not seem to be affected by air pollution, and variations in the epiphytic lichen communities presumably depended on other environmental factors (Loppi et al. 1995).

Other quantitative methods were also proposed for monitoring the effects of air pollution in Germany. The VDI method (1995) is based on the calculation of the frequency of species within the ladder of 10 units (each 10 × 10 cm) positioned on the trunks of trees. In Italy, the biodiversity of epiphytic lichens was calculated as the sum of frequencies within the sampling ladder of 10 quadrats (each 15 × 10 cm) (Nimis et al. 1990). A standardized methodology was proposed by Asta et al. (2002) which provides a general picture of environmental quality. This method supplies information on the long-term effects of air pollutants, eutrophication, anthropization and climatic changes on lichens. The objective method (sampling design, procedure and data interpretation) allows comparing results throughout the different studies (Asta et al. 2002).

The qualitative scale for estimating SO_2 pollution was developed by Hawksworth and Rose (1970). The basic element is the species, each of them having a range of tolerance to pollution. The tolerance can be expressed as ecological indicator value. This scale has been widely applied in different countries (Bates et al. 1990; Khalil and Asta 1998; Svoboda 2007; Lisowska 2011). However, such scale designed to assess SO_2 may no longer be accurate indicator of atmospheric pollution because SO_2 levels in many areas have declined meanwhile nitrogen deposition in urban and agricultural areas is the main driver of lichen communities changes (Larsen et al. 2007; Mayer et al. 2009; Rogers et al. 2009; Grandin 2011; Johansson et al. 2012; Evju and Bruteig 2013; Sparrius et al. 2013). Zone scales as the biomonitoring approach allow predicting the level of pollution by sulphur dioxide from the assemblages of lichen species typical for unaltered habitat. Scales of lichen sensitivities have been published by Hawksworth and Rose (1970), de Wit (1976), and Wirth (1991).

Index of lichen diversity. ILD is calculated as the sum of frequencies of epiphytic lichens in a sampling grid (Giordani 2007). To interpret the IDL values in terms of air pollution, Loppi (2004) suggested using derivative ILD values. For example, ILD_{NN} value without the frequency of nitrophytic species as reflection of air pollution by phytotoxic gases (SO_2, NO_x); ILD_N calculated only with frequency of nitrophytic species, expected to reflex eutrophication and ILD_{SN} calculated only with the frequency of strictly nitrophytic species as reflection of ammonia pollution (Loppi and Frati 2006). For each index, a different interpretative scale of environmental alteration was devised (Table 4.1).

Nitrophytic species are those which has an indicator value for eutrophication of four (rather high eutrophication) or five (very high eutrophication) according to Nimis (2003). Strictly nitrophytic species are those with an indicator value for eutrophication from four to five (Loppi and Frati 2006). Lichen diversity values with only the frequency of nitrophytic species could be taken as rough estimates of eutrophication, mainly nitrogen deposition.

Despite the decrease in the main lichen abundance affecting pollutant SO_2, decreased lichen diversity over time and space are still detected (Slaby and Lisowska 2012; Hauck et al. 2013b; Mayer et al. 2013). Various anthropogenic variables are responsible for lichen decline along with successive processes of lichen communities. In the complex ecosystems with the high climatic variability and sharp environmental gradients, the factors are particularly complex and hard to detect (Brunialti and Giordani 2003; Giordani 2007). Bates et al. (2001) proposed that methods based totally on the total lichen diversity may not be appropriate for future monitoring urban air pollution. According to Giodani (2007), the relationship between lichen diversity and environmental variable changes under different ecological conditions, air pollutants are the main limiting factor in non-forested areas, whereas harvesting or fires predominate in forested areas.

The transplant method. In areas with low pollution levels where lichens are not killed by contaminants, the biomonitoring studies could be performed using native lichen species. When biomonitors are insufficient in the sites to be monitored or when the aim is to assess pollution deposition and accumulation rates, the use of transplants is a very useful tool in terrestrial and aquatic environment (Augusto et al. 2013a).

The lichen thalli are collected from relatively non-polluted site and transplanted to the study sites to be monitored under the exposure of pollution. The use of transplanted lichens allows assessing the response of organisms exposed to different atmospheric conditions including pollutants over fixed exposure period. This method also allows to determine threshold levels of toxic substances in a thallus (Mikhailova 2002). The transplant method is used in bioaccumulation

Table 4.1 Interpretative scale of environmental alteration using different ILD (Loppi and Frati 2006)

ILD_{NN}	ILD_N	ILD_{SN}	Environmental alteration
0	>36	>12	Very high
1–18	25–36	9–12	High
19–36	13–24	5–8	Moderate
37–54	1–12	1–4	Low
>54	0	0	Negligible

studies and analysing morphological, physiological and ultrastructural damage. Samples can be taken periodically in order to evaluate vitality of lichen thalli. The transplantation studies are used often in order to assess air pollution in urban and industrial areas (Garty et al. 1996; Gonzalez and Pignata 1997; Canas and Pignata 1998; Zambrano and Nash 2000; Gonzalez et al. 2003; Carreras et al. 2005; Cloquet et al. 2009; Oztetik and Cicek 2011; Aslan et al. 2013).

Physiological and biochemical parameters to assess the vitality of lichens exposed to air pollution. Due to different sensitivity of lichens to air pollution, they are potentially useful for air monitoring. In order to assess pollution induced stress effects in lichens, the methodologies have to be sensitive and reliable (Garty et al. 2000). The physiological reactions of lichens to atmospheric contamination have been evaluated by means of changes in photosynthesis and fluorescence rates, the degradation of chlorophylls and decrease in the cell membrane integrity.

The measurement of modulated chlorophyll *a*, fluorescence in lichens is one of several valid tools in biomonitoring studies (Jensen and Kricke 2000; Tretiach et al. 2007b). The chlorophyll fluorescence parameter ratio of variable fluorescence to maximal fluorescence (Fv/Fm) has been widely used to assess the state of the photosynthetic apparatus in response air pollution (Bjerke et al. 2005; Fernandez-Salegui et al. 2006a; Tretiach et al. 2007a; Paoli et al. 2010; Piccotto et al. 2011; Karakoti et al. 2014).

It has been shown that air pollution reduces both photosynthesis and respiration rates in lichens (Haffner et al. 2001; Nash and Gries 2002; Zambrano and Nash 2000; Boonpragob and Nash 1991). This parameter may provide diagnostic information on the state of photosystems, being related to the environmental factors and pollution. The results of recent transplantation studies showed that the pulse amplitude modulation fluorimetry applied may become an important tool in biomonitoring studies. In such cases, particular care should be taken in sample collection, selection and exposure (Tretiach et al. 2007b).

With increasing levels of air pollution, lichens show a transformation of chlorophyll *a* into phaeophytin *a*, followed by a reduction of chlorophyll content (Gonzalez and Pignata 1994; Zambrano et al. 1999; Gauslaa and Solhaug 2000; Zambrano and Nash 2000; Hauck et al. 2003; Backor et al. 2004; Godinho et al. 2004; Pignata et al. 2004; Shukla and Upreti 2007; Frati et al. 2011; Bajpai and Upreti 2012; Majumder et al. 2013; Karakoti et al. 2014). A loss of chlorophyll finally could lead to extinction of lichen species.

Significant damage to lichen cell membranes has also been shown as a result of exposure to pollutants (Adamo et al. 2003; Alebic-Juretic and Arko-Pijevac 2005; Spagnuolo et al. 2011; Majumder et al. 2013; Sujetoviene and Sliumpaite 2013). In the case of mycobiont, content of ergosterol, the main sterol of fungal membranes, is the marker of the response to pollution in the environment (Backor et al. 2006; Vantova et al. 2013).

Cellular injury may be caused by free radicals which are induced upon exposure to air pollution. Exposure to pollution increases the production of reactive oxygen species (ROS), including superoxide radicals, hydroxyl radicals and hydrogen peroxide along with disbalance of cellular redox status. Under the high pollution, algal cells are injured because the capacity of cellular antioxidant protection system is not able to cope with high ROS levels. ROS can react with cellular components (lipids, proteins, pigments and nucleic acids) to cause lipid peroxidation (del Hoyo et al. 2011). To protect themselves from potentially damaging effects of the oxidative stress induced by ROS, living organisms have enhanced some antioxidant enzymes [superoxide dismutases (SOD), catalases (CAT) , peroxidases (POD) and glutathione S-transferases (GSTs)] and non-enzymatic antioxidants (glutathione (GHS), ascorbate (APX) and phenolic compounds) as defence mechanisms (Oztetik and Cicek 2011). Superoxide dismutase (SOD) catalyses the dismutase of superoxide radicals ($O_2^{\cdot-}$) to yield hydrogen peroxide (H_2O_2) and oxygen (O_2). The enzyme is important as defence mechanism against oxygen radical-mediated toxicity. Thus, the induction of SOD can be as indicator of enhanced $O_2^{\cdot -}$ production. It has been shown the good correlation between SOD activity and atmospheric O_3 and SO_2 concentrations in transplants of *Hypogymnia physodes* (Egger et al. 1994). An increase in the activity of SOD by exposure to pollutants has been documented in *H. physodes*, *Xanthoria parietina* and *Ramalina farinacea* (Egger et al. 1994; Silberstein et al. 1996b). Declines in the activity of SOD, POD and CAT in pollution sensitive *Ramalina duriaei* were documented (Silberstein et al. 1996b). The activity of all the antioxidant enzymes assayed was also decreased in *Evernia prunastri* by SO_2 exposure (Deltoro et al. 1999). It is possible that pollutant accumulation exceeded the rate of detoxification processes in sensitive lichen species with the result in H_2O_2 accumulation (Hippeli and Elstner 1996). The increase in glutathione (GSH) concentrations and glutathione S-transferase (GSTs) activities was observed under the exposure of air pollutants (Oztetik and Cicek 2011).

The peroxidation of membrane lipid is one of the most important aspects of oxidative stress induced by the reactivity of oxygen free radicals. Malonedialdehyde (MDA) is one of the final products of polyunsaturated fatty acids peroxidation. An increase in free radicals causes the surplus production of MDA. Therefore, the content of MDA can be used as indicator of the degree of oxidative stress. This is one the most frequently used indicator of oxidative stress in lichens (Gonzalez et al. 1996; Gonzalez and Pignata 1997; Carreras et al. 1998; Gonzalez et al. 1998; Kong et al. 1999; Cuny et al. 2004; Monnet et al. 2005; Carreras and Pignata 2007; Sujetoviene and Sliumpaite 2013).

During lipid peroxidation conjugation of the ethylenic groups of polyunsaturated fatty acids may be observed which results in an increase in hydroperoxy conjugated dienes (HPDC). HPDC as peroxidation product seems to be a better estimator of damage because no changes were detected in the MDA content under the exposure of urban atmospheric pollutants (Rodriguez et al. 2007). HPDC was successfully applied in the studies of transplanted lichens to urban areas (Canas et al. 1997; Gonzalez and Pignata 1997, 2000; Carreras et al. 1998; Gonzalez et al. 1998, 2012; Carreras and Pignata 2002; Pignata et al. 2004).

For the evaluation of air quality under the mix of contaminants (e.g. urban areas), the use of pollution index (PI) was proposed by Gonzalez and Pignata (1994) and is used in the biomonitoring studies (Levin and Pignata 1995; Gonzalez and Pignata 1997; Carreras et al. 1998; Pignata et al. 2007). The index includes such changes in the transplanted lichens parameters as content of chlorophyll *a* and *b* and sulphur content in lichen thallus, and the concentration of HPDC compared with the control. The PI was proposed as a good indicator of global damage to lichen and of different air quality (Levin and Pignata 1995; Gonzalez et al. 1996; Gonzalez and Pignata 1997). The PI values were highest in lichens transplanted to sites with high industrial activity and traffic.

Despite used changes in physiological and biochemical parameters as biomarkers determining the level of environmental stress on lichens, extracellular and intracellular content of nitrogen compounds or heavy metals is also used. It is suggested that intracellular rather than total concentrations must be considered (Backor and Loppi 2009). In the intracellular spaces accumulated particles can remain unaltered for a long time. Because the negatively charged anionic sites of the cell walls of symbionts in lichens present, the extracellular content also plays important part of compound uptake in lichens. These binding sites are carboxyl, phosphate, amine and hydroxyl groups (Backor and Loppi 2009). Because elements bounded to cell wall are readily exchangeable, it is suggested that extracellular amounts and proportions reflects environmental input (Brown 1987).

Lichens can accumulate and retain many heavy metals in quantities above their physiological requirements. The lichens continuously accumulates pollutants from the air until equilibrium is reached; thus, transplanted lichens are useful for monitoring air pollution concentrations overtime (Kularatne and de Freitas 2013). The more polluted area, the longer the equilibrium time. Transplantation studies showed that lichens respond to changes in heavy metals concentrations within a period of several months (Kularatne and de Freitas 2013; Cloquet et al. 2009), after one year (Mikhailova and Sharunova 2008) or several years (Sondergaard 2013).

4.4 Evidence from Experiments and Filed Studies: Lichen Diversity in Response Pollution and Physiological Response of Lichens

4.4.1 Nitrogen Compounds

Large quantities of nitrogen (N) compounds produced naturally exceed anthropogenic emissions. Anthropogenic inputs from power generation, industrial activity, transport and agriculture have disrupted this balance. Other atmospheric contributions of nitrogen compounds come from non-combustion processes, for example nitric acid manufacture, welding processes and the use of explosives. The main two categories of nitrogen compounds are oxidized [nitric oxide (NO) and nitrogen dioxide (NO_2)] and reduced nitrogen forms [ammonia (NH_3)]. The atmospheric deposition of nitrogen has increased about threefold from pre-industrial times in 1995, and by 2050, it is projected to redouble (Galloway et al. 2004). Nowadays, the rates of nitrogen deposition are steadily increasing and becoming a big problem along with increase in world population and its demands (Galloway et al. 2008).

Nitrogen deposition can eutrophy and acidify the ecosystems (Stevens et al. 2011). The impact of N deposition on a species depends on several factors, including the duration of exposure, total amount and form of nitrogen and the sensitivity of species (Bobbink et al. 2010; Erisman et al. 2013). In order to monitor atmospheric pollution by nitrogen compounds with lichens, three approaches are used: (1) changes in lichen species composition, (2) changes in physiological parameters and (3) accumulation of nitrogen in lichens.

Species composition. Unlike most other pollutants, nitrogen is an essential nutrient. Increased N availability usually leads to increase growth by investing more in their photosynthetic capacity. As nitrogen concentration in lichen thalli increase in order to balance their carbon-to-nitrogen ratio, the growth rate and carbon assimilation capacity increase (Palmqvist and Dahman 2006; Johansson et al. 2011). The higher growth rate could be characteristic to fast-growing species which may out compete the slow-growing ones. Thus, changes in nitrogen availability could induce changes in lichen species composition. Johansson et al. (2012) proposed that those community responses could be due to physiological responses of the individual species rather than changes in competitive interactions.

Adding nitrogen to the environment tends to shift lichen community composition. Lichen responses to nitrogen are different due to their sensitivity to their ionic environment. Increase in nitrate nitrogen can acidify the substratum while enrichment in the form of ammonium results in higher pH levels. NO_x has a strong effect on lichen diversity (Davies et al. 2007), its community composition and frequency and dispersal (Larsen et al. 2007). In addition, lichen population declines under high NO_x concentration (van Dobben et al. 2001; Giordani 2007). A number of studies have observed a positive relationship between the proximity to livestock farms and abundance of nitrophilous lichen species (Sanchez-Hoyos and Manrique 1995; van Dobben and DeBakker 1996; van Dobben and ter Braak 1998; van Herk 1999; Ruoss 1999). Despite the negative effects, some lichen species such as *Lecanora dispersa* and *Phaeophyscia orbicularis* are NO_x tolerant (Davies et al. 2007). Phytotoxic effect of NO_2 on lichens that were fumigated with 4 ppm (7,520 μg m^{-3}) for six hours in a laboratory experiment was confirmed by Nash (1976) and suggested that the pollutant would probably not be harmful to lichens since the NO_2 concentration detected in natural environment was usually less than 1 ppm. Thus, the effect of NO_x on lichens seems controversial and unclear.

Some N tolerant species are favoured by N addition improving their physiological status and a luxury consumption of the extra N supplied (Ochoa-Hueso and Manrique 2011). It was concluded that the tolerant lichen *Cladonia foliacea* could be used as biomarker of responses to nitrogen. In contrast, other species adapted to nutrient poor habitats and acidophytic (sensitive to eutrophication) species abundance and distribution are negatively correlated with nitrogen enrichment (Gaio-Oliveira et al. 2004; Friedel and Muller 2004; Davies et al. 2007; Franzen-Reuter and Frahm 2007; Pilkington et al. 2007; Pinho et al. 2008). High levels of nitrogen cause a decrease in lichen abundance and overall homogenization of the lichen species with remaining the most tolerant species (Liska and Herben 2008). The loss of ground lichen species adapted to nutrient poor conditions were observed when studying eutrophication effects in pine forests over 45 years period (Reinecke et al. 2014).

Direct nitrogen fertilization highlights the potential threat to lichens. The disappearance of lichen *H. physodes* and *Lepraria incana* following 21 months of fertilization was observed by Franzen-Reuter and Frahm (2007). The lichen community composition responded to fertilization even at the lowest nitrogen application level —6 kg N ha^{-1} $year^{-1}$ (Johansson et al. 2012). The data suggest that those community responses could be used as sensitive bioindicators in order to detect low levels of N pollution. As most studies have focused on the total amount of N deposition, it is very important to investigate the effects of different ratios of oxidised and reduced N forms on lichen communities. The form of N deposition has important consequences on lichen community response to N enrichment.

Changes in physiological parameters. Green algae–lichens are solely depended on atmosphere nitrogen directly deposited on their thallus surfaces to meet their nitrogen demand. Lichen pigment concentration can reflect N deposition. The lichen transplanted from an environment relatively low in nitrogen, used nitrogen as a nutrient, and in response chlorophyll *a* content could increase. It was suggested that chlorophyll synthesis could be stimulated at the urban sites due to higher concentrations of nitrogen compounds, which are beneficial to lichens (Boonpragob and Nash 1991). Such trend was found in studies with transplants in urban areas—with *P. sulcata* transplanted to the urban area of Biel city (von Arb and Brunold 1990), with *Usnea amblyoclada* transplanted at the centre of Cordoba city (Carreras et al. 2005; 2009) and with *E. prunastri* exposed in the urban parks and residential sites (Sujetoviene and Sliumpaite 2013). The concentration of chlorophyll was also positively affected by the traffic level: the content of chlorophylls increased in parallel with the level of traffic density and pollutants emitted by transport (Carreras et al. 1998; Satya and Upreti 2009) and under the impact of agricultural activities (Fig. 4.1) (Frati et al. 2011). Increase in chlorophyll *a* concentration was recorded along with accumulated nitrogen under the fertilization experiment (Dahlman et al. 2003; Gaio-Oliveira et al. 2005a). Short-term (24-, 48- and 96 h) exposure of N excess do not cause chlorophyll degradation in *X. parietina* (Munzi et al. 2009).

The increase in photosynthetic pigments was not always characteristic because the algal partner could suffer under the high nitrogen deposition. Degradation of photosynthetic pigments is commonly used as an indicator of lichen vitality. The rate of degradation of chlorophyll *a* into phaeophytin is closely parallel to environmental pollution (Garty et al. 2000), which may be indicated by decreased concentrations of chlorophylls content, carotenoids and chlorophyll degradation to phaeophytin as effects of NO_2 and NH_3 from agricultural activities and road traffic (Fig. 4.1) (Frati et al. 2006, 2011). Samples of nitrophytic *X. parietina* also showed loss of chlorophyll when exposed to increasing ammonium concentrations (Gaio-Oliveira et al. 2004). Experimental evidence supports the hypothesis that high concentrations of nitrogen have toxic effects on lichens (Munzi et al. 2010; Meysurova et al. 2011). Different physiological responses were observed nitrophytic and oligotrophic lichens treating with combination of oxidised and reduced nitrogen (Gaio-Oliveira et al. 2005a). Chlorophyll *a* and ergosterol concentrations

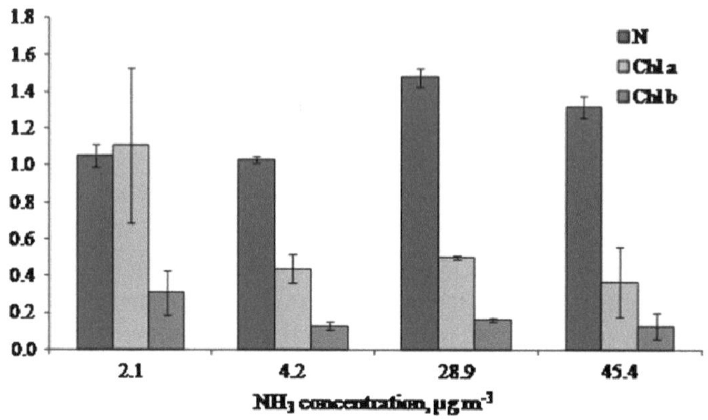

Fig. 4.1 The concentrations of total nitrogen (N, % dry weight), chlorophyll *a* (Chl *a*, mg g^{-1}) and chlorophyll *b* (Chl *b*, mg g^{-1}) in *Evernia prunastri* thalli transplanted for 2 months at plots with different levels of agricultural activities (Frati et al. 2011)

remained unaltered in nitrophytic *X. parietina* irrespective of N dosage, while ergosterol decreased with increasing N uptake in acidophytic *E. prunastri*. The latter species had accumulated a large pool of ammonium at the highest N dosage, while in *X. parietina* showed a significant nitrate concentration.

At the molecular level, Tretiach et al. (2007b) reported that a large amount of NO_x can damage the photobionts of transplanted *Flavoparmelia caperata*, hindering photosynthesis probably due to the increased reactive oxygen species (ROS), as high concentration of NO_2 in the cells forms nitrous and nitric acid, which acidifies cytoplasm and results in protein denaturation, deamination of amino acids and nucleic acid. When exposed to excessive nitrogen supply, the higher damage was observed for mycobiont than the photobiont (Dahlman et al. 2002, 2003; Gaio-Oliveira et al. 2004).

Ergosterol a component of the fungal plasma membrane is used as a marker of mycobiont respiration. The significant decrease in ergosterol with increasing N uptake was observed in *E. prunastri* (Gaio-Oliveira et al. 2005b), suggesting that increasing nitrogen uptake has caused more damage to the fungal partner than to the photobiont. Chitin a component of the fungal cell wall is being correlated with amount of mycobiont biomass. Increase in photobiont investment took parallel by a subsequent decrease of investments in the mycobiont was proved by the increased ratio chlorophyll to ergosterol and chlorophyll to chitin (Dahlman et al. 2003).

Different degrees of lichen sensitivity are related with different mechanisms of avoiding excess of nitrogen uptake. Some nitrogen sensitive species are able to avoid toxicity and are able to avoid the excess of nitrogen uptake while others are able to neutralize it once when it entered the cell. The latter is related with lichen ability to assimilate N into non-toxic forms such as arginine (Silberstein et al. 1996a; Dahlman et al. 2003). The accumulation of arginine could be a way to avoid toxic effects of surplus NH_4^+ assimilation. The lichens may be able to adapt to the increased N supply either through uptake regulation or through increased growth (Johansson et al. 2010). Recent studies have shown that some species have repairing systems involved in its metabolism. The neutralization of nitrogen excess was observed when species produce or use polyamines for that (Pirintsos et al. 2009). Studies have indicated some species (e.g. *E. prunastri*) could not avoid nitrogen entrance into the cell due to its high cation exchange capacity (Gaio-Oliveira et al. 2001). Lower cation exchange capacity prevents ion absorption within lichen thallus, and this could explain why some species (e.g. *X. parietina*) are more adaptable and react different under high N availability (Gaio-Oliveira et al. 2005a).

Cell membrane damage. A common sign of N compounds toxicity is the decline of essential cations in the lichen thallus. Ammonium supply resulted in a decrease in intracellular K^+ and Mg^{2+} of *X. parietina* (Munzi et al. 2011). An excessive ammonium load affected the cell membrane integrity until rupture and as a consequence, a loss of cytoplasmatic material occurred that can be confirmed by intracellular K^+ leakage. Electrical conductivity indicated cell membrane damage in the cases of acute, high N stress episodes (Munzi et al. 2009) but no significant differences were detected during the four weeks experimental conditions (Munzi et al. 2012).

Photosynthesis and fluorescence. Under the non-toxic increase in N availability, lichens showed the response by investing more N in their photosynthetic and carbon assimilation capacity (Palmqvist et al. 2002; Ra et al. 2005) and finally to increase in growth rate (Palmqvist and Dahlman 2006). The increased N supply favours photosynthesis with loads of N as high as 20 kg N ha^{-1} $year^{-1}$ (Ochoa-Hueso and Manrique 2011) and above this threshold, the improved photosynthesis was not observed.

It is thus clear that the excess nitrogen might negatively affect the N pathway and the photosynthesis process of lichens. High NO_x levels induce deleterious effects in the photosynthetic apparatus of the photobionts of transplanted lichens (Tretiach et al. 2007b). It is suggested that nitrogen oxides interrupt the formation of the trans-thylakoid proton gradient as showed heavily affected non-photochemical quenching.

According to Paoli et al. (2010) samples of *E. prunastri* transplanted to close proximity of a point source of NH_3 showed a marked decrease in the potential quantum yield of PSII after 15 days. Nitrogen sensitive lichen *E. prunastri* exposed to NH_3 showed decrease in Fv/Fm above 3 µg m^{-3}, suggesting direct adverse effects of NH_3 on its photosynthetic performance (Munzi et al. 2014). By contrast, nitrogen tolerant *X. parietina* showed decreased capacity of fluorescence above 50 µg m^{-3} NH_3. Smaller chlorophyll *a* fluorescence variation in pollution-resistant lichen *X. parietina* was also observed (Piccotto et al. 2011). The results of effect of ammonia on lichens clearly indicate that the ecological success and species-specific tolerance of *X. parietina* at ammonia-rich sites or urban areas might be related to indirect effects of increased nitrogen availability.

Accumulation. Lichens are largely dependent on the atmosphere for acquisition of major nutrients including nitrogen. Nitrogen concentration in lichen thallus can respond to N from the atmosphere and can be a good indicator which reflects the content of it in the atmosphere (Hyvarinen and Crittenden 1998; Olsen et al. 2010; Boltersdorf and Werner 2013). Nitrogen concentrations in *Usnea lapponica* were strongly correlated with nitrogen deposition (McMurray et al. 2013). Nitrogen accumulation was observed in transplants as a response of anthropogenic N enrichment by traffic activity (Gombert et al. 2003; Frati et al. 2006; Sujetoviene 2010). Lichen N accumulation patterns matched throughfall N deposition patterns (Fenn et al. 2007). Experimentally nitrogen deposition was simulated by irrigating trees over a 3 yr period with NH_4NO_3, providing N loads ranging from ambient to 50 kg N ha^{-1} $year^{-1}$ (Johansson et al. 2010). The experiment results showed that the thallus showed increased N concentration in *Alectoria sarmentosa* and *Platismatia glauca* with increasing N load.

Lichens when transplanted in an agricultural area with high atmospheric NH_3 concentrations have responded to NH_3 air pollution accumulating nitrogen in its thalli (Søchting 1995; Ruoss 1999; Frati et al. 2011) and showed maximum content of nitrogen in *X. parietina* after one month of exposure but at sites with background concentrations, it took several months to reach statistically significant uptake (Olsen et al. 2010). The excessive nitrogen supply applying fertilizers also induced the increase in lichen thallus N concentration (Dahlman et al. 2003).

Root et al. (2013) presented a new monitoring approach that estimates throughfall inorganic N deposition from N concentration in lichens. It was concluded that lichen N concentration is a cost-effective way to monitor N deposition, and its N concentrations can be used to estimate throughfall N deposition and such lichen monitoring allows identification of areas that may exceed N critical loads.

The identification of early stages of biological effects of nitrogen pollution is essential to prevent consequences for a whole ecosystem. Lichen physiological changes are successfully used as an early indicator of environmental stress (Vingiani et al. 2004). Composition or cover changes in lichen communities indicate exceedance of critical loads for terrestrial ecosystems.

Using these approaches, a threshold value was suggested for nitrogen compounds. A threshold value was defined as a concentration of atmospheric NH_3 above which changes in species diversity or physiology occurs. Concentration of 3 μg m^{-3} NH_3 in the atmosphere was suggested which cause adverse effects on lichens photosynthetic performance—a decrease in lichen species frequency and Fv/Fm of oligotrophic *E. prunastri* (Munzi et al. 2014). Empirical data suggest that the critical load for epiphytic lichens in boreal forests is c. 10–15 kg N ha^{-1} $year^{-1}$, although lower critical loads (as low as 3 kg ha^{-1} $year^{-1}$) have been suggested (Fenn et al. 2008).

4.4.2 Sulphur Compounds

Sulphur dioxide originates from both natural and anthropogenic sources in the environment. Natural sources of the sulphur in the atmosphere are volcanic activity, fires and bacterial activity, and the sea and ocean. The main anthropogenic source of SO_2 is industrial activity that contains sulphur, e.g. the fossil fuel combustion, smelting, manufacture of sulphuric acid, incineration of refuse and production of elemental sulphur. Anthropogenic emissions constituted about 90 % of the total sulphur input to the atmosphere (Whelpdale et al. 1997). Sulphur emissions have been increasing steadily since the year 1850, now mainly in Asian countries by developing industries. According to Europe's Environment Agency assessment (2012), in Europe, sulphur emissions have decreased by 74 % between the years 1990 and 2011.

Sulphur dioxide (SO_2) was formerly viewed as the most important phytotoxic pollutant and until the early 1980s was the subject of the greatest research in this field. Now, it has attained less attention because of its declining concentrations as emissions have been reduced and also because nitrogen gases and ozone have been found to be of increasing significance.

Sensitivity of lichens to sulphur compounds has been investigated for a long time and is the basis of the use of lichens as bioindicators. High sensitivity of lichens to sulphur compounds lies on the acidic characteristics of sulphur dioxide which is the main source of sulphur in the atmosphere (Batic 2002). Damaging effects of sulphur compounds are related with proton formation along with oxidation of sulphite to sulphate and free radical generation during the oxidation process.

The main tools for assessing the response of lichens to sulphur pollution is changes in their distribution and frequency and changes in physiological parameters. It was proved that lichens are very suitable bioindicators as they are sensitive to atmospheric sulphur dioxide, and sulphur dioxide considered as a primary factor causing death of lichens in most urban and industrial areas (Giordani 2007). Fruticose species are more sensitive to sulphur compounds than many foliose and crustose species (Owczarek et al. 1999; Tretiach and Ganis 1999).

The summarized historical evidence of lichen decline with increasing burning of fossil and rising concentrations of SO_2 along with lichen species decline are provided by Bates (2002). The disappearance of lichens from polluted urban and industrialized areas was recorded under high pollution with sulphur dioxide (LeBlank and De Sloover 1970; LeBlanc et al. 1972). In recent decades, concentrations in the atmosphere decreased markedly and as consequence, the air quality improved quite rapidly and remarkable in some polluted areas. Recent trends towards lower concentrations of sulphur dioxide lead to changes in lichen occurrence and recolonization of lichen desert (Rose and Hawksworth 1981; Hawksworth and Mcmanus 1989; Letrouitgalinou et al. 1992;

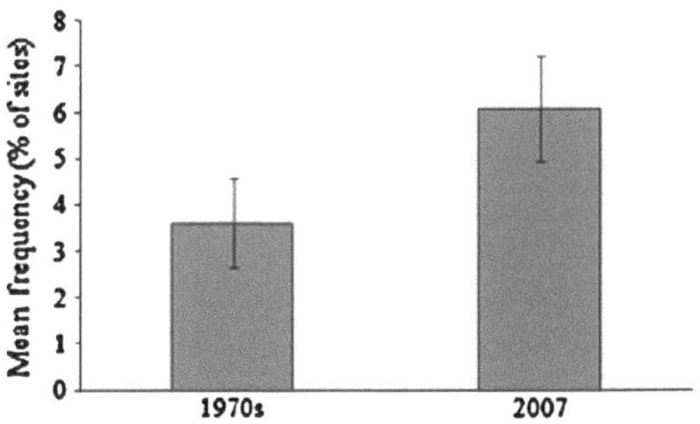

Fig. 4.2 Changes in number of epiphytic species observed in Skawina, in the 1970s and in 2007, in the urban-industrial area of southern Poland. The increase in species number showed lichen recolonisation process as a result of air quality improvement (Lisowska 2011)

van Dobben and De Bakker 1996; Ranta 2001; Loppi and Corsini 2003; Loppi et al. 2003; Hultengren et al. 2004; Adamska 2011; Lisowska 2011; Slaby and Lisowska 2012; Evju and Bruteig 2013; Fig. 4.2). However, in the case of heavy polluted areas ,species recolonisation cannot be expected and lichens in such area shows growth of toxitolerant species (Upreti et al. 2005). Modest SO_2 levels, interacting with low bark pH, still inhibit recolonization of sensitive lichens in urban areas (Bates et al. 2001; Batty et al. 2003).

Individual lichen species differ in their sensitivities to sulphur dioxide. Such differences in sensitivities lead to biomonitoring approach to divide polluted areas into zone according to lichen species assemblages. According to measurements of ambient sulphur dioxide and the IAP the surveyed area may be divided into pollution zones (Gilbert 1968, 1970; Hawksworth and Rose 1970).

Sulphur dioxide may be adsorbed on a thallus and then absorbed when the thallus are wet. The sensitivity of lichens to sulphur dioxide depends on their ability to absorb and retain water and on morphological characteristics of a lichen thallus such as a thick cortex that prevent or impedes the penetration of sulphur dioxide into the thallus (Vernon 1993). Because of the high solubility of SO_2 in water, lichen absorb SO_2 even when not metabolically active and absorption leads to higher total sulphur levels in lichen thallus (Nash 2010). Accumulation of sulphur appeared to be independent of growth form (St Clair et al. 2002). There is a positive relationship between the concentration of sulphur dioxide in the atmosphere and content of sulphur found in lichen thallus (LeBlanc and Rao 1973; Pyatt 1973; LeBlanc et al. 1974; Pakarinen 1981; Takala et al. 1985; Garty et al. 1988; Showman and Long 1992; Oztetik and Cicek 2011). The sulphur content in lichens has been associated quite well with the exposure to sulphur dioxide time (Carreras et al. 1998; Gonzalez and Pignata 2000; Wiseman and Wadleigh 2002; Vingiani et al. 2004; Tretiach et al. 2007a). In some cases, it is difficult to correlate SO_2 exposure and sulphur content in lichens because injured lichens lose their ability to accumulate sulphur due to increased membrane permeability.

The severity of injuries of SO_2 has been related to the concentrations of SO_2: the higher is concentration, the greater symptoms of injuries. Over time lichens could be affected under the SO_2 exposure and the morphological changes occur. Sulphur dioxide affects the growth of lichens: the smaller, more compacted thalli and decrease in cover were recorded (Massara et al. 2009). Sulphur compounds caused reduction or absence of fruiting bodies. Under the exposure, the colour of the lichen thallus could also be changed. For example some lichens could be pigmented or turn to brown. More sensitive species degrade more or less rapidly. Degradation includes discolourations, mostly bleaching due to damage of the photobionts. Some of lichen could become white and detached from the substrate (Saunders 1970).

A destructive effect of SO_2 on lichens was indicated by a statistically significant negative correlation between $\delta^{13}C$ values in the whole lichen and atmospheric sulphur dioxide concentrations (Batts et al. 2004). $\delta^{13}C$ values in lichens in wet environments, in the presence of high SO_2 concentrations, were more negative than that in samples from drier sites. An explanation of this effect may relate to an increase in carboxylation resistance resulting from damage to chlorophyll and cell structure by SO_2 (Batts et al. 2004).

The observed effects of sulphur dioxide on lichens also involve decline in nitrogen fixation (in cyanolichens lichens with bacteria and actinomycetes), decreases in photosynthesis rate, increase in membrane permeability and ultrastructural changes. Correlation between the S content and physiological markers of SO_2 injuries were observed in lichens under the sulphur pollution (Boonpragob and Nash 1990; von Arb et al. 1990; Gonzalez and Pignata 1994). The nitrogen fixation proved to be more susceptible than photosynthesis (Hällgren and Huss 1975; Henriksson and Pearson 1981). Nitrogen fixation rates of *Peltigera canina* decreased when the lichen was exposed to sulphur dioxide (Henriksson and Pearson 1981). Photosynthesis is one of the first processes to be affected by SO_2 pollution (Malhotra and Hocking 1976). Photosynthesis was found to be up to five times more sensitive than respiration to sulphur dioxide (Beekley and Hoffman 1981). Since under the moderate pollution conditions, the opposite results were observed—SO_2 showed the highest positive correlation with net photosynthesis along with increase in chlorophyll content (von Arb et al. 1990). Under the SO_2 fumigation, net photosynthesis rates were more negatively affected in *E. prunastri* than in *R. farinacea* (Deltoro et al. 1999). At low concentration, inhibition of photosynthesis occurs and a prolonged exposition to hydrogen sulphide seems to cause permanent damage to the photosynthetic apparatus, with a decrease in photosynthetic capacity, causing a drastic decrease in thallus growth (Tretiach and Baruffo 2001).

An application of pulse amplitude method (PAM) to measure fluorescence in order to monitor the status of photosystem II (PSII) in lichens exposed to gaseous SO_2. The results indicated rapid induction kinetics of fluorescence upon fumigation, which indicated the inhibition of PSII by SO_2 (Gries et al. 1995). Fluorescence was negatively correlated with S content (Garty et al. 2003).

One of the direct effects of sulphur dioxide on the metabolism of lichens is destruction of photosynthesis pigments. The loss of chlorophyll finally could lead to extinction. With the increasing sulphur dioxide concentrations, a transformation of chlorophyll *a* into phaeophytin along with a reduction of chlorophyll content was observed (Hällgren and Huss 1975; Kong et al. 1999). Chlorophyll reduction in fruticose lichen *E. prunastri* was observed only after severe damage to photosynthesis (Sanz et al. 1992). The experimental results were in accordance with pigment losses reported under the field conditions (Gonzalez and Pignata 1994; Gonzalez et al. 1998; Silberstein et al. 1996a; Garty et al. 1998). The field evidence has suggested that chlorophyll degradation as a result of SO_2 could be affected under the moderate stress conditions (Satya and Upreti 2009).

Lichens exposed to a range of SO_2 concentrations display ultrastructural alterations. Ultrastructural changes in *Cladina stellaris* cells which was affected by SO_2 included lipid accumulation and membrane thickening (Plakunova and Plakunova 1987). Symptoms are characteristic when thallus is ageing. Ultrastructural injuries in exposed *H. physodes* involve swelling of the mitochondria and stretching of thylakoids (Holopainen and Kauppi 1989). At the biochemical level, reduction of protein and lipid synthesis was recorded (Bychek-Guschina et al. 1999). The main effect of SO_2 on cells is acidification, which influence enzymatic reactions (Grube 2010). Lichen sensitivity to acidification could be related to the buffering capacity of secondary metabolites produced by lichens. Usnic acid producing species makes lichens vulnerable to acidity and such species would be rare in areas of high pollution levels (Hauck and Jürgens 2008).

Damage to cell membranes. Lichens under the sulphur stress have been experienced loss of electrolytes (Pearson and Rodgers 1982; Grace et al. 1985). Leakage of electrolytes is a result of damage to cell membranes of the bionts. Exposures of lichens to SO_2 have induced ion efflux, indicating changes in membrane integrity under laboratory conditions (Puckett et al. 1977; Pearson and Henriksson 1981; Fields and St. Clair 1984) or in field studies (Fuchs and Garty 1983; Alebic-Juretic and Arko-Pijevac 1989; Boonpragob and Nash 1990; Garty et al. 1993). The increase in conductivity values in *P. sulcata* was positively related to the increases in SO_2

values in the atmosphere (Marques et al. 2005). The highest cell membrane damage was observed in lichen thallus collected at sites with highest winter SO_2 levels (Alebic-Juretic and Arko-Pijevac 2005). The conductivity of leachates was positively correlated with modelled atmospheric SO_2 concentrations and with bark S (Tarhanen et al. 1996). On the basis of sensitivity, low cost, portability and rapidity, the membrane permeability test has been recommended for evaluation of air pollution damage to lichens (Garty et al. 1993).

Oxidative stress. The response to SO_2 in lichens is an increase in activities of enzymes as a defence against increased oxidant formation. The tolerant lichen *X. parietina* showed increase in peroxidise activity more than that of the sensitive *Ramalina duriaei,* and the activity was higher at polluted than at unpolluted sites (Silberstein et al. 1996b). SO_2-induced lipid peroxidation in biomembranes as indicated by its positive correlation with the MDA content (Egger et al. 1994).

In polluted areas, sulphur dioxide is often in conjunction with other pollutants. Therefore, it is difficult to distinguish the effects of sulphur along with mixture of other pollutants. As it was proved that the direct toxic effects to lichens occur at much lower concentrations than those causing damage to higher plants; thus, lichens is valuable indicator setting critical level for sulphur in the environment. Lichens are useful bioindicators of sulphur pollution because they can absorb and accumulate toxic sulphur compounds and have different sensitivities to them.

4.4.3 Ozone

Ground level ozone is not emitted into the atmosphere but is created by chemical reactions between oxides of nitrogen and volatile organic compounds. Ozone is the most important air oxidant causing injuries to vegetation (Fuhrer 2002). In general, lichens are not considered as sensitive as vascular plants to ozone (Nash and Sigal 1998; Bertuzzi et al. 2013). Comparing historical and recent lichen sampling data, different sensitivity lichen species against ozone were listed by Sigal and Nash (1983). Sensitive species such as *E. prunastri*, *Peltigera* spp. *Pseudocyphellaria* spp. and *Ramalina* spp. disappeared as consequence of this oxidant impact in the Los Angeles. Sensitive species (*Usnea* spp., *Collema nigrescens*) are found in protected areas, and moderately tolerant Hypogymnia enteromorpha are in reduced abundance and show morphological anomalies (thallus bleaching and decay).

Physiological studies showed conflicting results in relation to ozone toxicity. A number of recent experiments with O_3 confirm the statements that it negatively affects lichens. The experimental fumigation with O_3 revealed a significant decline in the efficiency of PSII photochemistry, the collapse of a high number of photobiont cells and as a consequence—bleaching of the thallus (Scheidegger and Schroeter 1995). The decrease in photosynthesis rate was also observed (Ross and Nash 1983; Zambrano and Nash 2000). Under the field conditions among other potential factors, regional ozone levels during the course of the transplant period were sufficiently high to have contributed to the adverse effects on lichens, i.e. reduced photosynthesis and chlorophyll *b* degradation (Zambrano et al. 1999; Zambrano and Nash 2000). Ozone has direct, deleterious effect on lichens through the fast formation of reactive oxygen species—O_3 concentration showed a positive correlation with the contents malondialdehyde and superoxide dismutase) in *H. physodes* (Egger et al. 1994), which is a product of lipid peroxidation and are indicators of oxidative damage to membranes and to the enzymes that protect against oxidation.

By contrast, Calatayud et al. (2000) found that exposure to O_3 for 4 h per day for 14 days did not induced changes in the efficiency of PSII as measured by fluorescence in all of the five lichen species studied. Additionally, neither photosynthetic CO_2 assimilation nor xanthophylls cycle activity was affected by high concentrations of ozone. Exposure of *P. sulcata* to ozone also did not have any effect on Fv/Fm (Balaguer et al. 1996). Under the field conditions, the content of chlorophyll in transplanted *H. physodes* was positively correlated with O_3 concentration (Egger et al. 1994). The results of the fumigation studies suggest that the functionality of photosystems was not impaired by ozone (Calatayud et al. 2000; Riddell et al. 2010, 2012; Bertuzzi et al. 2013).

It is well evident that the effects of ozone under experimental and field conditions included biophysical and physiological, as well as structural impairment in the lichens studied. Despite the evidence of O_3 effects, some researchers suggest that lichens are not good indicators of ozone as no significant consequence to O_3 exposure were detected (Ruoss and Vonarburg 1995; Riddell et al. 2010; Bertuzzi et al. 2013). It is suggested that lichens have significant capacities to withstand oxidative stress induced by high ozone concentrations (Calatayud et al. 2000; Bertuzzi et al. 2013).

4.4.4 Heavy Metals

Heavy metals originate from naturally and anthropogenic sources in the environment. The natural sources of these elements are volcanic eruptions, dusts, fires and salt spray. Anthropogenic sources include emissions from different anthropogenic activities—mining, industrial production (smelters, oil refineries, petrochemical plants and chemical industry) and untreated sewage sludge and from diffuse sources—coal combustion and traffic. The problem is that the amount naturally occurring heavy metals is relatively very small as compared with anthropogenic loads of these elements.

There are some detailed reviews which have referred to the entrapment, uptake and accumulation of heavy metals by lichen thalli, made apparent by parameters of lichen vitality and stress (Garty 2001; Wolterbeek 2002; Szczepaniak and Biziuk 2003; Backor and Loppi 2009).

The main approach identifying pollution by heavy metals with lichens has been the accumulation of metals in lichen thallus. The accumulation process depends on many factors such as environmental conditions (temperature, humidity, wind velocity etc.), characteristics and health of lichen and availability of heavy metal in the environment (Folkenson 1979). It is no doubt that lichens are excellent bioaccumulators of heavy metals because the concentrations found in their thalli have been directly correlated with concentrations in the environment (Pawlik-Skowronska et al. 2008; Conti et al. 2009; Aprile et al. 2010; LeGalley et al. 2013; Hauck et al. 2013a). The metal concentrations in lichens are depended primarily on what metal are in the environment and the chemical and physical properties of the metal containing particles (size and acidity) (Folkenson 1979). The different lichen uptake rates were observed for some heavy metals, and this evidenced the existence of a selective metal uptake. The differences observed in metal uptake are probably due to the composition and type of cell wall binding sites leading to a differential absorption and/or cation exchange (Carrearas and Pignata 2007). Metals with a higher affinity for binding sites than divalent cations could saturate the absorption capacity of lichen thalli and thus block the incorporation of other cations (Gailey and Lloyd 1993).

The tolerance of lichen to accumulate elements varies greatly between species and elements (Folkeson 1979). Species-specific information on accumulation capacity is scarce. One method for the evaluation of the accumulation capacity is the index of exposed-to-control ratio (E/C) (Frati et al. 2005; Bergamaschi et al. 2007; Aprile et al. 2010), which helps to compare on the same scale the responses of the different biomonitors after exposure. Some lichen species could be considered as "severe accumulators". Transplantation studies have highlighted that lichen could accumulate metals by several times comparing with the initial concentration in its thalli: the E/C ratios investigated in all sites show "severe accumulation" for *F. caperata,* while for *Parmotrema chinense* they indicate "accumulation" (Aprile et al. 2010). *H. physodes*, *P. furfuracea* and *Usnea hirta* transplanted to an urban site of Italy have a similar accumulation capacity, while that of *P. sulcata* is lower (Bergamaschi et al. 2007).

The accumulation capacity could be different for some elements. The accumulation capacities of two species—*X. parietina* and *Parmelia tiliacea*

—differed only for the elements Ca, Cl, Cu and Pb (Karakas and Tuncel 2004). A relative high concentration of K, Fe, Mg, Zn, Mn, Ph, and Cu was found in thalli of *H. physodes,* and Mg, Zn, Pb, Cu, and Cd in *U. hirta* transplanted to traffic sites (Garty et al. 1996). Other species could have low accumulation capacity and could be considered as weak accumulators. The concentrations of heavy metals in *Cladonia* spp. thalli were relatively low considering the frequently enormous substrate contamination and for example such species could have ecological importance as stable and resistant pioneer in anthropogenically affected sites (Osyczka and Rola 2013).

Some lichen species are considered to be more suitable for transplanting and heavy metal accumulation. The lichen species with finely divided or densely sorediate thalli usually showed a greater affinity for collecting particulate matter (Garty 2001; Aprile et al. 2010). Surface development strongly influenced accumulation capacity of the biomonitors (Adamo et al. 2007). The accumulating capacity of *P. tiliacea* was higher than that of *X. parietina* and this difference proposed would be due to morphological differences in lichen features (Sloof 1993; Karakas and Tuncel 2004). Other properties such as cation exchange capacity and amount of ligands could be also important (Folkenson 1979). A lower incorporation of some metallic cations could be due to the formation of weak complexes with cell wall ligands that can consequently be easily removed and replaced by cations with a higher affinity for the cell wall binding sites (Carreras and Pignata 2007).

Some heavy metals induce inhibition of chlorophyll synthesis. Chlorophyll *a* was sensitive to the presence of copper and mercury but chlorophyll *b* increased in response to copper and zinc exposure (Backor and Dzubaj 2004). As a result, a significant decrease of the chlorophyll *a/b* ratio was found, while the total chlorophyll content was almost unchanged. Thus, the total content of chlorophyll is not suitable indicator of physiological damage. Chlorophyll *a* was more sensitive to metals than chlorophyll *b*, and its decrease was attributable to conversion of chlorophyll *a* to chlorophyll *b* by copper (Chettri et al. 1998). It has been suggested that the Chl *b*/Chl *a* ratio is a useful parameter to determine the physiological conditions of lichens and photobionts subjected to air pollution and heavy metals (Chettri et al. 1998; Carreras and Pignata 2007).

High negative correlations between Al and pigment concentrations were found (Nakajima et al. 2013) which suggests that Al stress reduces pigment concentrations. Accumulation of Al, Cr, Fe, Pb and Zn in the thallus suppressed the concentrations of pigments such as chlorophyll *a*, chlorophyll *b* and total chlorophyll, however, enhanced the level of protein (Bajpai et al. 2010). Short-term exposure to high Cu^{2+} concentrations (40 and 400 mu mol g^{-1}) under non-complexing conditions caused a dose-dependent decrease in chlorophyll *a* content; chlorophyll *b* and total carotenoid contents remained constant (Pawlik-Skowronska et al. 2006). It has been observed that Pb and Cd are mainly complexed to the fungal cell wall (Di Toppi et al. 2005; Sarret et al. 1998) and thus do not have any detrimental effects on physiological parameters in lichens. These ions could be tightly bound to binding sites of the mycobiont cell wall forming stable complexes (Carreras and Pignata 2007).

Some studies showed that there is correlation between chlorophyll integrity and the concentrations of heavy metals. The chlorophyll integrity decreased with increase in amount of Cu, Pb and Zn (Shukla and Upreti 2008). The integrity of chlorophyll correlated with the concentration of K and correlated inversely with the concentration of Cr, Fe, Mn, Ni, Pb and B (Garty et al. 1998). The chlorophyll integrity was especially affected following treatment of the thalli with Cu, Zn, Mn, Pb and Fe salts under acidic conditions (Garty et al. 1992). The chlorophyll degradation as well as alteration in the pigment content was the most sensitive parameters to assess the vitality of lichen thallus against polluted environment (Karakoti et al. 2014).

The lichen vitality under the metal stress was examined using the potential quantum yield of photosynthesis expressed as fluorescence ratio Fv/Fm. The Fv/Fm ratios were correlated with Al, Cd, Cr and Cu elements in lichens transplanted or naturally growing at sites with different degree of pollution level (Garty et al. 2000; Karakoti et al. 2014). Significantly positive correlations between Fv/Fm and element contents (Pb, Cu, Fe and Cr) were found in naturally growing lichen *Pyxine cocoes* (Karakoti et al. 2014). Garty et al. (2000) found that Fv/Fm ratios in transplanted *Ramalina lacera* were correlated with Al, Cd, Cr and Cu elements. The decreased Fv/Fm values of experimentally treated lichens and their bionts with different metal concentrations were also reported (Backor et al. 2004; Unal et al. 2010; Vantova et al. 2013). Cadmium induced the decrease in Fv/Fm values indicating damage to PSII (Backor et al. 2007).

Calcium, Cu and Sr showed an inverse correlation with the Fv/Fm ratio expressing the potential quantum yield of PSII (Garty et al. 2002). In *Ramalina fastigiata* intracellular Cu concentrations above ca. 4.0 mu mol g^{-1} were linked to a decline in chlorophyll fluorescence. The fluorescence parameter Fv/Fm was shown to be useful in determining the sensitivity of the lichens to Cu uptake (Branquinho et al. 1997a).

Pb uptake, particularly in cyanobiont lichens, caused a decrease in photosystem II (PSII) photochemical reactions, measured as a change in the ratio Fv/Fm (Branquinho et al. 1997b). Lichen PSII photochemical reactions were more sensitive to Pb than cell membrane damage since, with increasing Pb concentrations, no clear alterations in lichen membrane permeability, as measured by the loss of intracellular K and Mg, were observed.

Regarding the effects of the cations studied on the physiological parameters measured in *U. amblyoclada*, the following sequence could explain their toxicity: $Cu^{2+} > Pb^{2+} > Ni^{2+} > Zn^{2+}$ (Carrearas and Pignata 2007). Metal cations appeared to be ionically bound within the cell wall in an exchangeable form with binding affinities of Pb > Cu > Zn (Chettri et al. 1998).

Among the two bionts, the algal partner appears to be more susceptible to Cd stress, probably because of the presence of delicate and sensitive components such as the chloroplast and photosynthetic pigments (Di Toppi et al. 2005). Heavy metals have induced oxidative stress as indicated enhancement of GSH concentrations and GST and SOD activities (Cuny et al. 2004; Oztetik and Cicek 2011; Fig. 4.3). Monnet et al. (2006) documented an increase in antioxidant activities (SOD, CAT and APX) with increased Cu concentrations. Concentrations of stress-ethylene correlated positively with Al, Ba, Pb, S and V content and negatively with Cu and Sn (Garty et al. 2003).

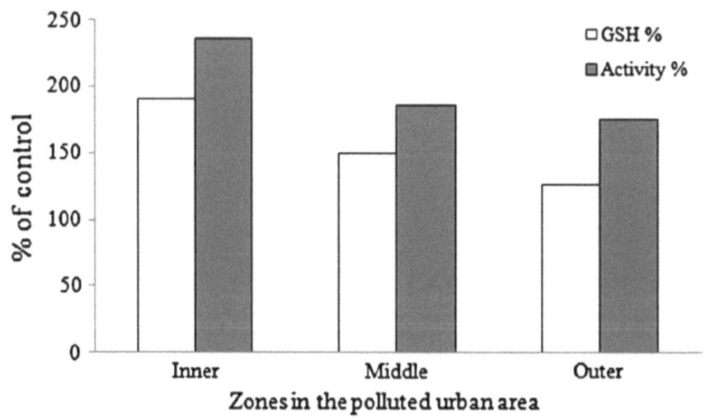

Fig. 4.3 Effects of air pollution on glutathione (*GSH*) concentrations and on the glutathione S-transferase (*GST*) enzyme activities in transplanted epiphytic lichen *Pseudevernia furfuracea*. The values represent the % difference from the control. The oxidative stress biomarkers were related with higher level of heavy metals in the inner zone as a result of heating houses, exhaust gases and industrial activities (Oztetik and Cicek 2011)

Various studies showed that cell membrane damage are related with the concentrations of elements. The extent of cell membrane integrity may be evaluated by measuring electrical conductivity. Electric conductivity values correlated positively with B, Fe, Mg, Mn, Na, Pb, Sn and Ti content (Garty et al. 2003). Calcium, Cr, Cu, Mg, Na and Sr showed a positive correlation with the electrical conductivity corresponding with cell membrane disintegration (Garty et al. 2002). High concentrations of Ni, Mg and B coincided with damage caused to cell membranes (Garty et al. 1998). Membrane integrity was highly correlated with the presence of Ca (Beckett and Brown 1984).

4.4.5 Persistent Organic Pollutants (POPs)

High toxicity, bioaccumulation and long persistence in the environment is characteristic for the POPs. The main pollution sources are related with human activities—industrial processes, domestic heating and traffic (Boström et al. 2002). Recently, a detail review of biomonitoring POPs using lichens was given by Augusto et al. (2013a).

The evaluation of the impact of POPs is receiving increasing attention because of their mutagenic and carcinogenic properties. Polychlorinated dibenzo-p-dioxins and polychlorinated dibenzofurans (PCDD/Fs) are important in environmental health studies due to their carcinogenic and potentially toxic effects, even in very low concentration (Augusto et al. 2007). PAHs persist in the environment for long periods due to their relatively stable molecular structure and tend to accumulate in organisms.

PAHs pollution present in the ecosystem can be efficiently detected by lichen biomonitoring techniques, and a large number of PAH accumulation in lichens from different regions of the world are available (Guidotti et al. 2003; Blasco et al. 2006, 2011; Augusto et al. 2009a; Shukla et al. 2010, 2013; Nascimbene et al. 2014; Protano et al. 2014).

Lichens accumulate PAHs over detection limits with highest values being during winter and the lowest—during summer (Augusto et al. 2013b). Interception and accumulation of POPs by lichens are dependent on the structure of the lichen thallus. Fruticose lichens accumulate higher concentrations of POPs than foliose lichens (Augusto et al. 2009a; Blasco et al. 2011). The results of the latter study showed that foliose lichens such as *E. prunastri*, *Usnea* sp. or *R. farinacea* accumulated more low molecular weight POPs (3- and 4-ringed PAHs) than foliose lichen *P. sulcata* which was enriched in high molecular weight POPs (5 and 6-ringed PAHs). Lichen biomonitoring studies indicated that PAHs with 2–4 rings are dominated in lichen thallus (Augusto et al. 2010; Blasco et al. 2011; Bajpai et al. 2013; Shukla et al. 2013; Nascimbene et al. 2014; Protano et al. 2014) while in the case of PCDD/Fs, the lichen accumulated mostly from TCDD/Fs to PeCDD/Fs (Augusto et al. 2007). Lichens accumulated higher PAH concentrations when compared with other environmental indicators (Fig. 4.4) and its concentrations were significantly and linearly correlated with concentrations of PAHs in soil (Augusto et al. 2010). The accumulation process was more intensive when lichens were hydrated than when they were desiccated (Kylin and Bouwman 2012). Accumulated airborne PAHs in lichens was proportional to traffic density (Guidotti et al. 2009; Shukla and Upreti 2009; Blasco et al. 2011).

Regarding the physiological changes, the chlorophyll degradation and protein content were found to be the most sensitive parameters to assess the vitality of lichen thallus against PAHs pollution (Bajpai et al. 2013).

It was argued that PCDD/F in lichens may be used as spatial estimators of the potential risk of inhalation by the population present in the area (Augusto et al. 2007). Using spatial models of PAHs accumulated in lichens, it was possible to fingerprint multiple sources of atmospheric PAHs at regional scale (Augusto et al. 2009b). When comparing several POPs in lichens, spruce needles and passive air sampling devices, it was found that lichens accumulated the highest

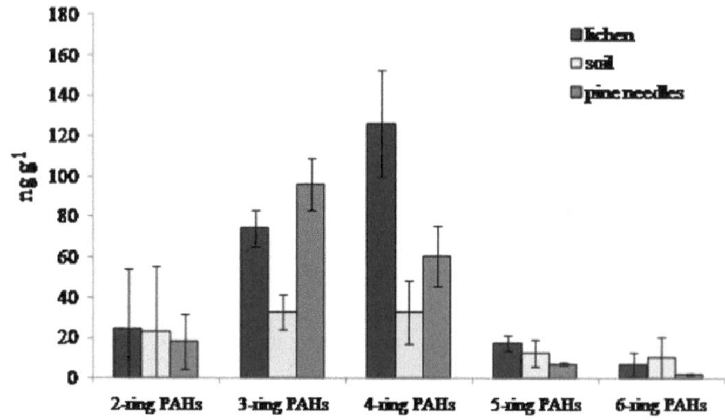

Fig. 4.4 PAH concentrations (ng g^{-1}) in lichens, soil and pine needles. Bars represent standard deviations (Augusto et al. 2010)

content (Ockenden et al. 1998; Migaszewski et al. 2002; Schrlau et al. 2011). Temporal variation of levels of PAHs in air and in lichens is in accordance (Augusto et al. 2013a).

4.5 Conclusions

In order to develop an early warning system and to avoid detrimental effects of pollutants on living organisms, lichens can be used as an excellent bioindicators to monitor atmospheric quality. Various studies have highlighted the relevance of both physiological and biochemical properties of lichens, which attributes to the biomonitoring potential of lichen species. In the present review, characteristic physiological attributes and lichen response to different pollutants have been discussed in detail.

The changes in physiological and biochemical parameters as a response to pollution are important for assessment of ambient air quality changes, but lichen vitality and diversity are influenced by number of other factors. Changes could be induced by microclimatic changes as different light regimes, temperature, drought or humidity, substrates and other environmental factors. Therefore, in order to draw any conclusion on the effect of pollution on lichens, these environmental parameters especially microclimatic changes should be taken into consideration.

Reproducibility of the data and comparison of various studies or datasets can be ensured by following standardized protocol for sampling and preservation of the lichen samples.

Lichens occur in all available substrata and geographical conditions; therefore, they are exposed to different ambient stresses too. In order to withstand such stresses, lichens have evolved complex mechanism of exclusion or avoidance which is still not well understood. Lichens respond differently to different pollutants/abiotic/biotic stresses depending on the type and nature of pollution and the lichen species involved. Therefore, it is very important to clarify and classify the causes of differential sensitivity of different lichen species towards various pollutants.

References

Adamo P, Giordano S, Vingiani S, Cobianchi RC, Violante P (2003) Trace element accumulation by moss and lichen exposed in bags in the city of Naples (Italy). Environ Pollut 122:91–103. doi:10.1016/S0269-7491(02)00277-4

Adamo P, Crisafulli P, Giordano S, Minganti V, Modenesi P, Monaci F, Pittao E, Tretiach M, Bargagli R (2007) Lichen and moss bags as monitoring devices in urban areas. Part II: trace element content in living and dead biomonitors and comparison with synthetic materials. Environ Pollut 146:392–399. doi:10.1016/j.envpol.2006.03.047

Adamska E (2011) Lichen recolonization in the city of Torun. Ecol Quest 15:119–125. doi:10.2478/v10090-011-0043-2

Alebic-Juretic A, Arko-Pijevac M (1989) Air pollution damage to cell membranes in lichens-results of simple biological test applied in Rijeka, Yugoslavia. Water Air Soil Pollut 47:25–33. doi:10.1007/BF00468994

Alebic-Juretic A, Arko-Pijevac M (2005) Lichens as indicators of air pollution in the city of Rijeka, Croatia. Fresen Environ Bull 14:40–43

Aprile GG, Di Salvatore M, Carratu G, Mingo A, Carafa AM (2010) Comparison of the suitability of two lichen species and one higher plant for monitoring airborne heavy metals. Environ Monit Assess 162:291–299. doi:10.1007/s10661-009-0796-x

Aslan A, Gurbuz H, Yazici K, Cicek A, Turan M, Ercisli S (2013) Evaluation of lichens as bio-indicators of metal pollution. J Elem 18:353–369. doi:10.5601/jelem.2013.18.3.01

Asta J, Rolley F (1999) Biodiversité et bioindication lichénique: qualité de l'air dans l'agglomération Grenobloise. Bull Int Assoc Fr Lichénol 3:121–126

Asta J, Erhardt W, Ferretti M, Fornasier F, Kirschbaum U, Nimis PL, Purvis O, Pirintsos S, Scheidegger C, van Haluwyn C, Wirth V (2002) Mapping lichen diversity as an indicator of environmental quality. In: Nimis PL, Scheidegger C, Wolseley P (eds) Monitoring with lichens—monitoring lichens. Kluwer Academic, Dordrecht

Augusto S, Pereira MJ, Soares A, Branquinho C (2007) The contribution of environmental biomonitoring with lichens to assess human exposure to dioxins. Int J Hyg Environ Health 210:433–438. doi:10.1016/j.ijheh.2007.01.017

Augusto S, Maguas C, Branquinho C (2009a) Understanding the performance of different lichen species as biomonitors of atmospheric dioxins and furans: potential for intercalibration. Ecotoxicol 18:1036–1042. doi:10.1007/s10646-009-0360-z

Augusto S, Maguas C, Matos J, Pereira MJ, Soares A, Branquinho C (2009b) Spatial modeling of PAHs in lichens for fingerprinting of multisource atmospheric pollution. Environ Sci Technol 43:7762–7769. doi:10.1021/es901024w

Augusto S, Maguas C, Matos J, Pereira MJ, Branquinho C (2010) Lichens as an integrating tool for monitoring PAH atmospheric deposition: a comparison with soil, air and pine needles. Environ Pollut 158:483–489. doi:10.1016/j.envpol.2009.08.016

Augusto S, Maguas C, Branquinho C (2013a) Guidelines for biomonitoring persistent organic pollutants (POPs), using lichens and aquatic mosses—a review. Environ Pollut 180:330–338. doi:10.1016/j.envpol.2013.05.019

Augusto S, Pereira MJ, Maguas C, Branquinho C (2013b) A step towards the use of biomonitors as estimators of atmospheric PAHs for regulatory. Chemosphere 92:626–632. doi:10.1016/j.chemosphere.2013.03.068

Backor M, Dzubaj A (2004) Short-term and chronic effects of copper, zinc and mercury on the chlorophyll content of four lichen photobionts and related alga. J Hattori Bot Lab 95:271–284

Backor M, Loppi S (2009) Interactions of lichens with heavy metals. Biol Plantarum 53:214–222. doi:10.1007/s10535-009-0042-y

Backor M, Fahselt D, Wu CT (2004) Free proline content is positively correlated with copper tolerance of the lichen photobiont *Trebouxia erici* (Chlorophyta). Plant Sci 167:151–157. doi:10.1016/j.plantsci.2004.03.012

Backor M, Pawlik-Skowronska B, Tomko J, Budova J, Di Toppi LS (2006) Response to copper stress in aposymbiotically grown lichen mycobiont *Cladonia cristatella*: uptake, viability, ergosterol and production of non-protein thiols. Mycol Res 110:994–999. doi:10.1016/j.mycres.2006.05.007

Backor M, Vaczi P, Bartak M, Budova J, Dzubaj A (2007) Uptake, photosynthetic characteristics and membrane lipid peroxidation levels in the lichen photobiont *Trebouxia erici* exposed to copper and cadmium. Bryologist 110:100–107. doi:10.1639/0007-2745(2007)110[100:UPCAML]2.0.CO;2

Bajpai R, Upreti DK (2012) Accumulation and toxic effect of arsenic and other heavy metals in a contaminated area of West Bengal, India, in the lichen *Pyxine cocoes* (Sw.) Nyl. Ecotoxicol Environ Saf 83:63–70. doi:10.1016/j.ecoenv.2012.06.001

Bajpai R, Upreti DK, Nayaka S, Kumari B (2010) Biodiversity, bioaccumulation and physiological changes in lichens growing in the vicinity of coal-based thermal power plant of Raebareli district, north India. J Hazard Mater 174:429–436. doi:10.1016/j.jhazmat.2009.09.071

Bajpai R, Karakoti N, Upreti DK (2013) Performance of a naturally growing Parmelioid lichen *Remototrachyna awasthii* against organic and inorganic pollutants. Environ Sci Pollut Res 20:5577–5592. doi:10.1007/s11356-013-1583-3

Balaguer L, Valladares F, Ascaso C, Barnes JD, de los Rios A, Manrique E, Smith EC (1996) Potential effects of rising tropospheric concentrations of CO_2 and O_3 on green-algal lichens. New Phytol 132:641–652. doi:10.1111/j.1469-8137.1996.tb01882.x

Bates JW (2002) Effects on bryophytes and lichens. In: Bell JNB, Threshow M (eds) Air pollution and plant life, 2nd edn. Wiley, Chrichester

Bates JW, Bell JNB, Farmer AM (1990) Epiphyte recolonization of oaks along a gradient of air-pollution in south-east England, 1979–1990. Environ Pollut 68:81–99. doi:10.1016/0269-7491(90)90014-4

Bates JW, Bell JNB, Massara AC (2001) Loss of *Lecanora conizaeoides* and other fluctuations of epiphytes on oak in SE England over 21 years with declining SO_2 concentrations. Atmos Environ 35:2557–2568. doi:10.1016/S1352-2310(00)00402-7

Batic F (2002) Bioindication of sulphur dioxide pollution with lichens. In: Kranner I, Beckett RP, Varma AK (eds) Protocols in lichenology. Culturing, biochemistry, ecophysiology and use in biomonitoring. Springer, Berlin

Batic F, Mayrhofer H (1996) Bioindication of air pollution by epiphytic lichens in forest decline studies in Slovenia. Phyton 36:85–90

Batts JE, Calder LJ, Batts BD (2004) Utilizing stable isotope abundances of lichens to monitor environmental change. Chem Geol 204:345–368. doi:10.1016/j.chemgeo.2003.11.007

Batty K, Bates JW, Bell JNB (2003) A transplant experiment on the factors preventing lichen colonization of oak bark in southeast England under declining

SO_2 pollution. Can J Bot 81:439–451. doi:10.1139/B03-039

Beckett RP, Brown DH (1984) The control of cadmium uptake in the lichen genus *Peltigera*. J Exp Bot 35:1071–1082

Beekley PK, Hoffman GR (1981) Effects of sulphur dioxide fumigation on photosynthesis, respiration and chlorophyll content of selected lichens. Bryologist 84:379–390

Bergamaschi L, Rizzio E, Giaveri G, Loppi S, Gallorini M (2007) Comparison between the accumulation capacity of four lichen species transplanted to a urban site. Environ Pollut 148:468–476. doi:10.1016/j.envpol.2006.12.003

Bermudez GMA, Rodriguez JH, Pignata ML (2009) Comparison of the air pollution biomonitoring ability of three *Tillandsia* species and the lichen *Ramalina celastri* in Argentina. Environ Res 109:6–14. doi:10.1016/j.envres.2008.08.014

Bertuzzi S, Davies L, Power SA, Tretiach M (2013) Why lichens are bad biomonitors of ozone pollution? Ecol Ind 34:391–397. doi:10.1016/j.ecolind.2013.05.023

Bjerke JW, Gwynn-Jones D, Callaghan TV (2005) Effects of enhanced UV-B radiation in the field on the concentration of phenolics and chlorophyll fluorescence in two boreal and arctic-alpine lichens. Environ Exp Bot 53:139–149. doi:10.1016/j.envexpbot.2004.03.009

Blasco M, Domeno C, Nerin C (2006) Use of lichens as pollution biomonitors in remote areas: comparison of PAHs extracted from lichens and atmospheric particles sampled in and around the Somport tunnel (Pyrenees). Environ Sci Technol 40:6384–6391. doi:10.1021/es0601484

Blasco M, Domeno C, Nerin C (2008) Lichens biomonitoring as feasible methodology to assess air pollution in natural ecosystems: combined study of quantitative PAHs analyses and lichen biodiversity in the Pyrenees Mountains. Anal Bioanal Chem 391:759–771. doi:10.1007/s00216-008-1890-6

Blasco M, Domeno C, Lopez P, Nerin C (2011) Behaviour of different lichen species as biomonitors of air pollution by PAHs in natural ecosystems. J Environ Monit 13:2588–2596. doi:10.1039/c0em00681e

Bobbink R, Hicks K, Galloway J, Spranger T, Alkemade R, Ashmore M, Bustamante M, Cinderby S, Davidson E, Dentener F, Emmett B, Erisman JW, Fenn M, Gilliam F, Nordin A, Pardo L, De Vries W (2010) Global assessment of nitrogen deposition effects on terrestrial plant diversity: a synthesis. Ecol Appl 20:30–59. doi:10.1890/08-1140.1

Boltersdorf S, Werner W (2013) Source attribution of agriculture-related deposition by using total nitrogen and N-15 in epiphytic lichen tissue, bark and deposition water samples in Germany. Isot Environ Health Stud 49:197–218. doi:10.1080/10256016.2013.748051

Boonpragob K, Nash TH (1990) Seasonal variation of elemental status in the lichen *Ramalina menziesii* Tayl from two sites in Southern California: evidence for dry deposition accumulation. Environ Exp Bot 30:415–428

Boonpragob K, Nash TH (1991) Physiological responses of the lichen *Ramalina-menziesii* Tayl to the Los Angeles urban environment. Environ Exp Bot 31:229–238. doi:10.1016/0098-8472(91)90075-Y

Boström CE, Gerde P, Hanberg A, Jernstrom B, Johansson C, Kyrklund T, Rannug A, Tornqvist M, Victorin K, Westerholm R (2002) Cancer risk assessment, indicators, and guidelines for polycyclic aromatic hydrocarbons in the ambient air. Environ Health Perspect 110:451–488

Branquinho C, Brown DH, Catarino F (1997a) The cellular location of Cu in lichens and its effects on membrane integrity and chlorophyll fluorescence. Environ Exp Bot 38:165–179. doi:10.1016/S0098-8472(97)00015-4

Branquinho C, Brown DH, Maguas C, Catarino F (1997b) Lead (Pb) uptake and its effects on membrane integrity and chlorophyll fluorescence in different lichen species. Environ Exp Bot 37:95–105. doi:10.1016/S0098-8472(96)01038-6

Brown DH (1987) the location of mineral elements in lichens implications for metabolism. Bibl Lichenol 25:361–376

Brunialti G, Giordani P (2003) Variability of lichen diversity in a climatically heterogeneous area (Liguria, NW Italy). Lichenologist 35:55–69. doi:10.1006/lich.2002.0417

Bychek-Guschina IA, Kotlova ER, Heipieper H (1999) Effects of sulfur dioxide on lichen lipids and fatty acids. Biochem 64:61–65

Calatayud A, Temple PJ, Barreno E (2000) Chlorophyll *a* fluorescence emission, xanthophyll cycle activity, and net photosynthetic rate responses to ozone in some foliose and fruticose lichen species. Photosynthetica 38:281–286. doi:10.1023/A:1007214915785

Calvelo S, Liberatore S (2004) Applicability of in situ or transplanted lichens for assessment of atmospheric pollution in Patagonia, Argentina. J Atmos Chem 49:199–210. doi:10.1007/s10874-004-1225-8

Calvelo S, Baccala N, Liberatore S (2009) Lichens as bioindicators of air quality in distant areas in Patagonia (Argentina). Environ Bioindic 4:123–135. doi:10.1080/15555270902963459

Canas MS, Pignata ML (1998) Temporal variation of pigments and peroxidation products in the lichen *Parmotrema uruguense* (Kremplh.) Hale transplanted to urban and non-polluted environments. Symbiosis 24:147–161

Canas MS, Orellana L, Pignata ML (1997) Chemical response of the lichens *Parmotrema austrosinense* and *P. conferendum* transplanted to urban and non-polluted environments. Ann Bot Fenn 34:27–34

Carreras HA, Pignata ML (2002) Biomonitoring of heavy metals and air quality in Cordoba City, Argentina, using transplanted lichens. Environ Pollut 117:77–87. doi:10.1016/S0269-7491(01)00164-6

Carreras HA, Pignata ML (2007) Effects of the heavy metals Cu^{2+}, Ni^{2+}, Pb^{2+}, and Zn^{2+} on some physiological parameters of the lichen *Usnea amblyoclada*. Ecotoxicol Environ Saf 67:59–66. doi:10.1016/j.ecoenv.2006.05.005

Carreras HA, Gudino GL, Pignata ML (1998) Comparative biomonitoring of atmospheric quality in five zones of Cordoba city (Argentina) employing the transplanted lichen *Usnea* sp. Environ Pollut 103:317–325. doi:10.1016/S0269-7491(98)00116-X

Carreras HA, Wannaz ED, Perez CA, Pignata ML (2005) The role of urban air pollutants on the performance of heavy metal accumulation in *Usnea amblyoclada*. Environ Res 97:50–57. doi:10.1016/j.envres.2004.05.009

Carreras HA, Rodriguez JH, Gonzalez CM, Wannaz ED, Garcia Ferreyra F, Perez CA, Pignata ML (2009) Assessment of the relationship between total suspended particles and the response of two biological indicators transplanted to an urban area in central Argentina. Atmos Environ 43:2944–2949. doi:10.1016/j.atmosenv.2009.02.060

Chettri MK, Cook CM, Vardaka E, Sawidis T, Lanaras T (1998) The effect of Cu, Zn and Pb on the chlorophyll content of the lichens *Cladonia convoluta* and *Cladonia rangiformis*. Environ Exp Bot 39:1–10

Cloquet C, De Muynck D, Signoret J, Vanhaecke F (2009) Urban/Peri-Urban Aerosol Survey by Determination of the Concentration and Isotopic Composition of Pb Collected by Transplanted Lichen *Hypogymnia physodes*. Environ Sci Technol 43:623–629. doi:10.1021/es801739p

Conti ME, Cecchetti G (2001) Biological monitoring: lichens as bioindicators of air pollution assessment—a review. Environ Pollut 114:471–492. doi:10.1016/S0269-7491(00)00224-4

Conti ME, Pino A, Botre F, Bocca B, Alimonti A (2009) Lichen *Usnea barbata* as biomonitor of airborne elements deposition in the Province of Tierra del Fuego (southern Patagonia, Argentina). Ecotoxicol Environ Saf 72:1082–1089. doi:10.1016/j.ecoenv.2008.12.004

Cuny D, Van Haluwyn C, Shirali P, Zerimech F, Jerome L, Haguenoer JM (2004) Cellular impact of metal trace elements in terricolous lichen *Diploschistes muscorum* (Scop.) R. Sant.—Identification of oxidative stress biomarkers. Water Air Soil Pollut 152:55–69. doi:10.1023/B:WATE.0000015332.94219.ff

Dahlman L, Nasholm T, Palmqvist K (2002) Growth, nitrogen uptake, and resource allocation in the two tripartite lichens Nephroma arcticum and Peltigera aphthosa during nitrogen stress. New Phytol 153:307–315. doi:10.1046/j.0028-646X.2001.00321.x

Dahlman L, Persson J, Nasholm T, Palmqvist K (2003) Carbon and nitrogen distribution in the green algal lichens *Hypogymnia physodes* and *Platismatia glauca* in relation to nutrient supply. Planta 217:41–48. doi:10.1007/s00425-003-0977-8

Das P, Joshi S, Rout J, Upreti DK (2013) Lichen diversity for environmental stress study: Application of index of atmospheric purity (IAP) and mapping around a paper mill in Barak Valley, Assam, northeast India. Trop Ecol 54:355–364

Davies L, Bates JW, Bell JNB, James PW, Purvis OW (2007) Diversity and sensitivity of epiphytes to oxides of nitrogen in London. Environ Pollut 146:299–310. doi:10.1016/j.envpol.2006.03.023

De Wit A (1976) Epiphytic lichens and air pollution in the Netherlands. Bibl Lichenol 5:1–115

Del Hoyo A, Álvarez R, Del Campo EM, Gasulla F, Barreno E, Casano LM (2011) Oxidative stress induces distinct physiological responses in the two *Trebouxia* phycobionts of the lichen *Ramalina farinacea*. Ann Bot 107:109–118

Deltoro VI, Gimeno C, Calatayud A, Barreno E (1999) Effects of SO_2 fumigations on photosynthetic CO_2 gas exchange, chlorophyll afluorescence emission and antioxidant enzymes in the lichen *Evernia prunastri* and *Ramalina farinacea*. Physiol Plant 105:648–654. doi:10.1034/j.1399-3054.1999.105408.x

Di Toppi LS, Musetti R, Vattuone Z, Pawlik-Skowronska B, Fossati F, Bertoli L, Badiani M, Favali MA (2005) Cadmium distribution and effects on ultrastructure and chlorophyll status in photobionts and mycobionts of *Xanthoria parietina*. Microsc Res Tech 66:229–238. doi:10.1002/jemt.20166

Egger R, Schlee D, Turk R (1994) Changes of physiological and biochemical parameters in the lichen *Hypogymnia-physodes* (L) Nyl due to the action of air-pollutants—a field-study. Phyton 34:229–242

Erisman JW, Galloway JN, Seitzinger S, Bleeker A, Dise NB, Petrescu AMR, Leach AM, de Vries W (2013) Consequences of human modification of the global nitrogen cycle. Philos Trans R Soc 368:1–9. doi:10.1098/rstb.2013.0116

European Environment Agency (2012) Emissions of acidifying substances (CSI 001)—assessment published December 2011 (www.eea.europa.eu/data-and-maps/indicators/emmisions-of-acidifying-substances-version2/assessment-1)

Evju M, Bruteig IE (2013) Lichen community change over a 15-year time period: effects of climate and pollution. Lichenologist 45:35–50. doi:10.1017/S0024282912000539

Fenn ME, Geiser L, Bachman R, Blubaugh TJ, Bytnerowicz A (2007) Atmospheric deposition inputs and effects on lichen chemistry and indicator species in the Columbia River Gorge, USA. Environ Pollut 146:77–91

Fenn ME, Jovan S, Yuan F, Geiser L, Meixner T, Gimeno BS (2008) Empirical and simulated critical loads for nitrogen deposition in California mixed conifer forests. Environ Pollut 155:492–511. doi:10.1016/j.envpol.2008.03.019

Fernandez-Salegui AB, Calatayud A, Terron A, Barreno EM (2006a) Chlorophyll a fluorescence in transplants of *Parmelia sulcata* Taylor near a power station (La Robla, Leon, Spain). Lichenologist 38:457–468. doi:10.1017/S0024282906005226

Fernandez-Salegui AB, Terron AA, Barreno E (2006b) Bioindicadores de la calidad del aire en La Robla (Leon, noroeste de Espana) diez anos despues. Lazaroa 27:29–41

Fields RD, St Clair LI (1984) A comparison of methods for evaluating SO_2 impact on selected lichen species—*Parmelia chlorochroa*, *Collema polycarpon* and *Lecanora muralis*. Bryologist 87:297–301. doi:10.2307/3242947

Folkeson L (1979) Interspecies calibration of heavy-metal concentrations in nine mosses and lichens: applicability to deposition measurements. Water Air Soil Pollut 11:253–260

Franzen-Reuter I, Frahm JP (2007) Effects of nitrogen additions on epiphytic lichen and bryophyte cover on permanent observation plots (Rhineland-Palatinate, Germany). Herzogia 20:61–75

Frati L, Brunialti G, Loppi S (2005) Problems related to lichen transplants to monitor trace element deposition in repeated surveys: a case study from Central Italy. J Atmos Chem 52:221–230

Frati L, Caprasecca E, Santoni S, Gaggi C, Guttova A, Gaudino S, Pati A, Rosamilia S, Pirintsos SA, Loppi S (2006) Effects of NO_2 and NH_3 from road traffic on epiphytic lichens. Environ Pollut 142:58–64. doi:10.1016/j.envpol.2005.09.020

Frati L, Brunialti G, Gaudino S, Pati A, Rosamilia S, Loppi S (2011) Accumulation of nitrogen and changes in assimilation pigments of lichens transplanted in an agricultural area. Environ Monit Assess 178:19–24. doi:10.1007/s10661-010-1667-1

Friedel A, Muller F (2004) Bryophytes and lichens as indicators for changes of air pollution in the Serrahn Natural Forest Reserve (Mueritz National Park). Herzogia 17:279–286

Fuchs C, Garry J (1983) Elemental content in the lichen *Ramalina duriaei* (De Not.) Jatta at air quality biomonitoring stations. Environ Exp Bot 23:29–43

Fuhrer J (2002) Ozone impacts on vegetation. Ozone Sci Eng 24:69–74. doi:10.1080/01919510208901597

Gailey F, Lloyd OL (1993) Spatial and temporal patterns of airborne metal pollution—the value of low technology sampling to an environmental epidemiology study. Sci Total Environ 133:201–219. doi:10.1016/0048-9697(93)90245-2

Gaio-Oliveira G, Branquinho C, Máguas C, Martins-Loução MA (2001) The concentration of nitrogen in nitrophilous and non-nitrophilous lichen species. Symbiosis 31:187–199

Gaio-Oliveira G, Dahlman L, Palmqvist K, Máguas C (2004) Ammonium uptake in the nitrophytic lichen *Xanthoria parietina* and its effects on vitality and balance between symbionts. Lichenologist 36:75–86

Gaio-Oliveira G, Dahlman L, Palmqvist K, Máguas C (2005a) Responses of the lichen *Xanthoria parietina* (L.) Th. Fr. to varying thallus nitrogen concentrations. Lichenologist 37:171–179

Gaio-Oliveira G, Dahlman L, Palmqvist K, Martins-Loução MA, Máguas C (2005b) Nitrogen uptake in relation to excess supply and its effects on the lichen *Evernia prunastri* (L.) Ach. and *Xanthoria parietina* (L.) Th. Fr. Planta 220:794–803

Galloway JN, Dentener FJ, Capone DG, Boyer EW, Howarth RW, Seitzinger SP, Asner GP, Cleveland CC, Green PA, Holland EA, Karl DM, Michaels AF, Porter JH, Townsend AR, Vorosmarty CJ (2004) Nitrogen cycles: past, present, and future. Biogeochem 70:153–226. doi:10.1007/s10533-004-0370-0

Galloway JN, Townsend AR, Erisman JW, Bekunda M, Cai Z, Freney JR, Martinelli LA, Seitzinger SP, Sutton MA (2008) Transformation of the nitrogen cycle: Recent trends, questions, and potential solutions. Science 320:889–892. doi:10.1126/science.1136674

Garty J (2001) Biomonitoring atmospheric heavy metals with lichens: theory and application. Crit Rev Plant Sci 20:309–371. doi:10.1016/S0735-2689(01)80040-X

Garty J, Kardish N, Hagemeyer J, Ronen R (1988) Correlations between the concentration of adenosine triphosphate, chlorophyll degradation and the amounts of airborne heavy metals and sulphur in a transplanted lichen. Arch Environ Contam Toxicol 17:601–611

Garty J, Karary Y, Harel J (1992) Effect of low pH, heavy metals and anions on chlorophyll degradation in the lichen *Ramalina duriaei* (De Not) Bagl. Environ Exp Bot 32:229–241. doi:10.1016/0098-8472(92)90006-N

Garty J, Karary Y, Harel J (1993) The impact of air pollution on the integrity of cell membranes and chlorophyll in the lichen *Ramalina duriaei* (Denot) Bagl transplanted to industrial sites in Israel. Arch Environ Contam Toxicol 24:455–460

Garty J, Kauppi M, Kauppi A (1996) Accumulation of airborne elements from vehicles in transplanted lichens in urban sites. J Environ Qual 25:265–272

Garty J, Cohen Y, Kloog N (1998) Airborne elements, cell membranes, and chlorophyll in transplanted lichens. J Environ Qual 27:973–979

Garty J, Weissman L, Tamir O, Beer S, Cohen Y, Karnieli A, Orlovsky L (2000) Comparison of five physiological parameters to assess the vitality of the lichen *Ramalina lacera* exposed to air pollution. Physiol Plant 109:410–418

Garty J, Tamir O, Hassid I, Eshel A, Cohen Y, Karnieli A, Orlovsky L (2001) Photosynthesis, chlorophyll integrity, and spectral reflectance in lichens exposed to air pollution. J Environ Qual 30:884–893

Garty J, Tamir O, Cohen Y, Lehr H, Goren AI (2002) Changes in the potential quantum yield of photosystem II and the integrity of cell membranes relative to the elemental content of the epilithic desert lichen *Ramalina maciformis*. Environ Toxicol Chem 21:848–858. doi:10.1897/1551-5028(2002)021<0848:CITPQY>2.0.CO;2

Garty J, Tomer S, Levin T, Lehr H (2003) Lichens as biomonitors around a coal-fired power station in Israel. Environ Res 91:186–198. doi:10.1016/S0013-9351(02)00057-9

Gauslaa Y, Solhaug KA (2000) High-light-intensity damage to the foliose lichen Lobaria pulmonaria within a natural forest: the applicability of chlorophyll fluorescence methods. Lichenologist 32:271–289. doi:10.1006/lich.1999.0265

Gibson MD, Heal MR, Li Z, Kuchta J, King GH, Hayes A, Lambert S (2013) The spatial and seasonal variation of nitrogen dioxide and sulfur dioxide in Cape Breton Highlands National Park, Canada, and the association with lichen abundance. Atmos Environ 64:303–311. doi:10.1016/j.atmosenv.2012.09.068

Gilbert OL (1968) Bryophytes as indicators of air pollution in the Tyne valley. New Phytol 67:15–30

Gilbert OL (1970) biological scale for the estimation of sulfur dioxide air pollution. New Phytol 69:629–634

Giordani P (2007) Is the diversity of epiphytic lichens a reliable indicator of air pollution? A case study from Italy. Environ Pollut 146:317–323. doi:10.1016/j.envpol.2006.03.030

Giordani P, Brunialti G, Alleteo D (2002) Effects of atmospheric pollution on lichen biodiversity (LB) in a Mediterranean region (Liguria, northwest Italy). Environ Poll 118:53–64. doi:10.1016/S0269-7491(01)00275-5

Giordano S, Sorbo S, Adamo P, Basile A, Spagnuolo V, Cobianchi RC (2004) Biodiversity and trace element content of epiphytic bryophytes in urban and extraurban sites of southern Italy. Plant Ecol 170:1–14. doi:10.1023/B:VEGE.0000019025.36121.5d

Godinho RM, Freitas MC, Wolterbeek HT (2004) Assessment of lichen vitality during a transplantation experiment to a polluted site. J Atmos Chem 49:355–361. doi:10.1007/s10874-004-1251-6

Gombert S, Asta J, Seaward MRD (2003) Correlation between the nitrogen concentration of two epiphytic lichens and the traffic density in an urban area. Environ Pollut 123:281–290

Gombert S, Asta J, Seaward MRD (2004) Assessment of lichen diversity by index of atmospheric purity (IAP), index of human impact (IHI) and other environmental factors in an urban area (Grenoble, southeast France). Sci Total Environ 324:183–199. doi:10.1016/j.scitotenv.2003.10.036

Gonzalez CM, Pignata ML (1994) The influence of air pollution on soluble proteins, chlorophyll degradation, MDA, sulphur and heavy metals in a transplanted lichen. Chem Ecol 9:105–113

Gonzalez CM, Pignata ML (1997) Chemical response of the lichen *Punctelia subrudecta* (Nyl.) Krog transplanted close to a power station in an urban-industrial environment. Environ Pollut 97:195–203. doi:10.1016/S0269-7491(97)00102-4

Gonzalez CM, Pignata ML (2000) Chemical response of transplanted lichen *Canomaculina pilosa* to different emission sources of air pollutants. Environ Pollut 110:235–242. doi:10.1016/S0269-7491(99)00300-0

Gonzalez CM, Casanovas SS, Pignata ML (1996) Biomonitoring of air pollutants from traffic and industries employing *Ramalina ecklonii* (Spreng) Mey and Flot in Cordoba, Argentina. Environ Pollut 91:269–277. doi:10.1016/0269-7491(95)00076-3

Gonzalez CM, Orellana LC, Casanovas SS, Pignata ML (1998) Environmental conditions and chemical response of a transplanted lichen to an urban area. J Environ Manag 53:73–81. doi:10.1006/jema.1998.0194

Gonzalez CM, Pignata ML, Orellana L (2003) Applications of redundancy analysis for the detection of chemical response patterns to air pollution in lichen. Sci Total Environ 312:245–253. doi:10.1016/S0048-9697(03)00253-5

Gonzalez MC, Lingua M, Gudino GL (2012) Estimate the air quality along a section in the suquia river (cordoba, argentina) using Usnea amblyoclada as active biomonitor. Rev Int Contam Ambient 28:311–322

Grace B, Gillespie TJ, Puckett KJ (1985) Uptake of gaseous sulfur dioxide by the lichen *Cladina rangiferina*. Can J Bot 63:797–805

Grandin U (2011) Epiphytic algae and lichen cover in boreal forests—a long-term study along a N and S deposition gradient in Sweden. Ambio 40:857–866. doi:10.1007/s13280-011-0205-x

Gries C, Sanz M-J, Nash TH III (1995) The effect of SO_2 fumigation on CO_2 gas exchange, chlorophyll fluorescence and chlorophyll degradation in different lichen species from western North America. Cryptogam Bot 5:239–246

Grube M (2010) Die hard: lichens. In: Seckback J, Grube M (eds) Symbioses and stress. Joint ventures in biology. Springer, Dordrecht. doi:10.1007/978-90-481-9449-0

Guidotti M, Stella D, Owczarek M, De Marco A, De Simone C (2003) Lichens as polycyclic aromatic hydrocarbon bioaccumulators used in atmospheric pollution studies. J Chromatogr 985:185–190. doi:10.1016/S0021-9673(02)01452-8

Guidotti M, Stella D, Dominici C, Blasi G, Owczarek M, Vitali M, Protano C (2009) Monitoring of traffic-related pollution in a province of central Italy with transplanted lichen *Pseudovernia furfuracea*. Bull Environ Contam Toxicol 83:852–858. doi:10.1007/s00128-009-9792-7

Haffner E, Lomsky B, Hynek V, Hallgren JE, Batic F, Pfanz H (2001) Air pollution and lichen physiology. Physiological responses of different lichens in a transplant experiment following an SO_2 gradient. Water Air Soil Pollut 131:185–201. doi:10.1023/A:1011907530430

Hällgren JE, Huss K (1975) Effects of SO_2 on photosynthesis and nitrogen fixation. Physiol Plant 34:171–176. doi:10.1111/j.1399-3054.1975.tb03815.x

Hauck M (2008) Susceptibility to acidic precipitation contributes to the decline of the terricolous lichens *Cetraria aculeata* and *Cetraria islandica* in central Europe. Environ Pollut 152:731–735. doi:10.1016/j.envpol.2007.06.046

Hauck M, Jürgens SR (2008) Usnic acid controls the acidity tolerance of lichens. Environ Pollut 156:115–122. doi:10.1016/j.envpol.2007.12.033

Hauck M, Paul A, Gross S, Raubuch M (2003) Manganese toxicity in epiphytic lichens: chlorophyll degradation and interaction with iron and phosphorus. Environ Exp Bot 49:181–191. doi:10.1016/S0098-8472(02)00069-2

Hauck M, Boening J, Jacob M, Dittrich S, Feussner I, Leuschner C (2013a) Lichen substance concentrations in the lichen *Hypogymnia physodes* are correlated with heavy metal concentrations in the substratum. Environ Exp Bot 85:58–63. doi:10.1016/j.envexpbot.2012.08.011

Hauck M, de Bruyn U, Leuschner C (2013b) Dramatic diversity losses in epiphytic lichens in temperate broad-leaved forests during the last 150 years. Biol Conserv 157:136–145. doi:10.1016/j.biocon.2012.06.015

Hawksworth DL, Mcmanus PM (1989) Lichen recolonization in London under conditions of rapidly falling sulfur-dioxide levels, and the concept of zone skipping. Bot J Linn Soc 100:99–109. doi:10.1111/j.1095-8339.1989.tb01712.x

Hawksworth DL, Rose F (1970) Qualitative scale for estimating sulphur dioxide air pollution in England and wales using epiphytic lichens. Nature 227:145–148. doi:10.1038/227145a0

Henriksson E, Pearson LC (1981) Nitrogen fixation rate and chlorophyll content of the lichen *Peltigera canina* exposed to sulfur dioxide. Am J Bot 68:680–684

Hippeli S, Elstner EF (1996) Mechanisms of oxygen activation during plant stress: biochemical effects of air pollutants. J Plant Physiol 148:249–257

Holopainen T, Kauppi M (1989) Comparison of light, fluorescence and electron-microscopic observations in assessing the SO_2 injury of lichens under different moisture conditions. Lichenologist 21:119–134. doi:10.1017/S0024282989000241

Hultengren S, Gralen H, Pleijel H (2004) Recovery of the epiphytic lichen flora following air quality improvement in south-west Sweden. Water Air Soil Pollut 154:203–211. doi:10.1023/B:WATE.0000022967.35036.ca

Hyvarinen M, Crittenden PD (1998) Relationships between atmospheric nitrogen inputs and the vertical nitrogen and phosphorus concentration gradients in the lichen *Cladonia portentosa*. New Phytol 140:519–530. doi:10.1046/j.1469-8137.1998.00292.x

Jensen M, Kricke R (2000) Chlorophyll fluorescence measurements in the field: assessment of the vitality of large numbers of lichen thalli. In: Nimis PL, Scheidegger C, Wolseley A (eds) Monitoring with lichens—monitoring lichens, NATO science series IV: earth and environmental sciences 7. Kluwer, Dordrecht

Jeran Z, Jacimovic R, Batic F, Mavsar R (2002) Lichens as integrating air pollution monitors. Environ Pollut 120:107–113. doi:10.1016/S0269-7491(02)00133-1

Johansson O, Nordin A, Olofsson J, Palmqvist K (2010) Responses of epiphytic lichens to an experimental whole-tree nitrogen-deposition gradient. New Phytol 188:1075–1084. doi:10.1111/j.1469-8137.2010.03426.x

Johansson O, Olofsson J, Giesler R, Palmqvist K (2011) Lichen responses to nitrogen and phosphorus additions can be explained by the different symbiont responses. New Phytol 191:795–805. doi:10.1111/j.1469-8137.2011.03739.x

Johansson O, Palmqvist K, Olofsson J (2012) Nitrogen deposition drives lichen community changes through differential species responses. Glob Change Biol 18:2626–2635. doi:10.1111/j.1365-2486.2012.02723.x

Kaeffer MI, de Azevedo Martins S M, Alves C, Pereira VC, Fachel J, Ferrao Vargas VM (2011) Corticolous lichens as environmental indicators in urban areas in southern Brazil. Ecol Indic 11:1319–1332. doi:10.1016/j.ecolind.2011.02.006

Karakas SY, Tuncel SG (2004) Comparison of accumulation capacities of two lichen species analyzed by instrumental neutron activation analysis. J Radioanal Nucl Chem 259:113–118

Karakoti N, Bajpai R, Upreti DK, Mishra GK, Srivastava A, Nayaka S (2014) Effect of metal content on chlorophyll fluorescence and chlorophyll degradation in lichen *Pyxine cocoes* (Sw.) Nyl.: a case study from Uttar Pradesh. India. Environ Earth Sci 71:2177–2183. doi:10.1007/s12665-013-2623-5

Khalil K, Asta J (1998) Les lichens, bioindicateurs de pollution atmosphérique dans la Région Lyonnaise. Ecologie 29:467–472

Kong FX, Hu W, Chao SY, Sang WL, Wang LS (1999) Physiological responses of the lichen *Xanthoparmelia mexicana* to oxidative stress of SO_2. Environ Exp Bot 42:201–209. doi:10.1016/S0098-8472(99)00034-9

Kricke R, Loppi S (2002) Bioindication: the I.A.P approach. In: Nimis PL, Scheidegger C, Wolseley A (eds) Monitoring with lichens—monitoring lichens, NATO science series IV: earth and environmental sciences 7. Kluwer, Dordrecht

Kularatne KIA, de Freitas CR (2013) Epiphytic lichens as biomonitors of airborne heavy metal pollution. Environ Exp Bot 88:24–32. doi:10.1016/j.envexpbot.2012.02.010

Kylin H, Bouwman H (2012) Hydration state of the moss *Hylocomium splendens* and the lichen *Cladina stellaris* governs uptake and revolatilization of airborne alpha- and gamma-hexachlorocyclohexane. Environ Sci Technol 46:10982–10989. doi:10.1021/es302363g

Larsen RS, Bell JNB, James PW, Chimonides PJ, Rumsey FJ, Tremper A, Purvis OW (2007) Lichen and bryophyte distribution on oak in London in relation to air pollution and bark acidity. Environ Pollut 146:332–340. doi:10.1016/j.envpol.2006.03.033

LeBlanc F, De Sloover J (1970) Relation between industrialization and the distribution and growth of epiphytic lichens and mosses in Montreal. Can J Bot 48:1485–1496

LeBlanc F, Rao DN (1973) Effects of sulfur dioxide on lichen and moss transplants. Ecology 54:612–617. doi:10.2307/1935347

LeBlanc F, Rao D, Comeau G (1972) The epiphytic vegetation of *Populus balsamifera* and its significance as an air pollution indicator in Sudbury, Ontario. Can J Bot 50:519–528

LeBlanc F, Robitaille G, Rao DN (1974) Biological responses of lichens and bryophytes to environmental pollution in the Murdochville copper mine area, Quebec. J Hattori Bot Lab 38:405–433

LeGalley E, Widom E, Krekeler MPS, Kuentz DC (2013) Chemical and lead isotope constraints on sources of metal pollution in street sediment and lichens in southwest Ohio. Appl Geochem 32:195–203. doi:10.1016/j.apgeochem.2012.10.020

Letrouitgalinou MA, Seaward MRD, Deruelle S (1992) On the return of epiphytic lichens to the Jardin-du-Luxembourg (Paris). Bull Soc Bot Fr 139:115–126

Levin AG, Pignata ML (1995) *Ramalina ecklonii* as a bioindicator of atmospheric-pollution in argentina. Can J Bot 73:1196–1202

Liska J, Herben T (2008) Long-term changes of epiphytic lichen species composition over landscape gradients: an 18 year time series. Lichenologist 40:437–448. doi:10.1017/S0024282908006610

Lisowska M (2011) Lichen recolonisation in an urban-industrial area of southern Poland as a result of air quality improvement. Environ Monit Assess 179:177–190. doi:10.1007/s10661-010-1727-6

Loppi S (1996) Lichens as bioindicators of geothermal air pollution in central Italy. Bryologist 99:41–48. doi:10.2307/3244436

Loppi S (2004) Mapping the effects of air pollution, nitrogen deposition, agriculture and dust by the diversity of epiphytic lichens in central Italy. In: Lambley P, Wolseley P (eds) Lichens in a changing pollution environment. English Nature Research Reports, London

Loppi S, Corsini A (2003) Diversity of epiphytic lichens and metal contents of *Parmelia caperata* thalli as monitors of air pollution in the town of Pistoia (C Italy). Environ Monit Assess 86:289–301. doi:10.1023/A:1024017118462

Loppi S, DeDominicis V (1996) Effects of agriculture on epiphytic lichen vegetation in Central Italy. Isr J Plant Sci 44:297–307

Loppi S, Frati L (2006) Lichen diversity and lichen transplants as monitors of air pollution in a rural area of central Italy. Environ Monit Assess 114:361–375. doi:10.1007/s10661-006-4937-1

Loppi S, Nascimbene J (1998) Lichen bioindication of air quality in the Mt. Amiata geothermal area (Tuscany, Italy). Geotherm 27:295–304. doi:10.1016/S0375-6505(98)00003-0

Loppi S, Putorti E, De Dominicis V, Barbaro A (1995) Lichens as bioindicators of air quality near a municipal solid waste incinerator (Central Italy). Allionia 33:121–129

Loppi S, Bosi A, Signorini C, De Dominicis V (2003) Lichen recolonization of *Tilia* trees in Arezzo (Tuscany, central Italy) under conditions of decreasing air pollution. Cryptogam Mycol 24:175–185

Majumder S, Mishra D, Ram SS, Jana NK, Santra S, Sudarshan M, Chakraborty A (2013) Physiological and chemical response of the lichen, *Flavoparmelia caperata* (L.) Hale, to the urban environment of Kolkata. India. Environ Sci Pollut Res 20:3077–3085. doi:10.1007/s11356-012-1224-2

Malhotra SS, Hocking D (1976) Biochemical and cytological effects of sulphur dioxide on plant metabolism. New Phytol 76:227–237

Marques AP, Freitas MC, Wolterbeek HT, Steinebach OM, Verburg T, De Goeij JJM (2005) Cell-membrane damage and element leaching in transplanted *Parmelia sulcata* lichen related to ambient SO_2, temperature, and precipitation. Environ Sci Technol 39:2624–2630. doi:10.1021/es0498888

Massara AC, Bates JW, Bell JNB (2009) Exploring causes of the decline of the lichen Lecanora conizaeoides in Britain: effects of experimental N and S applications. Lichenologist 41:673–681. doi:10.1017/S0024282909990119

Mayer AL, Vihermaa L, Nieminen N, Luomi A, Posch M (2009) Epiphytic macrolichen community correlates with modeled air pollutants and forest conditions. Ecol Ind 9:992–1000. doi:10.1016/j.ecolind.2008.11.010

Mayer W, Pfefferkorn-Dellali V, Tuerk R, Dullinger S, Mirtl M, Dirnboeck T (2013) Significant decrease in epiphytic lichen diversity in a remote area in the European Alps, Austria. Basic Appl Ecol 14:396–403. doi:10.1016/j.baae.2013.05.006

McMurray JA, Roberts DW, Fenn ME, Geiser LH, Jovan S (2013) Using epiphytic lichens to monitor nitrogen deposition near natural gas drilling operations in the Wind River Range, WY. USA. Water Air Soil Pollut 224:1487. doi:10.1007/s11270-013-1487-3

Meysurova AF, Khizhnyak SD, Pakhomov PM (2011) Toxic effect of nitrogen and sulfur dioxides on the chemical composition of *Hypogymnia physodes* (L.) Nyl.: IR spectroscopic analysis. Contemp Probl Ecol 4:186–194. doi:10.1134/S199542551102010X

Migaszewski ZM, Galuszka A, Paslawski P (2002) Polynuclear aromatic hydrocarbons, phenols, and trace metals in selected soil profiles and plant bioindicators in the Holy Cross Mountains, South-Central Poland. Environ Int 28:303–313. doi:10.1016/S0160-4120(02)00039-9

Mikhailova I (2002) Transplanted lichens for bioaccumulation studies. In: Nimis PL, Scheidegger C, Wolseley A (eds) Monitoring with lichens—monitoring lichens, NATO science series IV: earth and environmental sciences 7. Kluwer, Dordrecht

Mikhailova IN, Sharunova IP (2008) Dynamics of heavy metal accumulation in thalli of the epiphytic lichen *Hypogymnia physodes*. Russ J Ecol 39:346–352. doi:10.1134/S1067413608050068

Monaci F, Bargagli R, Gasparo D (1997) Air pollution monitoring by lichens in a small medieval town of central Italy. Acta Bot Neerl 46:403–412

Monnet F, Bordas F, Deluchat V, Chatenet P, Botineau M, Baudu M (2005) Use of the aquatic lichen *Dermatocarpon luridum* as bioindicator of copper pollution: accumulation and cellular distribution tests. Environ Pollut 138:455–461. doi:10.1016/j.envpol.2005.04.019

Monnet F, Bordas F, Deluchat V, Baudu M (2006) Toxicity of copper excess on the lichen *Dermatocarpon luridum*: Antioxidant enzyme activities. Chemosphere 65:1806–1813. doi:10.1016/j.chemosphere.2006.04.022

Munzi S, Pirintsos SA, Loppi S (2009) Chlorophyll degradation and inhibition of polyamine biosynthesis in the lichen *Xanthoria parietina* under nitrogen stress. Ecotoxicol Environ Saf 72:281–285. doi:10.1016/j.ecoenv.2008.04.013

Munzi S, Pisani T, Paoli L, Loppi S (2010) Time- and dose-dependency of the effects of nitrogen pollution on lichens. Ecotoxicol Environ Saf 73:1785–1788. doi:10.1016/j.ecoenv.2010.07.042

Munzi S, Loppi S, Cruz C, Branquinho C (2011) Do lichens have "memory" of their native nitrogen environment? Planta 233:333–342. doi:10.1007/s00425-010-1300-0

Munzi S, Paoli L, Fiorini E, Loppi S (2012) Physiological response of the epiphytic lichen *Evernia prunastri* (L.) Ach. to ecologically relevant nitrogen concentrations. Environ Pollut 171:25–29. doi:10.1016/j.envpol.2012.07.001

Munzi S, Cruz C, Branquinho C, Pinho P, Leith ID, Sheppard LJ (2014) Can ammonia tolerance amongst lichen functional groups be explained by physiological responses? Environ Pollut 187:206–209. doi:10.1016/j.envpol.2014.01.009

Nakajima H, Yamamoto Y, Yoshitani A, Itoh K (2013) Effect of metal stress on photosynthetic pigments in the Cu-hyperaccumulating lichens *Cladonia humilis* and *Stereocaulon japonicum* growing in Cu-polluted sites in Japan. Ecotoxicol Environ Saf 97:154–159. doi:10.1016/j.ecoenv.2013.07.026

Nascimbene J, Tretiach M, Corana F, Lo Schiavo F, Kodnik D, Dainese M, Mannucci B (2014) Patterns of traffic polycyclic aromatic hydrocarbon pollution in mountain areas can be revealed by-lichen biomonitoring: a case study in the Dolomites (Eastern Italian Alps). Sci Total Environ 475:90–96. doi:10.1016/j.scitotenv.2013.12.090

Nash TH III (1976) Sensitivity of lichens to nitrogen dioxide fumigations. Bryologist 79:103–106

Nash TH III (2010) Lichen sensitivity to air pollution. In: Nash TH III (ed) Lichen Biology, 2nd edn. Cambridge University Press, Cambridge

Nash TH, Gries C (2002) Lichens as bioindicators of sulfur dioxide. Symbiosis 33:1–21

Nash TH III, Sigal LL (1998) Epiphytic lichens in the San Bernardino mountains in relation to oxidant gradients. In: Miller PR, McBride JR (eds) Oxidant air pollution impacts in the Montane Forests of Southern California: a case study of the San Bernardino Mountains. Springer, New York

Nimis PL (2003) Checklist of Italian lichens 3.0. University of Trieste, Department of Biology, IN3.0/2. http://dbiodbs.univ.trieste.it. Accessed 16 April 2014

Nimis PL, Castello M, Perotti M (1990) Lichens as biomonitors of sulfur dioxide pollution in La Spezia (Northern Italy). Lichenologist 22:333–344. doi:10.1017/S0024282990000378

Nimis PL, Scheidegger C, Wolseley PA (2002) Monitoring with lichens–monitoring lichens. An introduction. In: Nimis PL, Scheidegger C, Wolseley A (eds) Monitoring with lichens—monitoring lichens, NATO science series IV: earth and environmental sciences 7. Kluwer, Dordrecht

Ochoa-Hueso R, Manrique E (2011) Effects of nitrogen deposition and soil fertility on cover and physiology of *Cladonia foliacea* (Huds.) Willd. a lichen of biological soil crusts from Mediterranean Spain. Environ Pollut 159:449–457. doi:10.1016/j.envpol.2010.10.021

Ockenden WA, Steinnes E, Parker C, Jones KC (1998) Observations on persistent organic pollutants in plants: implications for their use as passive air samplers and for POP cycling. Environ Sci Technol 32:2721–2726. doi:10.1021/es980150y

Olsen HB, Berthelsen K, Andersen HV, Søchting U (2010) *Xanthoria parietina* as a monitor of ground-level ambient ammonia concentrations. Environ Pollut 158:455–461. doi:10.1016/j.envpol.2009.08.025

Osyczka P, Rola K (2013) Response of the lichen *Cladonia rei* Schaer. to strong heavy metal contamination of the substrate. Environ Sci Pollut Res 20:5076–5084. doi:10.1007/s11356-013-1645-6

Owczarek M, Spadoni M, De Marco A, De Simone C (1999) Lichens as indicators of air pollution in urban and rural sites of Rieti (Central Italy). Fresen Environ Bull 8:288–295

Oztetik E, Cicek A (2011) Effects of urban air pollutants on elemental accumulation and identification of oxidative stress biomarkers in the transplanted lichen *Pseudovernia furfuracea*. Environ Toxicol Chem 30:1629–1636. doi:10.1002/etc.541

Pakarinen P (1981) Regional variation of sulphur concentration in *Sphagnum* mosses and *Cladonia* lichens in Finnish bogs. Ann Bot Fennici 18:275–279

Palmqvist K, Dahlman L (2006) Responses of the green algal foliose lichen *Platismatia glauca* to increased nitrogen supply. New Phytol 171:343–356. doi:10.1111/j.1469-8137.2006.01754.x

Palmqvist K, Dahlman L, Valladares F, Tehler A, Sancho LG, Mattsson JE (2002) CO_2 exchange and thallus nitrogen across 75 contrasting lichen associations from different climate zones. Oecologia 133:295–306. doi:10.1007/s00442-002-1019-0

Paoli L, Pirintsos SA, Kotzabasis K, Pisani T, Navakoudis E, Loppi S (2010) Effects of ammonia from livestock farming on lichen photosynthesis. Environ Pollut 158:2258–2265. doi:10.1016/j.envpol.2010.02.008

Paoli L, Pisani T, Guttova A, Sardella G, Loppi S (2011) Physiological and chemical response of lichens transplanted in and around an industrial area of south Italy: relationship with the lichen diversity. Ecotoxicol Environ Saf 74:650–657. doi:10.1016/j.ecoenv.2010.10.011

Pawlik-Skowronska B, Purvis OW, Pirszel J, Skowronski T (2006) Cellular mechanisms of Cu-tolerance in the epilithic lichen *Lecanora polytropa* growing at a copper mine. Lichenologist 38:267–275. doi:10.1017/S0024282906005330

Pawlik-Skowronska B, Wojciak H, Skowronski T (2008) Heavy metal accumulation, resistance and physiological status of some epigeic and epiphytic lichens inhabiting Zn and Pb polluted areas. Pol J Ecol 56:195–207

Pearson LC, Henriksson E (1981) Air pollution damage to cell membranes in lichens. 2. Laboratory experiments. Bryologist 84:515–520. doi:10.2307/3242559

Pearson LC, Rodgers GA (1982) Air pollution damage to cell membranes in lichens. III. Field experiments. Phyton 22:329–337

Piccotto M, Bidussi M, Tretiach M (2011) Effects of the urban environmental conditions on the chlorophyll a fluorescence emission in transplants of three ecologically distinct lichens. Environ Exp Bot 73:102–107. doi:10.1016/j.envexpbot.2010.09.010

Pignata ML, Gonzalez CM, Wannaz ED, Carreras HA, Gudino GL, Martinez MS (2004) Biomonitoring of air quality employing in situ *Ramalina celastri* in Argentina. Int J Environ Pollut 22:409–429. doi:10.1504/IJEP.2004.005678

Pignata ML, Pla RR, Jasan RC, Martinez MS, Rodriguez JH, Wannaz ED, Gudino GL, Carreras HA, Gonzalez CM (2007) Distribution of atmospheric trace elements and assesment of air quality in Argentina employing the lichen, *Ramalina celastri*, as a passive biomonitor: detection of air pollution emission sources. Int J Environ Health 1:29–46. doi:10.1504/IJENVH.2007.012223

Pilkington MG, Caporn SJM, Carroll JA, Cresswell N, Lee JA, Emmett BA, Bagchi R (2007) Phosphorus supply influences heathland responses to atmospheric nitrogen deposition. Environ Pollut 148:191–200. doi:10.1016/j.envpol.2006.10.034

Pinho P, Augusto S, Martins-Loucao MA, Pereira MJ, Soares A, Maguas C, Branquinho C (2008) Causes of change in nitrophytic and oligotrophic lichen species in a Mediterranean climate: Impact of land cover and atmospheric pollutants. Environ Pollut 154:380–389. doi:10.1016/j.envpol.2007.11.028

Pirintsos SA, Munzi S, Loppi S, Kotzabasis K (2009) Do polyamines alter the sensitivity of lichens to nitrogen stress? Ecotoxicol Environ Saf 72:1331–1336. doi:10.1016/j.ecoenv.2009.03.001

Plakunova OV, Plakunova VG (1987) Ultrastructure of components of the lichen *Cladina stellaris* in health and disease during SO_2 environmental pollution. Biol Bull Acad Sci USSR 14:223–230

Protano C, Guidotti M, Owczarek M, Fantozzi L, Blasi G, Vitali M (2014) Polycyclic aromatic hydrocarbons and metals in transplanted lichen (*Pseudovernia furfuracea*) at sites adjacent to a solid waste landfill in central Italy. Arch Environ Contam Toxicol 66:471–481. doi:10.1007/s00244-013-9965-6

Puckett KJ, Tomassini FD, Nieboer E, Richardson DHS (1977) Potassium efflux of lichen thalli following exposure to aqueous sulphur dioxide. New Phytol 179:135–145

Pyatt FB (1973) Plant sulphur content as air pollution gauge in the vicinity of a steelworks. Environ Pollut 5:103–115

Ra HSY, Geiser LH, Crang RFE (2005) Effects of season and low-level air pollution on physiology and element content of lichens from the US Pacific Northwest. Sci Total Environ 343:155–167. doi:10.1016/j.scitotenv.2004.10.003

Ranta P (2001) Changes in urban lichen diversity after a fall in sulphur dioxide levels in the city of Tampere, SW Finland. Ann Bot Fennici 38:295–304

Reinecke J, Klemm G, Heinken T (2014) Vegetation change and homogenization of species composition in temperate nutrient deficient Scots pine forests after 45 yr. J Veg Sci 25:113–121. doi:10.1111/jvs.12069

Riddell J, Padgett PE, Nash TH III (2010) Responses of the lichen *Ramalina menziesii* Tayl. to ozone fumigations. In: Nash TH, Geiser L, McCune B, Triebel D, Tomescu AMF, Sanders WB (eds) Biology of lichens—symbiosis, ecology, environmental monitoring, systematics and cyber applications. Bibl Lichenol 105:113–123

Riddell J, Padgett PE, Nash TH III (2012) Physiological responses of lichens to factorial fumigations with nitric acid and ozone. Environ Pollut 170:202–210. doi:10.1016/j.envpol.2012.06.014

Rodriguez JH, Carreras HA, Pignata ML, Gonzalez CM (2007) Nickel exposure enhances the susceptibility of lichens *Usnea amblyoclada* and *Ramalina celastri* to urban atmospheric pollutants. Arch Environ Contam Toxicol 53:533–540. doi:10.1007/s00244-006-0034-2

Rogers PC, Moore KD, Ryel RJ (2009) Aspen succession and nitrogen loading: a case for epiphytic lichens as bioindicators in the Rocky Mountains, USA. J Veg Sci 20:498–510. doi:10.1111/j.1654-1103.2009.01064.x

Romeralo C, Diez JJ, Santiago NF (2012) Presence of fungi in Scots pine needles found to correlate with air quality as measured by bioindicators in northern Spain. Forest Pathol 42:443–453. doi:10.1111/efp.12002

Root HT, Geiser LH, Fenn ME, Jovan S, Hutten MA, Ahuja S, Dillman K, Schirokauer D, Berryman S, McMurray JA (2013) A simple tool for estimating throughfall nitrogen deposition in forests of western North America using lichens. Forest Ecol Manage 306:1–8. doi:10.1016/j.foreco.2013.06.028

Rose CI, Hawksworth DI (1981) Lichen recolonization in London's cleaner air. Nature 289:289–292. doi:10.1038/289289a0

Ross LJ, Nash TH III (1983) Effects of ozone on gross photosynthesis of lichens. Environ Exp Bot 23:71–77

Ruoss E (1999) How agriculture affects lichen vegetation in central Switzerland. Lichenologist 31:63–73

Ruoss E, Vonarburg C (1995) Lichen diversity and ozone impact in rural areas of Central Switzerland. Cryptogam Bot 5:252–263

Sanchez-Hoyos MA, Manrique E (1995) Effect of nitrate and ammonium ions on the pigment content (xanthophylls, carotenes and chlorophylls) of *Ramalina capitata*. Lichenologist 27:155–160. doi: 10.1017/S0024282995000168

Sanz MJ, Gries C, Nash TH III (1992) Dose-response relationship for SO_2 fumigations in the lichen *Evernia prunastri* (L.) Ach. New Phytol 122:313–319

Sarret G, Manceau A, Cuny D, Van Haluwyn C, Deruelle S, Hazemann JL, Soldo Y, Eybert-Berard L, Menthonnex JJ (1998) Mechanisms of lichen resistance to metallic pollution. Environ Sci Technol 32:3325–3330. doi:10.1021/es970718n

Satya, Upreti DK (2009) Correlation among carbon, nitrogen, sulphur and physiological parameters of *Rinodina sophodes* found at Kanpur city, India. J Hazard Mater 169:1088–1092. doi:10.1016/j.jhazmat.2009.04.063

Saunders PJW (1970) Air pollution in relation to lichens and fungi. Lichenologist 4:337–349. doi:10.1017/S0024282970000439

Scheidegger C, Schroeter B (1995) Effects of ozone fumigation on epiphytic macrolichens—ultrastructure, CO_2 gas-exchange and chlorophyll fluorescence. Environ Pollut 88:345–354. doi:10.1016/0269-7491(95)93449-A

Schrlau JE, Geiser L, Hageman KJ, Landers DH, Simonich SM (2011) Comparison of lichen, conifer needles, passive air sampling devices, and snowpack as passive sampling media to measure semi-volatile organic compounds in remote atmospheres. Environ Sci Technol 45:10354–10361. doi:10.1021/es202418f

Seaward MRD (2004) The use of lichens for environmental impact assessment. Symbiosis 37:293–305

Showman RE, Long RP (1992) Lichen studies along wet sulfate deposition gradient. Bryologist 95:166–170

Shukla V, Upreti DK (2007) Physiological response of the lichen *Phaeophyscia hispidula* (Ach.) Essl. to the urban environment of Pauri and Srinagar (Garhwal), Himalayas. India. Environ Pollut 150:295–299. doi:10.1016/j.envpol.2007.02.010

Shukla V, Upreti DK (2008) Effect of metallic pollutants on the physiology of lichen, *Pyxine subcinerea* Stirton in Garhwal Himalayas. Environ Monit Assess 141:237–243. doi:10.1007/s10661-007-9891-z

Shukla V, Upreti DK (2009) Polycyclic aromatic hydrocarbon (PAH) accumulation in lichen, *Phaeophyscia hispidula* of DehraDun City, Garhwal Himalayas. Environ Monit Assess 149:1–7. doi:10.1007/s10661-008-0225-6

Shukla V, Upreti DK, Patel DK, Ranu T (2010) Accumulation of polycyclic aromatic hydrocarbons in some lichens of Garhwal Himalayas, India. Int J Environ Waste Manage 5:104–113. doi:10.1504/IJEWM.2010.029695

Shukla V, Upreti DK, Patel DK, Yunus M (2013) Lichens reveal air PAH fractionation in the Himalaya. Environ Chem Lett 11:19–23. doi:10.1007/s10311-012-0372-4

Sigal LL, Nash TH III (1983) Lichen communities on conifers in southern California mountains—an ecological survey relative to oxidant air pollution. Ecology 64:1343–1354. doi:10.2307/1937489

Silberstein L, Siegel BZ, Siegel SM, Mukhtar A, Galun M (1996a) Comparative studies on *Xanthoria parietina*, a pollution-resistant lichen, and *Ramalina duriaei*, a sensitive species. I. Effects of air pollution on physiological processes. Lichenologist 28:355–365

Silberstein L, Siegel BZ, Siegel SM, Mukhtar A, Galun M (1996b) Comparative studies on *Xanthoria parietina*, a pollution-resistant lichen, and *Ramalina duriaei*, a sensitive species. II. Effects of possible air pollution-protection mechanisms. Lichenologist 28:367–383

Sim-Sim M, Carvalho P, Sergio C (2000) Cryptogamic epiphytes as indicators of air quality around an industrial complex in the Tagus valley, Portugal. Factor analysis and environmental variables. Cryptogam Bryol 21:153–170. doi:10.1016/S1290-0796(00)00110-3

Slaby A, Lisowska M (2012) Epiphytic lichen recolonization in the centre of Cracow (southern Poland) as a result of air quality improvement. Pol J Ecol 60:225–240

Sloof JE (1993) Environmental lichenology: biomonitoring trace element air pollution. PhD thesis, Delft University of Technology, Delft

Søchting U (1995) Lichens as monitors of nitrogen deposition. Cryptogam Bot 5:264–269

Sondergaard J (2013) Accumulation dynamics and cellular locations of Pb, Zn and Cd in resident and transplanted *Flavocetraria nivalis* lichens near a former Pb-Zn mine. Environ Monit Assess 185:10167–10176. doi:10.1007/s10661-013-3321-1

Spagnuolo V, Zampella M, Giordano S, Adamo P (2011) Cytological stress and element uptake in moss and lichen exposed in bags in urban area. Ecotoxicol Environ Saf 74:1434–1443. doi:10.1016/j.ecoenv.2011.02.011

Sparrius LB, Kooijman AM, Sevink J (2013) Response of inland dune vegetation to increased nitrogen and phosphorus levels. Appl Veg Sci 16:40–50. doi:10.1111/j.1654-109X.2012.01206.x

St Clair SB, St Clair LL, Weber DJ, Mangelson NF, Eggett DL (2002) Element accumulation patterns in foliose and fruticose lichens from rock and bark substrates in Arizona. Bryologist 105:415–421. doi:10.1639/0007-2745(2002)105[0415:EAPIFA]2.0.CO;2

Stamenkovic S, Cvijan M (2010) Determination of air pollution zones in Knjazevac (south eastern Serbia) by using epiphytic lichens. Biotechnol Biotechnol Equip 24:278–283

Stamenkovic S, Cvijan M, Arandjelovic M (2010) Lichens as bioindicators of air quality in Dimitrovgrad (south-eastern Serbia). Arch Biol Sci 62:643–648. doi:10.2298/ABS1003643S

Stamenkovic SM, Ristic SS, Dekic TL, Mitrovic TU, Baosic RM (2013) Air quality indication in Blace (southeastern Serbia) using lichens as bioindicators. Arch Biol Sci 65:893–897. doi:10.2298/ABS1303893S

Stevens C, Manning P, van den Berg LJL, de Graaf MCC, Wamelink GWW, Boxman AW, Bleeker A, Vergeer P, Arroniz-Crespo M, Limpens J, Lamers LPM, Bobbink R, Dorland E (2011) Ecosystem responses to reduced and oxidised nitrogen inputs in European terrestrial habitats. Environ Pollut 159:665–676. doi:10.1016/j.envpol.2010.12.008

Sujetoviene G (2010) Road traffic pollution effects on epiphytic lichens. Ekologija 56:64–71. doi:10.2478/v10055-010-0009-5

Sujetoviene G, Sliumpaite I (2013) Response of *Evernia prunastri* transplanted to an urban area in central Lithuania. Atmos Pollut Res 4:222–228

Svoboda D (2007) Evaluation of the European method for mapping lichen diversity (LDV) as an indicator of environmental stress in the Czech Republic. Biologia 62:424–431. doi:10.2478/s11756-007-0085-5

Szczepaniak K, Biziuk M (2003) Aspects of the biomonitoring studies using mosses and lichens as indicators of metal pollution. Environ Res 93:221–230

Takala K, Olkkonen H, Ikonen J, Jääskeläinen J, Puumalainen P (1985) Total sulphur contents of epiphytic and terricolous lichens in Finland. Ann Bot Fennici 22:91–100

Tarhanen S, Holopainen T, Poikolainen J, Oksanen J (1996) Effect of industrial emissions on membrane permeability of epiphytic lichens in northern Finland and the Kola Peninsula industrial areas. Water Air Soil Pollut 88:189–201

Trass H (1973) Lichen sensitivity to air pollution and index of poleotolerance (I.P.). Folia Cryptog Estonica 3:19–22

Tretiach M, Baruffo L (2001) Effects of H_2S on CO_2 gas exchanges and growth rates of the epiphytic lichen *Parmelia sulcata* Taylor. Symbiosis 31:35–46

Tretiach M, Ganis P (1999) Hydrogen sulphide and epiphytic lichen vegetation: a case study on Mt. Amiata (Central Italy). Lichenologist 31:163–181. doi:10.1006/lich.1998.0173

Tretiach M, Adamo P, Bargagli R, Baruffo L, Carletti L, Crisafulli P, Giordano S, Modenesi P, Orlando S, Pittao E (2007a) Lichen and moss bags as monitoring devices in urban areas. Part I: influence of exposure on sample vitality. Environ Pollut 146:380–391. doi:10.1016/j.envpol.2006.03.046

Tretiach M, Piccotto M, Baruffo L (2007b) Effects of ambient NO_x on chlorophyll *a* fluorescence in transplanted *Flavoparmelia caperata* (lichen). Environ Sci Technol 41:2978–2984

Unal D, Isik NO, Sukatar A (2010) Effects of chromium VI stress on photosynthesis, chlorophyll integrity, cell viability, and proline accumulation in lichen *Ramalina farinacea*. Russ J Plant Physiol 57:664–669. doi:10.1134/S1021443710050092

Upreti DK, Nayaka S, Bajpai A (2005) Do lichens still grow in Kolkata city? Curr Sci 88:338–339

van Dobben HF, De Bakker AJ (1996) Re-mapping epiphytic lichen biodiversity in The Netherlands: effects of decreasing SO_2 and increasing NH_3. Acta Bot Neerl 45:55–71

van Dobben HF, ter Braak CJF (1998) Effects of atmospheric NH3 on epiphytic lichens in the Netherlands: the pitfalls of biological monitoring. Atmos Environ 32:551–557. doi:10.1016/S1352-2310(96)00350-0

van Dobben HF, Wolterbeek HT, Wamelink GWW, Ter Braak CJF (2001) Relationship between epiphytic lichens, trace elements and gaseous atmospheric pollutants. Environ Pollut 112:163–169. doi:10.1016/S0269-7491(00)00121-4

van Herk CM (1999) Mapping of ammonia pollution with epiphytic lichens in the Netherlands. Lichenologist 31:9–20

Vantova I, Backor M, Klejdus B, Backorova M, Kovacik J (2013) Copper uptake and copper-induced physiological changes in the epiphytic lichen *Evernia prunastri*. Plant Growth Regul 69:1–9. doi:10.1007/s10725-012-9741-z

VDI (1995) Richtlinie 3799, blatt 1: Ermittlung und Beurteilung phytotoxischer Wirkungen von Immissionen mit Flechten: Flechtenkartierung. VDI, Dusseldorf

Vernon A (1993) The lichen symbiosis. Wiley, New York

Vingiani S, Adamo P, Giordano S (2004) Sulphur, nitrogen and carbon content of *Sphagnum capillifolium* and *Pseudevernia furfuracea* exposed in bags in the Naples urban area. Environ Pollut 129:145–158. doi:10.1016/j.envpol.2003.09.016

Vokou D, Pirintsos SA, Loppi S (1999) Lichens as bioindicators of temporal variations in air quality around Thessaloniki, northern Greece. Ecol Res 14:89–96. doi:10.1046/j.1440-1703.1999.00294.x

von Arb C, Brunold C (1990) Lichen physiology and air pollution i. Physiological responses of in-situ *Parmelia sulcata* among air pollution zones within Biel Switzerland. Can J Bot 68:35–55

von Arb C, Mueller C, Ammann K, Brunold C (1990) Lichen physiology and air pollution II. Statistical analysis of the correlation between SO_2, NO_2, NO and O_3, and chlorophyll content, net photosynthesis, sulphate uptake and protein synthesis of *Parmelia sulcata* Taylor. New Phytol 115:431–437

Washburn SJ, Culley TM (2006) Epiphytic macrolichens of the greater Cincinnati metropolitan area—part II: distribution, diversity and urban ecology. Bryologist 109:516–526. doi:10.1639/0007-2745(2006)109[516:EMOTGC]2.0.CO;2

Weissman L, Fraiberg M, Shine L, Garty J, Hochman A (2006) Responses of antioxidants in the lichen *Ramalina lacera* may serve as an early-warning bioindicator system for the detection of air pollution stress. FEMS Microbiol Ecol 58:41–53. doi:10.1111/j.1574-6941.2006.00138.x

Whelpdale DM, Summers PW, Sanhueza E (1997) A global overview of atmospheric acid deposition fluxes. Environ Monit Assess 48:217–247. doi:10.1023/A:1005708821454

Wirth V (1991) Zeigerwerte von Flechten. Scr Geobot 18:215–237

Wiseman RD, Wadleigh MA (2002) Lichen response to changes in atmospheric sulphur: isotopic evidence. Environ Pollut 116:235–241. doi:10.1016/S0269-7491(01)00133-6

Wolterbeek B (2002) Biomonitoring of trace element air pollution: principles, possibilities and perspectives. Environ Pollut 120:11–21

Wolterbeek HT, Garty J, Reis MA, Freitas MC (2003) Biomonitors in use: lichens and metal air pollution. In: Markert BA, Breure AM, Zechmeister HG (eds) Trace metals and other contaminants in the environment, vol 6: Bioindicators and Biomonitors. Elsevier, Amsterdam

Yule FA, Lloyd OLL (1984) An index of atmospheric-pollution survey in Armadale, central Scotland. Water Air Soil Pollut 22:27–45. doi:10.1007/BF00587462

Zambrano A, Nash TH (2000) Lichen responses to short-term transplantation in Desierto de los Leones, Mexico City. Environ Pollut 107:407–412. doi:10.1016/S0269-7491(99)00169-4

Zambrano A, Nash TH, Gries C (1999) Physiological effects of the Mexico city atmosphere on lichen transplants on oaks. J Environ Qual 28:1548–1555

Zechmeister HG, Hohenwallner D (2006) A comparison of biomonitoring methods for the estimation of atmospheric pollutants in an industrial town in Austria. Environ Monit Assess 117:245–259. doi:10.1007/s10661-006-0991-y

Bio-monitoring in Western North America: What Can Lichens Tell Us About Ecological Disturbances?

5

Steven D. Leavitt and Larry L. St. Clair

Contents

Abstract
Throughout modern history, many habitats across western North America have faced continuous anthropogenic disturbances. Mining, forestry, agriculture, grazing, industrial and residential development, and air pollution have altered natural habitats to varying degrees. Lichens have been considered 'a canary in the coalmine' and can provide important insights into the biological impact of human-related disturbances. Here, we discuss the role of lichens as bio-monitors in the intermountain region of western North America.

Keywords
Air pollution • Bio-accumulation • Bio-indicator • Climate change • Eutrophication • Mining

5.1 Introduction

Lichens are listed among a suite of ecosystem health indicators, which also includes some terrestrial and aquatic invertebrates (Hodkinson and Jackson 2005), nematodes (Bongers and Ferris 1999), bryophytes (Frego 2007; Pesch and Schroeder 2006), vascular plants (Coulston et al. 2003), and other sensitive species and/or communities. Rather than abiotic metrics, bio-indicators are likened to canaries in a coalmine, serving as a direct surrogate for assessing disturbances on biological communities. Lichens

S.D. Leavitt (✉)
Committee on Evolutionary Biology, University of Chicago, Chicago, IL, USA
e-mail: sleavitt@fieldmuseum.org

L.L. St. Clair
Department of Biology and M.L. Bean Life Science Museum, Brigham Young University, 2103C MLBM, Provo, UT 84602, USA
e-mail: larry_stclair@byu.edu

D.K. Upreti et al. (eds.), *Recent Advances in Lichenology*,
DOI 10.1007/978-81-322-2181-4_5

are particularly useful as bio-indicators due to the fact that many live and grow continuously for decades, or even hundreds of years, showing cumulative responses to changes in atmospheric pollution levels, land management practices, and climate change. Others are inherently sensitive to some particularly important pollutants, and their presence or absence can be indicative of current ecological conditions. Measurable responses of individual lichen thalli (e.g., differential accumulation of atmospheric pollutants) and lichen communities (e.g., changes in community composition and population density) can provide a means to quantitatively assess ecosystem health (McCune 2000) and provide a record of the kinds, and relative quantities, of air pollutants (Loppi and Pirintsos 2003; St. Clair et al. 2002b; Hawksworth 1971).

As lichens have no vascular system, they obtain mineral nutrients and co-occurring contaminants through either wet or dry deposition, rather than from their substrates (Agnan et al. 2014). Dry deposition involves the deposition of airborne gases or other particles due to increasing influence of gravity with decreasing wind speed. In contrast, wet deposition involves airborne aerosols and particles that are deposited on lichen thalli during precipitation events (i.e., fog, dew, snow, and rain) (Fenn et al. 2003; St. Clair et al. 2002b).As a result, the general indiscriminant accumulation of atmospheric outwash by lichens also makes them useful as bio-monitors of air quality (Hawksworth 1971; Henderson and Seaward 1979; Geiser and Neitlich 2007; Jovan 2008).

Monitoring air pollutant concentrations in lichen thalli can thus be used as an indirect measure of the potential impact of air pollution on other less sensitive biological systems. Patterns of element accumulation by lichens can also be used to document the occurrence and potential biological impact of specific sources of air pollution (Table 5.1). While in some cases lichens may simply accumulate and concentrate atmospheric pollutants with little noticeable physiological impact, in other cases, pollutants may reach toxic concentrations and the most sensitive lichens tend to disappear from highly polluted areas (Nash 1975; Branquinho et al. 1999). For example, the lichen *Hypogymnia physodes* was completely absent in close proximity to a heavily polluted smelter town in the Ural Mountains, Karabash, Russia, and occurrences of the lichen in nearby regions consistently showed elevated concentrations of heavy metals (Purvis et al. 2004). Furthermore, healthy lichen transplants set up near the smelter showed visible damage within three months, correlated to high concentrations of K, S, and toxic metals (Purvis et al. 2004).

Documenting the impact of air pollution on biological systems using lichens as bio-indicators generally involves four basic approaches: (1) elemental analysis of potential pollutants accumulated by lichen thalli; (2) assessing the presence or absence of pollution-sensitive, or conversely, pollution-tolerant, indicator species; (3) transplant studies assessing changes in appearance, physiology, growth rates, etc.; and (4) ecological assessments of lichen community structure (Nimis et al. 2002; St. Clair 1999; McCune and Geiser 2009). Lichens have been used to monitor a variety of atmospheric pollutants, including sulfur dioxide, ozone, nitrous oxides, peroxyacetyl nitrates (PAN), fluorides,

Table 5.1 A list of major sources of pollution and associated pollutants that are commonly monitored using lichens

Pollutant source	Characteristic pollutants
Coal-fired power plants	Hg, S, $\delta^{34}S$, SO_2, NO_X
Mining	Ore metals
Agriculture	N (NH_3), S, and phosphate
Vehicle exhaust	Cr, Ni, N (NO and NO_2), V, and T
Tire and brake wear	Ti, Fe, Cu, Sb, Zr, ZnO, and Ba
Nuclear radiation	^{106}Ru, ^{137}Cs, and ^{144}Ce
Aluminum manufacturing	F

Sources (Adachi and Tainosho 2004; Thorpe and Harrison 2008; McCune and Geiser 2009; Nimis et al. 2002)

aromatic compounds, various metals, and radioactive nuclides (St. Clair 1999; Nash 2008; Nimis et al. 2002; McCune and Geiser 2009).

While bio-monitoring studies evaluating airborne element accumulation patterns relative to a specific pollution source are common (Rose and Hawksworth 1981; Seaward 1993; Naeth and Wilkinson 2008; Aznar et al. 2008; Perlmutter 2010), directly correlating element concentrations in the air with lichen-accumulated elements are confounded by a number of factors (Jovan 2008; Riddell et al. 2011; Will-Wolf et al. 2006; Will-Wolf 2010). For example, although data suggest that the lichen genus *Parmelia* may be a useful indicator of atmospheric particulates in the city of Sendai, Japan, attempts to definitively correlate element air concentrations and lichen-accumulated elements were largely unsuccessful (Saeki et al. 1977). Not surprisingly, lichen morphology can influence, to some degree, the rate at which lichens accumulate elements from the atmosphere, and at least in some cases, it appears that foliose lichens may more effectively accumulate airborne pollutants than fruticose species (St. Clair et al. 2002a, b; Lawrey and Hale 1981). Similarly, different lichen species respond differently to air pollution. Some lichens are sensitive to specific pollutants and relatively insensitive to others. Differential sensitivity is likely related to several factors, including growth form, substrate chemistry and pH, and other lichen-specific physiological differences. Ultimately, attempts to correlate air pollution effects with a specific response are often difficult (Riddell et al. 2011), and data from lichen air quality bio-monitoring surveys require cautious interpretation and limited extrapolation (Nash 2008; Will-Wolf et al. 2006).

In this chapter, we discuss the roles of lichens in assessing air quality and ecological health in the intermountain region of western North America. Geographically, the Intermountain West is located between the rocky mountains on the east and Cascade and Sierra Nevada ranges on the west and extends north from Arizona and New Mexico into British Columbia, Canada. In the USA, the region contains a diverse array of habitats, ranging from low elevation arid deserts (e.g., the Mojave Desert) to inland temperate rainforests (e.g., portions of the Cabinet Mountains Wilderness Area in western Montana) to alpine tundra habitats (e.g., High Uinta Wilderness Area in northeastern Utah) (Dellasala et al. 2011). Throughout modern history, many habitats across western North America have faced continuous anthropogenic disturbances, and bio-monitoring with lichens has the potential to offer valuable insights into the impact of historic, contemporary, and projected disturbances in the region.

5.2 Bio-monitoring with Lichens in the Intermountain West

Throughout western North America, forestry, grazing, mining, energy development and processing, and recreation have played a major role in determining land-use strategies and economic development of federal, state, and private lands. To varying degrees, these land-use practices have had a substantial ecological impact, including habitat and community alteration, increased levels of pollutants, and the introduction and spread of non-native species (Fleischner 1994; Miller et al. 1989). Additionally, air pollution, altered fire regimes, and other broadscale disturbances have been shown to impact ecological health throughout the intermountain region (Smith et al. 2000; Miller et al. 1989). In spite of historic and ongoing disturbances, large tracks of quality habitat have been protected and managed by federal, state, and private agencies. Public lands —including state-owned, Bureau of Land Management (BLM), United States Department of Agriculture, Forest Service (USFS), National Park Service (NPS), National Wildlife Refuge (NWR), and military lands—comprise a substantial percentage of the total land area in the Intermountain West, ranging from ca. 35 % in Montana to over 81 % in Nevada. Effectively monitoring the ecological impact of anthropogenic disturbances can play an important role in assessing general ecological health while providing informed direction for future land-use strategies in western North America (Dillman 1996; Hyvärinen and Crittenden 1998).

Altitudinal, geological, and climatic variation create diverse habitats supporting distinct and varied lichen communities throughout the intermountain region of western North America. For over thirty years, lichens have been used to assess spatial and temporal patterns of atmospheric pollution in the Intermountain West. As part of a long-term bio-monitoring program, we have established over 350 reference sites throughout the intermountain region. Reference sites are selected to monitor both broadscale and more localized disturbances. The majority of our reference sites are found on land managed by the USFS, but the program also includes reference sites on lands managed by the BLM and NPS. For each reference site, we compile a comprehensive checklist of all lichens occurring at the site, document the presence (or absence) of pollution-sensitive indicator species, and assess mean concentrations of over 25 potential pollutant elements accumulated by documented sensitive lichen species. Due to the fact that certain lichen species are inherently sensitive to airborne contaminants, air pollution-related changes can be effectively documented by periodic reviews of reference sites and comparisons to original baseline data. Overall, our goal is to establish valuable baseline data and periodically review reference sites to effectively monitor ecological health throughout the Intermountain West using lichens. Generally, we plan to re-evaluate vulnerable reference sites on a three- to five-year cycle and more stable sites every 5–10 years; and the vast majority of our sites have been resampled at least once.

In recent decades, a number of policy changes have been implemented to mitigate the environmental impact of atmospheric pollution and other anthropogenic disturbances. In spite of these changes, both historic and contemporary disturbances continue to impact lands throughout western North America. While contemporary data and onetime assessments can provide critical insights into the general ecological health of a region or site, the strength of any bio-monitoring program is related to consistent reviews and long-term assessments of subsequent reviews compared back to the baseline data. Our ongoing lichen air quality bio-monitoring program at sites throughout the intermountain region offers a unique and valuable perspective into the ecological health across the region and affords opportunity for future long-term research.

5.2.1 Lichens as Air Quality Bio-indicators in the Intermountain West

A striking example of the role of lichens as bio-indicators of air quality in western North America is reflected in the major changes observed in lichen communities in southern California's South Air Basin, which includes the heavily urbanized Los Angeles and Orange Counties (Riddell et al. 2011). Historic records from the early 1900s and more recent surveys in the 1970s and then again in 2008 reveal an ongoing pattern of declining species ranges and community shifts. For example, high levels of N deposition have been documented in the Los Angeles Basin over the last 30 years, and several nitrophilous lichen genera, *Physcia*, *Physconia*, and *Xanthomendoza*, have steadily increased in abundance. Conversely, one of the most pollution-sensitive lichens remaining in the basin region, *Melanohalea subolivacea*, showed significant decline between 1976 and 2008, and no sensitive indicator species have returned since the 1976 inventory (Riddell et al. 2011).

In the intermountain region, the most common air pollution-sensitive indicator lichens used for elemental analyses include the following: *Alectoria sarmentosa* (epiphyte), *Bryoria fremontii* (epiphyte), *Letharia vulpina* (epiphyte), *Rhizoplaca melanophthalma* (saxicolous), and various *Xanthoparmelia* species (saxicolous and terricolous). *Xanthomendoza montana* (epiphyte) has also been collected and analyzed from a number of bio-monitoring reference sites; however, this species actively accumulates nitrogen and sulfur and consistently shows unusually elevated levels of these two elements. Additional research may ultimately provide important insights into varying elemental accumulation patterns among acidophytes, neutrophytes, and nitrophytes.

Various analytical techniques, including atomic absorption spectrometry, inductively coupled plasma emission spectroscopy (ICP), X-ray fluorescence spectrometry, and proton-induced X-ray emission technology (PIXE), have been used to determine pollutant element concentrations in lichen thalli (St. Clair 1999; Nimis et al. 2002). Care must be taken when collecting, preparing, and analyzing lichen material for elemental analysis. Samples should be collected with a ceramic knife to avoid metal contamination, and additional care must be taken to avoid other potential contaminants during collection, processing, and storage. Samples should be completely dried and stored in sterile plastic bags, avoiding the use of commonly used paper bags for more general collecting. Nasco Whirl-Pak 18 oz. plastic bags prevent contamination of samples and offer an effective and relatively inexpensive container for long-term storage of elemental analysis samples (St. Clair 1999).

An alternative, or often complementary, approach to elemental analysis for assessing air quality is tracking the presence or absence of sensitive indicator species—usually the larger, conspicuous macrolichens (McCune and Geiser 2009). A floristic survey at bio-monitoring reference sites, including collecting and identifying lichen species from as many habitats and substrates in the study area as possible, provides a general list of the lichens occurring at each site, and from the general list of lichens, a second list restricted to the documented sensitive indicator species and their relative abundance is compiled (St. Clair 1999). Periodic re-evaluations of the lichen flora, and particularly sensitive indicator lichens, at reference sites are central to documenting temporal changes in the relative abundance of indicator species and assessing the impact of changing air quality through time. Air quality ratings for many lichens occurring in western North America are found in McCune and Geiser (2009), St. Clair (1999), and the USDA Forest Service "National Lichens and Air Quality Database and Clearinghouse" (http://gis.nacse.org/lichenair/), providing important resources for qualitatively assessing the impact of air quality based on the occurrence of sensitive indicator lichens. Sensitive lichens should be assessed for air pollution-related changes, including bleaching of the upper cortex, thallus discoloration, necrotic spots, hyper-production of asexual reproductive propagules, or other abnormalities in thallus morphology. Consistent patterns in thallus damage can be indicative of air pollution-induced stress. The establishment of lichen air quality bio-monitoring references sites is limited by a number of practical constraints, most notably, the significant investment of time and the high level of training and expertise required. Given the environmental heterogeneity of lands in western North America and the species-specific microhabitat conditions required to support lichens, the absence of sensitive indicator species may not necessarily reflect poor air quality. Therefore, subjectively selecting reference sites that are likely to support sensitive indicator species, incorporating elemental analysis of lichen thalli to determine levels of atmospheric pollutants, and consistent periodic reviews are essential requirements for effectively monitoring air pollution-related impacts over time.

Assessing ecological health and recovery may be particularly important in areas that have been severely impacted by air pollution. However, impacted sites generally lack baseline floristic surveys and information about pollutant element concentrations in lichen thalli (McMurray et al. 2013; Riddell et al. 2011; Howe and Lendemer 2011). In these cases, transplant studies may provide important insights concerning the current health and recovery potential of damaged sites. Ideally, lichens thought to be native to the damaged site are transplanted from adjacent clean-air sites, and changes to physiological parameters (ex. photosynthetic activity, membrane leakage, etc.) are evaluated over time (Riddell et al. 2011; Nimis et al. 2002). Transplant samples may also be analyzed to evaluate patterns of atmospheric pollutant accumulation (Conti and Cecchetti 2001). Data obtained from transplant studies can then be used as an artificial baseline for assessing future air quality-related impact or recovery in habitats that have previously sustained significant air pollution-related damage (St. Clair 1999).

Thorough ecological surveys and community analyses likely yield the most comprehensive assessment of air pollution-related impact (McCune and Geiser 2009). This type of study generates data that can be directly used for evaluating air pollution-related effects on lichen species diversity, air quality, and climate, although they require high levels of training and expertise, a substantial investment of time and manpower, and the application of sophisticated statistical techniques to effectively separate the major factors influencing lichen community composition and structure. The USFS Forest Inventory and Analysis program (FIA) includes lichens among a suite of other forest health indicators (Stolte 2001; Will-Wolf 2010). The FIA program uses specially trained field crews to periodically survey epiphytic macrolichen communities to evaluate and monitor biodiversity, air quality, and climate (Will-Wolf 2010). In western North America, this approach has provided detailed insights into the responses of epiphytic lichens and lichen communities to air pollution and climate change (Geiser et al. 2010; Hyvärinen and Crittenden 1998; Jovan 2008; Jovan and Carlberg 2007). Although the FIA program provides an effective strategy for indicating particular environmental conditions in specific lichen communities (Will-Wolf 2010), interpretation of lichen species and community composition should not be generalized beyond the geographic and ecological scale of the observation (Will-Wolf et al. 2006). In the intermountain region of western North America, the FIA program may be difficult to implement for a number of reasons, including the dominance of arid and semiarid habitats with vegetation types that do not fit the FIA model. In addition, many intermountain area forests are relatively depauperate of epiphytic macrolichens due to climatic conditions rather than poor air quality or environmental health-related issues, and incorporating saxicolous sensitive indicator lichens is not a part of the FIA sampling protocol.

One of the best known examples of the use of lichens as bio-monitors of air pollution involves efforts to monitor sulfur dioxide (SO_2) levels (Nash and Gries 2002; Seaward 1993). In the Intermountain West, the major sources of SO_2 pollution include coal-burning power plants, diesel consumption, petroleum processing, and ore smelting. However, beginning in the 1970s, SO_2 emission levels have continuously declined in North America and Europe due to legislation regulating SO_2 emissions (Mylona 1996; Ellerman and Montero 1998). This dramatic decrease in SO_2 emissions has resulted in the recolonization of many SO_2—sensitive lichens into urban areas in Europe formerly impacted by poor air quality (Henderson and Seaward 1979; Rose and Hawksworth 1981; Seaward and Letrouit-Galinou 1991). However, recovery patterns associated with decreasing SO_2 emissions have not generally been observed in the intermountain region of western North America, although this may be due to a combination of factors including a more recent history of industrialization, a general lack of pre-industrialization baseline studies of lichen species diversity, and generally lower human population densities.

Anthropogenic-related increases in the levels of atmospheric sulfur (S) and nitrogen (N) are known to have potentially damaging effects on both terrestrial and aquatic ecosystems (Fowler et al. 2013). For example, even small increases in N deposition can alter ecosystem processes in N-limited environments, and previous studies suggest that current levels of N deposition are predicted to adversely affect a range of sensitive ecosystems in the western North America (Fenn et al. 2008; McMurray et al. 2013; Geiser et al. 2010; Glavich and Geiser 2008; Hyvärinen and Crittenden 1998; Jovan and Carlberg 2007). The primary nitrogen-containing atmospheric pollutants are nitrogen oxides and ammonia, but may also include further oxidized forms including, ammonium, nitrates, and nitric acid vapor

(McCune and Geiser 2009). While fossil fuel combustion, forest fires, and combustion of other forms of biomass release nitrogen oxides, ammonia waste is a common by-product of agriculture (i.e., fertilizers and animal waste). Nitrogen deposition rates are largely unknown for most areas in the Intermountain West, but the available data indicate that over much of the region, N deposition rates range from 1–4 to 30–90 kg/ha/year downwind of major air pollution sources, particularly agricultural and urban areas (Fenn et al. 2003). Rapid expansion of natural gas drilling throughout the intermountain region has raised important concerns about the potential ecological impact of increased N deposition associated with oil and gas extraction and processing. Analysis of lichen thalli for total nitrogen is known to accurately document N deposition (McMurray et al. 2013; Fenn et al. 2008; Hyvärinen and Crittenden 1998; Root et al. 2013).

Long-term bio-monitoring networks will be crucial for evaluating the impact of increased N deposition associated with oil and gas development, urban development, increased vehicular traffic, and large concentrations of domestic animal feeding operations. Root et al. (2013) proposed a simple model to monitor N deposition across broad spatial and topographical regions in western North America using widespread, easily recognized lichens. This approach offers particular utility to federal land managers for assessing N critical load exceedance through a cost-effective approach for assessing air quality in their management units, identifying vulnerable areas for more intensive monitoring, and directly assessing the impact of potential sources of air pollution (e.g., vehicular traffic).

5.2.2 Monitoring Past Disturbances

Pollutants associated with abandoned or uncontrolled areas pose significant risk to both ecosystems and human health at hundreds of sites throughout the Intermountain West. For example, heavy metals and arsenic waste from abandoned mining operations are artifacts of the prominent mining history of western North America. The impact of mining and smelting activities on lichens is well documented (Howe and Lendemer 2011; St. Clair et al. 2002a; Bargagli et al. 2002; Rusu et al. 2006). The US Federal Superfund program, operated under the management of the US Environmental Protection Agency, identifies, analyzes, and remediates hazardous waste sites in an attempt to improve the overall health of the region. In western Montana, 19 sites have been listed on the National Priorities List (NPL) and are eligible for long-term remedial action through the Superfund program (Fig. 5.1). In western North America, the majority of these NPL Superfund sites are the direct result of historic mining and smelting operations.

Beginning in 1992, a total of 49 lichen air quality bio-monitoring reference sites were established in or near eight wilderness areas in western Montana and the Idaho Panhandle (Fig. 5.1), including sites in the Anaconda-Pintler (including the Mount Haggin Wildlife Management Area), Bob Marshall, Cabinet Mountains, Gates of the Mountains, Gospel Hump, Mission Mountains, Scapegoat, and Selway Bitterroot wilderness areas (St. Clair et al. 2013). These bio-monitoring reference sites have been systematically reviewed and offer an ongoing perspective into the ecological health of western Montana. A number of air quality bio-monitoring reference sites have been established in and near the Mount Haggin Wildlife Management Area and the Gates of the Mountains Wilderness Area in western Montana. Several of these reference sites were established in close geographic proximity to decommissioned smelters which are now Superfund sites, including the Anaconda copper smelter, operated from 1919 until 1981, and the East Helena lead smelter (1888–2001) (Fig. 5.1).

At most bio-monitoring reference sites in western Montana, the lichen communities appear to be diverse and healthy, suggesting generally positive ecological conditions in the wilderness areas in western Montana (St. Clair et al. 2013). Additionally, the number of sensitive indicator species per reference site has remained high in

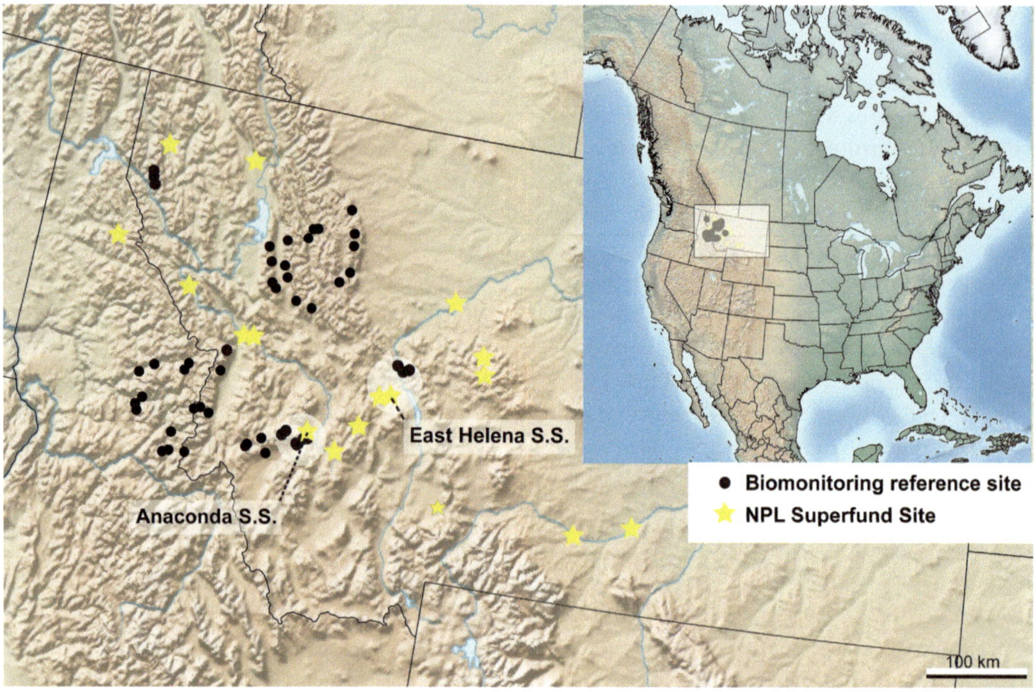

Fig. 5.1 Distribution of air quality bio-monitoring reference sites in western Montana and the Idaho Panhandle, USA. National Priority List (*NPL*). Superfund sites in Montana are indicated with *yellow stars*, and the Anaconda copper smelter and East Helena smelter Superfunds sites are highlighted on the map

comparison to other similar sites in the Intermountain western USA. While elemental analysis data over the entire study period have shown some minor fluctuations in the concentrations of some pollutant elements, the lichen flora of western Montana generally remains free from air pollution-related impact. However, the impact of historic mining operations and their subsequent decommissioning and remediation have been documented at a number of reference sites near the Anaconda and East Helena Superfund sites.

Distinct improvements in lichen communities have been observed at sites in close proximity (ca. 8 km) to the decommissioned Anaconda copper smelting operation (St. Clair et al. 2013). Closure of the copper smelter in the early 1980s, followed by remediation efforts, has resulted in significant improvements in local air quality, marked by an increase in overall species diversity as well as the number and abundance of sensitive indicator species. When the original bio-monitoring reference sites in the Anaconda-Pintler Wilderness Area and eastward sites in the Mount Haggin Wildlife Management Area were established in the early 1990s, a total of 143 lichen species, in 65 genera, were reported for the Anaconda-Pintler Wilderness Area, while only 21 species were reported for the sites between the Anaconda copper smelter and the eastern boundary of the wilderness area. Over the last 22 years, reference sites in and near the Mount Haggin Wildlife Management Area have shown significant improvement in response to declining levels of As, Cu, Ni, and Pb (Fig. 5.2). Likewise, Cu/Zn ratios from samples in the wildlife management area have also declined over the same time period (1.15 → 0.683) further documenting reduced levels of airborne Cu. The re-establishment and increase in relative abundance of a number of sensitive lichen species (*Bryoria fucescens*, *Letharia vulpina*, and *Usnea lapponica*), coinciding with the general decline in concentrations of atmospheric pollutants in the thalli of sensitive indicator species, suggests that

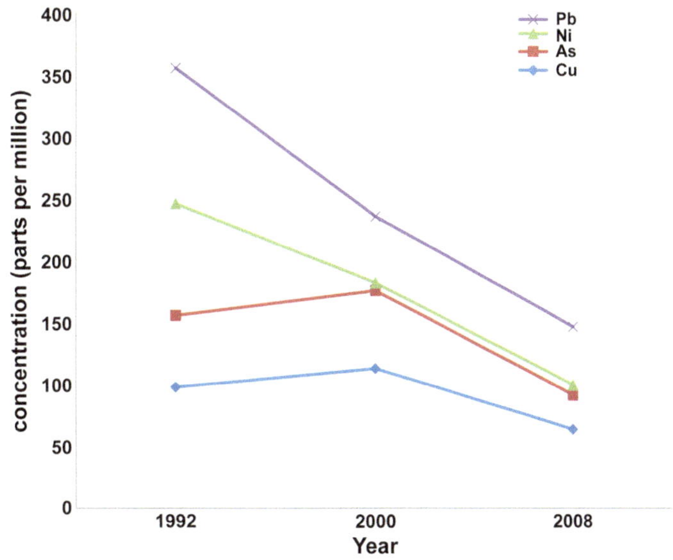

Fig. 5.2 Declining concentrations of four potential air pollutants, arsenic (As), copper (Cu), lead (Pb), and nickel (Ni), in lichen samples at a bio-monitoring reference site in Deerlodge County, Montana near the Anaconda copper smelter NPL Superfund site

decommissioning of the smelter operations and remediation of the slag debris and flu dust have had a significant positive impact on the adjacent biological communities. The average thallus size of several air pollution sensitive species at the eastern (closer to the Anaconda Superfund site) reference sites was smaller, suggesting recent re-establishment of these species at these sites. Continued monitoring is crucial to monitor recovery in this region. In other nearby reference sites in the Anaconda-Pintler Wilderness Area, 25–60 km west and upwind of the Anaconda Superfund site, pollutant element concentrations have generally remained within background levels and the number of sensitive indicator species per reference site has remained high throughout the sampling period. By all parameters, the lichen flora in the western portion of the Anaconda-Pintler Wilderness Area is healthy and free from air pollution-related impact, while the eastern reference sites between the eastern boundary of the wilderness area and the Anaconda Superfund site are showing evidence of significant recovery.

Three lichen air quality bio-monitoring reference sites were established during the summers of 2000–2001 in the Gates of the Mountains Wilderness Area, located ca. 25 km northeast of Helena, Montana (St. Clair et al. 2013). Elemental analysis data from the initial sampling period showed elevated levels of S, As, Ni, and Pb from all three reference sites (Table 5.2). The East Helena Superfund site represents a potential contributor to the elevated pollutant levels in the lichen samples collected from the Gates of the Mountains Wilderness Area. For over 100 years, the East Helena smelting operation deposited lead, arsenic, copper, zinc, cadmium, and other contaminants into the soil, surface water, and groundwater of Helena Valley. The East Helena smelting operation was decommissioned in 2001, and the area was subsequently listed as an NPL Superfund site. However, following closure of the smelter, concentrations of several critical pollutant elements continue to fluctuate at elevated levels. In spite of the high concentrations of pollutant elements, the overall species diversity as well as the number and abundance of sensitive indicator species per site in this wilderness area suggests that the lichen community is generally healthy. Ongoing monitoring of pollutant element concentrations in lichen samples from the Gates of the Mountains Wilderness Area will be essential for documenting full recovery of the biological systems following remediation of the East Helena Superfund site. Our data from bio-monitoring reference sites near the Anaconda superfund site suggest that a

Table 5.2 Mean concentrations of potential pollutant elements in sensitive indicator species from air quality bio-monitoring sites in the Gates of the Mountains Wilderness Area

Species and collection site	Elements (ppm except where indicated)													
	As		Cu		Ni		Pb		Zn		S %		N %	
	2000–2001	2009	2000–2001	2009	2000–2001	2009	2000–2001	2009	2000–2001	2009	2000-2001	2009	2000–2001	2009
Usnea sp. Meriwether Canyon, sample #808, 1192	**6.00**	**6.61**	8.00	6.30	2.00	1.62	**11.0**	**8.11**	**55.0**	**50.81**	**0.1752**	**0.148**	n.a.	**1.38**
Letharia vulpina, Meriwether Canyon, sample #809, 1191	**4.00**	**4.38**	6.00	6.71	2.00	1.30	**44.0**	**14.98**	**73.0**	**45.75**	**0.1278**	**0.148**	n.a.	**1.92**
Letharia vulpina, Refrigerator Canyon, sample #883, 1195	**4.29**	**7.85**	5.84	8.23	2.50	**4.26**	**43.6**	**14.55**	**66.9**	41.17	**0.1752**	**0.097**	n.a.	0.58
Rhizoplaca melanophthalma, Refrigerator Canyon, Sample #884, 1197	**10.40**	**9.59**	13.50	12.06	**3.85**	1.82	**41.9**	**106.10**	42.3	42.45	**0.1374**	**0.103**	n.a.	0.69
Usnea sp. Refrigerator Canyon, sample #885, 1196	**4.38**	**4.96**	5.26	9.06	1.31	3.21	12.8	**15.13**	56.7	42.03	**0.1594**	**0.087**	n.a.	**0.85**
Letharia vulpina, Hunter's Gulch, sample #886, 1194	**3.87**	**5.68**	7.16	7.50	1.86	**5.78**	**32.3**	**9.55**	**49.1**	42.03	**0.2142**	**0.149**	n.a.	**1.35**
Usnea sp. Hunter's Gulch, sample #887, 1193	**4.95**	**5.36**	7.47	6.55	2.40	1.30	**30.8**	9.11	57.1	39.13	**0.3213**	**0.106**	n.a.	**1.1**

Values for 2000–2001 and 2009 collections are shown; **bolded** values indicate values above provisional threshold of background levels (see: http://gis.nacse.org/lichenair/index.php?page=cleansite). All voucher material is housed at the Brigham Young University Herbarium of Nonvascular Cryptogams (BRY), Provo, Utah, USA

reduction in pollutant elements to background levels will likely require decades following the decommissioning and cleanup of the smelter operation.

Similar to the results from our study near the decommissioned Anaconda copper smelter, Howe and Lendemer (2011) documented the re-establishment of a formerly degraded lichen community downwind of the decommissioned Palmerton zinc smelter (also a NPL Superfund site) in Pennsylvania. A number of lichen bio-monitoring reference sites in the Gila National Forest in southwestern New Mexico and the Chiricahua National Monument in southeastern Arizona are located in close proximity to some of the most intensive copper mining and smelting operations in the world (St. Clair et al. 2002a). Lichen samples from these sites had copper concentrations >100 μg/g dry weight (background copper concentrations in other intermountain area locations are generally <20 μg/g dry weight) and Cu/Zi ratios ≥1 (Cu/Zn ratios > 0:6 suggest accumulation of airborne copper from emissions associated with copper processing activities) (St. Clair et al. 2002a). In southern Idaho, concentrations of heavy metal pollutants and P in rock posy lichens were found to be negatively correlated with the distance of sample sites from phosphate refineries (Dillman 1996).

Notwithstanding the potential use of lichens for monitoring degraded sites associated with mining activity, the appropriate sample design for accurately characterizing ecological health using lichen indicators remains challenging. For example, both the temporal impact and scale of disturbances may vary widely across specific sites, and effective sampling design is crucial to assessing the impact of the disturbance (Loppi et al. 2004; Lisowska 2011; Aznar et al. 2008). Establishing bio-monitoring reference sites along a gradient, relative to potential pollutant sources, and serial sampling can aid in characterizing the spatial and temporal impact at disturbed sites. For example, serial sampling of bio-monitoring reference sites near NPL Superfund sites in western Montana (Anaconda copper smelter) and eastern Pennsylvania (Palmerton zinc smelter) revealed a gradual recovery of lichen communities and decrease in elemental pollutants in sensitive lichen indicators approximately 30 years after the smelting operations were decommissioned (Howe and Lendemer 2011; St. Clair et al. 2013). In other cases, elevated heavy metal pollutants can be observed in otherwise health lichen communities, including references sites established in the Gates of the Mountains Wilderness Area, near the recently decommissioned East Helena smelter (St. Clair et al. 2013). In a separate case, one of six references sites established in the Flaming Gorge National Recreation Area in northeastern Utah during the summers of 2010 and 2011 showed elevated levels for As (2.8 ppm), Ni (3.39 ppm), and Pb (218.6 ppm) in the sensitive indicator species *Xanthoparmelia-coloradoënsis* (St. Clair and Leavitt 2013). The Pb concentrations were of particular concern—representing some of the highest values reported for the intermountain area. In spite of the elevated levels of heavy metal pollutants, the overall diversity of lichen species, and particularly the number of sensitive indicator species, at the site suggests that at some temporal scales lichens may effectively accumulate airborne pollutants without any apparent impact on community composition or species diversity.

In conclusion, a combined approach, including assessing overall lichen diversity, the abundance of sensitive indicator species, and elemental analysis of potential pollutants, is central to monitoring the impact of historic disturbances in long-term lichen bio-monitoring programs.

5.2.3 Monitoring Future Disturbances

Climate change is projected to promote major ecological shifts worldwide (Araújo and Rahbek 2006; Parmesan and Yohe 2003; Sykes et al. 1999). Shifting biomes due to climate change may lead to adjustments in species distributions and potential extinctions if populations are unable to tolerate or adapt in situ to altered environmental conditions or migrate to suitable habitats (Aitken et al. 2008; Jackson and Overpeck 2000; Beck et al. 2011; Bradshaw and Holzapfel 2006). Some components of cryptogamic communities have been shown to be particularly sensitive to climatic shifts (McCune 2000; Jägerbrand et al. 2006; Cornelissen et al. 2001; Bjerke 2011), and these communities, due to their sensitivity, may play an important role in monitoring the potential impacts of climate change. Although lichens are well-known indicators of air quality, recent studies indicate that they may also be useful in assessing ecological shifts related to climate change (Cornelissen et al. 2001; Rogers and Ryel 2008; Bjerke 2011).

On a global scale, macrolichens in climatically milder arctic ecosystems may decline if and where global changes cause vascular plants to increase in abundance (Cornelissen et al. 2001). In the UK, the southern elements of Britain's lichen flora, and other lichen species adapted to warmer climates, are projected to expand northward, while the montane species appear to be disproportionally threatened by climate change (Ellis et al. 2007). Other data suggest that a warmer, humid climate in Norway will likely be beneficial for the generalist species *H. physodes*, but detrimental to the subalpine birch specialist *Melanohalea olivacea* (Evju and Bruteig 2013). These studies highlight the importance of continued research on specific responses of lichen species and communities to changing climate. However, only by consistently coupling efficient quantitative methods with accurately characterized species- and/or community-specific responses to changing climatic conditions will we be able to effectively document climate change-related impact on vulnerable lichen communities.

In general, specific responses of most species, including non-vascular cryptogams, to rapid climate changes in vulnerable habitats remain uncertain, and detailed, long-term monitoring will be essential to accurately assessing biologically meaningful shifts in community composition and species distributions (Manel et al. 2010; Eaton and Ellis 2012).

Due to the limited number of experts, time constraints, funding, and the need for long-term monitoring, simple field methods that can be applied consistently over multiple years are crucial to adequately monitor shifts in population density and community composition resulting from climate change (Leavitt and St. Clair 2011; Nimis et al. 2002). While a number of effective methods have been developed for assessing community structure in corticolous lichen communities (Nimis et al. 2002), distance methods provide a scalable alternative for assessing population density and community composition in some types of terricolous and saxicolous lichen communities (Leavitt and St. Clair 2011). Distance methods are less prone to significant variation among observers, are generally less labor and time intensive, and generally minimize damage to the study area (Benavides and Sastre-De Jesús 2009; Kunze 1980; St. Clair et al. 2007; Dahdouh-Guebas and Koedam 2006). The population density of *Xanthoparmelia* communities in six long-term reference sites in subalpine habitats in the southwestern USA has been determined using distance methods, although the specific response of subalpine *Xanthoparmelia* communities to changing climate remains unknown (Leavitt and St. Clair 2011). Lichens and bryophytes are commonly important components of high-altitude/high-latitude ecosystems (Longton 1997; Jagerbrand et al. 2006) and Arctic/alpine lichens have been shown to be sensitive to a warmer and more fluctuating winter climate (Bjerke 2011). In addition, winter icing events may have a substantial effect on lichen-dominated ecosystems (Bjerke 2011). A number of studies on the impact of climate change on lichen communities in Greenland reveal the potential for dynamic shifts in lichen communities, where a warmer climate will likely result in more favorable conditions for some lichens in specific habitats and unfavorable conditions for others (Leavitt et al. 2014; Hansen 2010, 2011a, b, 2012). Non-vascular cryptogams are an important component of alpine tundra communities in western North America, and effective monitoring of alpine tundra lichen communities has the potential to make significant contributions to conservation and management of some of the most vulnerable biological communities in the Intermountain West (Conti et al. 2009; St. Clair et al. 2007; Ustin et al. 2009; Leavitt et al. 2014).

Alpine tundra habitat in the western USA is comprised of a variety of specialized vegetative communities adapted to severe climate. Of the estimated 3 million hectare of alpine tundra habitat in North America, nearly one-third lies in Colorado (Fig. 5.3). Historically, livestock grazing, mining, and recreation have resulted in significant disturbances in alpine habitats that may take decades to centuries to recover (St. Clair et al. 2007). Now global climate change provides an additional threat to sensitive alpine tundra communities through the encroachment of forest communities and increased variation in winter temperatures and precipitation.

Among the specialized Arctic-alpine lichens, two species from the genus *Flavocetraria* (Parmeliaceae), *F. cucullata* and *F. nivalis*, are important components of Colorado's alpine tundra community and have been collected historically from a number of alpine tundra communities throughout the state (Fig. 5.4). Evidence from recent field surveys suggests that lichen communities in Colorado's alpine tundra habitats have likely been impacted by climate change, along with other disturbances. Due to the specific habitat conditions required to support *F. cucullata* and *F. nivalis* and their restricted distribution in alpine tundra habitats in the continental USA, *Flavocetraria* spp. may potentially serve as effective bio-indicators for monitoring the impact of climate change on biological communities in alpine tundra habitats in the intermountain region.

During the summer of 2013, we established baseline estimates of population density for *F. nivalis* at four long-term, bio-monitoring sites in alpine tundra habitats in Colorado: (1) Niwot Ridge Long-Term Ecological Research Site, ca. 15 km northwest of Nederland in north-central Colorado; (2) vicinity of Loveland Pass, ca. 95 km west of Denver; (3) vicinity of Red Mountain Pass in the San Juan National Forest, along US Route 550; and (4) Cottonwood Pass in

Fig. 5.3 Alpine tundra habitat in Colorado, USA. The *panel on the left* shows Kiowa (4,047 m.a.s.l.) and Navajo (4,087 m.a.s.l.) peaks from the Niwot Ridge Long-term Ecological Research site; *panel on right* shows a habit shot of *Flavocetraria nivalis* at Loveland Pass, Colorado (scale bar = 1 cm)

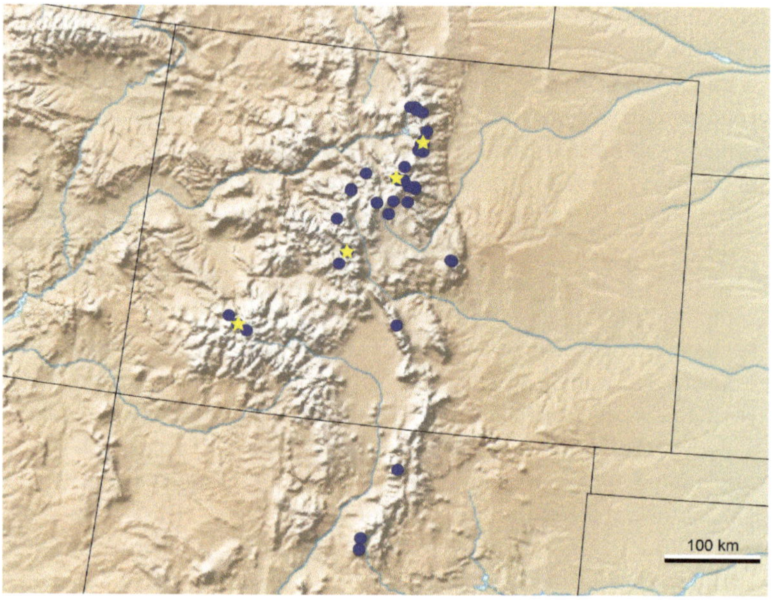

Fig. 5.4 Southernmost distribution of *Flavocetraria* species in North America: *blue circles* indicate all known occurrences of *Flavocetraria* species in Colorado and New Mexico, and *yellow stars* indicate sites established for long-term monitoring of *Flavocetraria nivalis* population densities (see Table 5.3)

south-central Colorado, ca. 30 km west of Buena Vista. *Flavocetraria nivalis* population density was estimated at the four sites following previously described methods, and baseline estimates are provided in Table 5.3 (Leavitt and St. Clair 2011; Zhu and Zhang 2009). Herbarium collections document that *F. nivalis* has historically occurred in abundance at each of these sites. In the recently established long-term plots, *F. nivalis* occurs in patchy, locally common populations at the Cottonwood Pass, Loveland Pass, and Niwot Ridge sites; however, *F. nivalis* was rare at the Red Mountain site (Table 5.3). *F. nivalis* populations at Red Mountain may be particularly vulnerable due occurrence near the southern-most extent of its range and low population density.

Recent records of *Flavocetraria* spp. are relatively sparse throughout Colorado. Only 13 collections made between 1990 and 2014 are reported in the Consortium of North American Lichen Herbaria, compared to 149 collections made between 1950 and 1990 (data accessed from http://lichenportal.org/portal/collections/index.php on March 11, 2014). Our baseline estimates of population density for *F. nivalis* populations and long-term bio-monitoring of these established plots will provide an important framework to help document the impact of climate change and other human-related disturbances on alpine tundra communities in the Intermountain West. Furthermore, revisiting additional sites where *Flavocetaria* spp. have historically been collected throughout the intermountain region may provide important perspectives about general patterns across the region. In particular, alpine tundra habitats in New Mexico represent the furthest south occurrences of *F. nivalis*, and these, along with occurrences at lower elevations further north, are likely vulnerable to a much higher probability of decline. However, in addition to observational studies, experimental manipulation studies will be essential to adequately characterize the response of *Flavocetraria* spp. to altered climate regimes. Other lichens restricted to alpine tundra habitats may also serve as important indicators of the impact of a warming climate on these sensitive habitats, and we advocate ongoing observational and experimental studies to develop effective strategies for monitoring climate change.

Table 5.3 Initial *Flavocetraria nivalis* population density estimates at four long-term reference sites in Colorado, USA

Site	Lat. lon. elevation	Abundance	Estimated population density 2013	Notes
Cottonwood Pass: Gunnison National Forest, Sawatch Range, just southwest of Cottonwood Pass on prominent hilltop	38° 49.441′N 106° 24.989′W 3,810 m.a.s.l.	Locally abundant	3.5 ± 0.02 individuals/m^2	Site receives heavy recreational traffic and moderate vehicle traffic during the summer months
Loveland Pass: Arapaho National Forest, Front Range, along US Route 6, vicinity of Loveland Pass	39° 39.875′N 105° 52.682′W 3,670 m.a.s.l.	Locally common	0.17 ± 0.09 individuals/m^2	Site is in close proximity to US Interstate 70 and receives heavy vehicle and recreational traffic year round
Niwot Ridge: Roosevelt National Forest, Niwot Ridge Long-Term Ecological Research Station, ca. 3 km west of Niwot Mountain	40° 03.190′N 105° 35.250′W 3,514 m.a.s.l.	Locally common	0.83 ± 0.10 individuals/m^2	Site is located within the Niwot Ridge Long-term Ecological Research Site, administered by the University of Colorado at Boulder
Red Mountain: San Juan National Forest, south of Red Mountain Pass on US Highway 550, east of County Road 16, between Big Horn and Browns Gulches	37° 52.422′N 107° 42.339′W 3,750 m.a.s.l.	Rare	≪0.1 individuals/m^2	Recreational and vehicle traffic are negligible

These sites are distributed across the southernmost distribution of *Flavocetraria* species in North America

5.3 Conclusions

Our lichen bio-monitoring program, including over 350 reference sites, represents the largest air quality bio-monitoring program in western North America and provides crucial baseline data across a wide range of habitats. In order to exploit the value of this research program, there is a critical need for ongoing periodic reviews of the existing reference sites. Ongoing reviews of the lichen flora, and particularly sensitive indicator lichens, and elemental analyses of potential pollutants at reference sites will be central to effectively assessing ecological health and the impact of changing air quality over time. Similarly, establishing lichen air quality bio-monitoring programs in and around important as of yet unprotected airsheds—including, national parks, wilderness areas (all classes), and other vulnerable undesignated wildland landscapes—should be a central focus of ongoing air quality bio-monitoring research. In light of expanding energy extraction and processing activity in western North America, there is a pressing need to establish additional lichen air quality bio-monitoring programs on lands that are most likely to be impacted by oil and natural gas development (McMurray et al. 2013). This is particularly important in light of the closing window relative to the opportunity for effectively establishing baseline, pre-impact programs. Additional research is needed, particularly in vulnerable and sensitive landscapes (alpine/Arctic tundra and warm deserts), to determine if sensitive indicators of climate change impact are in decline. It is imperative that we collect the data while it is still available—future opportunities for systematically exploring and characterizing the lichen flora of these rare and unique systems may be limited. In the case of *F. nivalis* in the Intermountain West, examination of historical distribution records, followed by determination of current population densities of *F. nivalis* across additional sites, would provide a more extensive baseline for monitoring climate change-related impact on alpine tundra systems. Furthermore, critical assessments of other potentially sensitive components of subalpine, alpine, and alpine tundra communities are desperately needed.

In arid regions of the Intermountain West, lichens are often important components in many biological soil crust communities. BSC play a major ecological role in arid and semiarid regions worldwide. BSC often provide 40–100 % of the ground cover in areas with relatively sparse vascular plant cover (St. Clair et al. 1993). They also effectively reduce wind and water erosion, significantly increasing soil fertility, and are generally important in ecosystem function (Belnap and Eldridge 2003). Soil crusts in general and the lichen component in particular are very sensitive to human-related perturbation. Over the last 150 years, a significant portion of the soil crust communities of the Great Basin and Colorado Plateau has been heavily damaged (Zhao et al. 2010; Chambers et al. 2014; St. Clair et al. 1993). Much of the disturbance is due to intensive grazing by cattle and sheep, urban and agricultural development, altered fire regimes, and off-road recreation (St. Clair et al. 1993; Belnap and Eldridge 2003). Today, novel combinations of disruptive factors, including climate change, invasive species, and changes in land use strategies, threaten these vulnerable soil crust communities [86]. Given the overall ecological characteristics of biological soil crust communities, including diversity and physiological status, these communities can also potentially serve as effective early indicators of ecosystem change or decline due to their rapid response to altered environmental conditions (Zhao et al. 2010; Chambers et al. 2014; Eldridge 2000). Both observational and experimental manipulation studies will be essential to appropriately characterizing the response of soil crust lichens to altered climate regimes and the development of effective bio-monitoring approaches.

Acknowledgments We wish to thank the dedicated employees of the USDA Forest Service whose knowledge, experience, insight, time, and kindness have played a crucial role in our research. The USDA Forest Service has provided long-term funding for this research, and additional funding sources include the Myrna Steinkamp Fund from the Colorado Native Plant Society, the Office of Research and Creative Activities at Brigham Young University, and the USDI Bureau of Land Management and National Park Service.

References

Adachi K, Tainosho Y (2004) Characterization of heavy metal particles embedded in tire dust. Environ Int 30 (8):1009–1017

Agnan Y, Séjalon-Delmas N, Probst A (2014) Origin and distribution of rare earth elements in various lichen and moss species over the last century in France. Sci Total Environ 487:1–12. doi:10.1016/j.scitotenv.2014.03.132

Aitken SN, Yeaman S, Holliday JA, Wang TL, Curtis-McLane S (2008) Adaptation, migration or extirpation: climate change outcomes for tree populations. Evol Appl 1(1):95–111. doi:10.1111/j.1752-4571.2007.00013.x

Araújo MB, Rahbek C (2006) How does climate change affect biodiversity? Science 313(5792):1396–1397. doi:10.1126/science.1131758

Aznar JC, Richer-Laflèche M, Cluis D (2008) Metal contamination in the lichen *Alectoria sarmentosa* near the copper smelter of Murdochville, Québec. Environ Pollut 156(1):76–81. doi:10.1016/j.envpol.2007.12.037

Bargagli R, Monaci F, Borghini F, Bravi F, Agnorelli C (2002) Mosses and lichens as biomonitors of trace metals. A comparison study on *Hypnum cupressiforme* and *Parmelia caperata* in a former mining district in Italy. Environ Pollut 116(2):279–287. doi:10.1016/S0269-7491(01)00125-7

Beck PSA, Juday GP, Alix C, Barber VA, Winslow SE, Sousa EE, Heiser P, Herriges JD, Goetz SJ (2011) Changes in forest productivity across Alaska consistent with biome shift. Ecol Lett 14(4):373–379. doi:10.1111/j.1461-0248.2011.01598.x

Belnap J, Eldridge D (2003) Disturbance and recovery of biological soil crusts. In: Belnap J, Lange O (eds) Biological soil crusts: structure, function, and management, vol 150. Ecological studies. Springer, Berlin, pp 363–383. doi:10.1007/978-3-642-56475-8_27

Benavides JC, Sastre-De Jesús I (2009) Digitized images provide more accuracy and efficiency to estimate bryophyte cover. Bryologist 112(1):12–18. doi:10.1639/0007-2745-112.1.12

Bjerke JW (2011) Winter climate change: ice encapsulation at mild subfreezing temperatures kills freeze-tolerant lichens. Environ Exp Bot 72(3):404–408. doi:10.1016/j.envexpbot.2010.05.014

Bongers T, Ferris H (1999) Nematode community structure as a bioindicator in environmental monitoring. Trends Ecol Evol 14(6):224–228. doi:10.1016/S0169-5347(98)01583-3

Bradshaw WE, Holzapfel CM (2006) Evolutionary response to rapid climate change. Science 312 (5779):1477–1478. doi:10.1126/science.1127000

Branquinho C, Catarino F, Brown DH, Pereira MJ, Soares A (1999) Improving the use of lichens as biomonitors of atmospheric pollution. Sci Total Environ 232 (1–2):67–77. doi:10.1016/S0048-9697(99)00111-4

Chambers J, Bradley B, Brown C, D'Antonio C, Germino M, Grace J, Hardegree S, Miller R, Pyke D (2014) Resilience to stress and disturbance, and resistance to Bromustectorum L. invasion in cold desert shrublands of western North America. Ecosystems 17 (2):360–375. doi:10.1007/s10021-013-9725-5

Conti ME, Cecchetti G (2001) Biological monitoring: lichens as bioindicators of air pollution assessment—a review. Environ Pollut 114(3):471–492. doi:10.1016/s0269-7491(00)00224-4

Conti ME, Pino A, Botrè F, Bocca B, Alimonti A (2009) Lichen *Usnea barbata* as biomonitor of airborne elements deposition in the Province of Tierra del Fuego (southern Patagonia, Argentina). Ecotoxicol Environ Saf 72(4):1082–1089. doi:10.1016/j.ecoenv.2008.12.004

Cornelissen JHC, Callaghan TV, Alatalo JM, Michelsen A, Graglia E, Hartley AE, Hik DS, Hobbie SE, Press MC, Robinson CH, Henry GHR, Shaver GR, Phoneix GK, Jones DG, Jonasson S, Iii FSC, Molau U, Neill C, Lee JA, Melillo JM, Sveinbjörnsson B, Aerts R (2001) Global change and Arctic ecosystems: is lichen decline a function of increases in vascular plant biomass? J Ecol 89(6):984–994. doi:10.1111/j.1365-2745.2001.00625.x

Coulston JW, Smith GC, Smith WD (2003) Regional assessment of ozone sensitive tree species using bioindicator plants. Environ Monit Assess 83 (2):113–127. doi:10.1023/A:1022578506736

Dahdouh-Guebas F, Koedam N (2006) Empirical estimate of the reliability of the use of the point-centred quarter method (PCQM): Solutions to ambiguous field situations and description of the PCQM + protocol. For Ecol Manage 228(1–3):1–18. doi:10.1016/j.foreco.2005.10.076

DellaSala D, Alaback P, Craighead L, Goward T, Paquet P, Spribille T (2011) Temperate and boreal rainforests of inland northwestern North America. In: Temperate and boreal rainforests of the world: ecology and conservation. Island Press/Center for Resource Economics, pp 82–110. doi:10.5822/978-1-61091-008-8_3

Dillman KL (1996) Use of the lichen *Rhizoplaca melanophthalma* as a biomonitor in relation to phosphate refineries near Pocatello Idaho. Environ Pollut 92(1):91–96. doi:10.1016/0269-7491(95)00084-4

Eaton S, Ellis CJ (2012) Local experimental growth rates respond to macroclimate for the lichen epiphyte *Lobaria pulmonaria*. Plant Ecol Divers 5(3):365–372. doi:10.1080/17550874.2012.728640

Eldridge D (2000) Ecology and management of biological soil crusts: recent developments and future challenges. Bryologist 103(4):742–747. doi:10.1639/0007-2745(2000)103[0742:EAMOBS]2.0.CO;2

Ellerman AD, Montero J-P (1998) The declining trend in sulfur dioxide emissions: implications for allowance prices. J Environ Econ Manage 36(1):26–45. doi:10.1006/jeem.1998.1033

Ellis CJ, Coppins BJ, Dawson TP, Seaward MRD (2007) Response of British lichens to climate change scenarios: trends and uncertainties in the projected impact for contrasting biogeographic groups. Biol Conserv 140 (3–4):217–235. doi:10.1016/j.biocon.2007.08.016

Evju M, Bruteig IE (2013) Lichen community change over a 15-year time period: effects of climate and pollution. Lichenologist 45(1):35–50. doi:10.1017/S0024282912000539

Fenn ME, Haeuber R, Tonnesen GS, Baron JS, Grossman-Clarke S, Hope D, Jaffe DA, Copeland S, Geiser L, Rueth HM, Sickman JO (2003) Nitrogen emissions, deposition, and monitoring in the western United States. Bioscience 53(4):391–403. doi:10.1641/0006-3568(2003)053[0391:nedami]2.0.co;2

Fenn ME, Jovan S, Yuan F, Geiser L, Meixner T, Gimeno BS (2008) Empirical and simulated critical loads for nitrogen deposition in California mixed conifer forests. Environ Pollut 155(3):492–511. doi:10.1016/j.envpol.2008.03.019

Fleischner TL (1994) Ecological costs of livestock grazing in western North America. Conserv Biol 8(3):629–644. doi:10.1046/j.1523-1739.1994.08030629.x

Fowler D, Coyle M, Skiba U, Sutton MA, Cape JN, Reis S, Sheppard LJ, Jenkins A, Grizzetti B, Galloway JN, Vitousek P, Leach A, Bouwman AF, Butterbach-Bahl K, Dentener F, Stevenson D, Amann M, Voss M (2013) The global nitrogen cycle in the twenty-first century. Philos Trans R Soc B: Biol Sci368(1621). doi:10.1098/rstb.2013.0164

Frego KA (2007) Bryophytes as potential indicators of forest integrity. Forest Ecol Manage 242(1):65–75. doi:10.1016/j.foreco.2007.01.030

Geiser LH, Neitlich PN (2007) Air pollution and climate gradients in western Oregon and Washington indicated by epiphytic macrolichens. Environ Pollut 145 (1):203–218. doi:10.1016/j.envpol.2006.03.024

Geiser LH, Jovan SE, Glavich DA, Porter MK (2010) Lichen-based critical loads for atmospheric nitrogen deposition in Western Oregon and Washington Forests. USA Environ Pollut 158(7):2412–2421. doi:10.1016/j.envpol.2010.04.001

Glavich DA, Geiser LH (2008) Potential approaches to developing lichen-based critical loads and levels for nitrogen, sulfur and metal-containing atmospheric pollutants in North America. Bryologist 111 (4):638–649. doi:10.1639/0007-2745-111.4.638

Hansen ES (2010) Lichens from five inland and coastal localities in south-west Greenland and their present climatic preferences in Greenland as regards oceanity and continentality. Bibl Lichenologica 104:143–154

Hansen ES (2011a) Qaanaaq og Siorapaluk—laver ognatur i et ændretklima. TidsskriftetGrønland 2011(2): 136–149

Hansen ES (2011b) Lichens from Qaanaaq and Siorapaluk, Northwest Greenland. Folia Cryptogamica Estonia 48:4–11

Hansen ES (2012) Lichens from five localities in southeast Greenland and their exposure to climate change. Bibl Lichenologica 108:123–134

Hawksworth DL (1971) Lichens as a litmus for air pollution: a historical review. Int J Environ Stud 1:281–296. doi:10.1080/00207237108709429

Henderson-Sellers A, Seaward MRD (1979) Monitoring lichen reinvasion of ameliorating environments. Environ Pollut (1970) 19(3):207–213. doi:10.1016/0013-9327(79)90042-9

Hodkinson ID, Jackson JK (2005) Terrestrial and aquatic invertebrates as bioindicators for environmental monitoring, with particular reference to mountain ecosystems. Environ Manage 35(5):649–666. doi:10.1007/s00267-004-0211-x

Howe NM, Lendemer JC (2011) The recovery of a simplified lichen community near the Palmerton Zinco Smelter after 34 years. Bibl Lichenologica 106:127–142

Hyvärinen M, Crittenden PD (1998) Relationships between atmospheric nitrogen inputs and the vertical nitrogen and phosphorus concentration gradients in the lichen *Cladonia portentosa*. New Phytol 140 (3):519–530. doi:10.1111/j.1469-8137.1998.00292.x

Jackson ST, Overpeck JT (2000) Responses of plant populations and communities to environmental changes of the late Quaternary. Paleobiology 26(4):194–220

Jägerbrand A, Lindblad KM, Björk R, Alatalo J, Molau U (2006) Bryophyte and lichen diversity under simulated environmental change compared with observed variation in unmanipulated alpine tundra. Biodivers Conserv 15(14):4453–4475. doi:10.1007/s10531-005-5098-1

Jovan S (2008) Lichen bioindication of biodiversity, air quality, and climate: baseline results from monitoring in Washington, Oregon, and California. General technical report, PNW-GTY-737, US Department of Agriculture, Forest Service

Jovan S, Carlberg T (2007) Nitrogen content of *Letharia vulpina* tissue from forests of the Sierra Nevada, California: geographic patterns and relationships to ammonia estimates and climate. Environ Monit Assess 129(1–3):243–251. doi:10.1007/s10661-006-9357-8

Kunze LM (1980) Distribution patterns of terricolous lichens occuring above treeline at Deer Park, Olympic National Park. University of Washington, Seattle, Washington

Lawrey JD, Hale ME (1981) Retrospective study of lichen lead accumulation in the northeastern United States. The Bryologist 84(4):449–456

Leavitt SD, St. Clair LL (2011) Estimating *Xanthoparmelia* (Parmeliaceae) population density in subalpine communities in southern Utah, U.S.A. using two distance methods, with implications for assessing community composition. Bryologist 114(3):625–636. doi:10.1639/0007-2745-114.3.625

Leavitt SD, Esslinger TL, Hansen ES, Divakar PK, Crespo A, Loomis BF, Lumbsch HT (2014) DNA barcoding of brown Parmeliae (Parmeliaceae) species: a molecular approach for accurate specimen identification, emphasizing species in Greenland. Org Divers Evol 14(1):11–20. doi:10.1007/s13127-013-0147-1

Lisowska M (2011) Lichen recolonisation in an urban-industrial area of southern Poland as a result of air quality improvement. Environ Monit Assess 179 (1–4):177–190. doi:10.1007/s10661-010-1727-6

Longton RE (1997) The role of bryophytes and lichens in polar ecosystems. In: Woodin SJ, Marquiss M (eds) The Arctic: environment, people, policy. Blackwell Science, Oxford, pp 69–96

Loppi S, Pirintsos SA (2003) Epiphytic lichens as sentinels for heavy metal pollution at forest ecosystems (central Italy). Environ Pollut 121(3):327–332. doi:10.1016/s0269-7491(02)00269-5

Loppi S, Frati L, Paoli L, Bigagli V, Rossetti C, Bruscoli C, Corsini A (2004) Biodiversity of epiphytic lichens and heavy metal contents of *Flavoparmelia caperatathalli* as indicators of temporal variations of air pollution in the town of MontecatiniTerme (central Italy). Sci Total Environ 326(1–3):113–122. doi:10.1016/j.scitotenv.2003.12.003

Manel S, Joost S, Epperson BK, Holderegger R, Storfer A, Rosenberg MS, Scribner KT, Bonin A, Fortin M-J (2010) Perspectives on the use of landscape genetics to detect genetic adaptive variation in the field. Mol Ecol 19:3760–3772. doi:10.1111/j.1365-294X.2010.04717.x

McCune B (2000) Lichen communities as indicators of forest health. Bryologist 103(2):353–356. doi:10.1639/0007-2745(2000)103[0353:LCAIOF]2.0.CO;2

McCune B, Geiser L (2009) Macrolichens of the Pacific Northwest. Oregon University Press, Corvalis

McMurray JA, Roberts DW, Fenn ME, Geiser LH, Jovan S (2013) Using epiphytic lichens to monitor nitrogen deposition near natural gas drilling operations in the Wind River Range, WY, USA. Water Air Soil Pollut 224(3):1–14. doi:10.1007/s11270-013-1487-3

Miller RR, Williams JD, Williams JE (1989) Extinctions of North American fishes during the past century. Fisheries 14(6):22–38. doi:10.1577/1548-8446(1989)014<0022:EONAFD>2.0.CO;2

Mylona S (1996) Sulphur dioxide emissions in Europe 1880–1991 and their effect on sulphur concentrations and depositions. Tellus B 48(5):662–689. doi:10.1034/j.1600-0889.1996.t01-2-00005.x

Naeth MA, Wilkinson SR (2008) Lichens as biomonitors of air quality around a diamond mine, Northwest Territories. Can J Environ Qual 37(5):1675–1684. doi:10.2134/jeq2007.0090

Nash TH (1975) Influence of effluents from a zinc factory on lichens. Ecol Monogr 45(2):183–198

Nash TH III (2008) Lichen biology. Cambridge University, Cambridge

Nash TH, Gries C (2002) Lichens as bioindicators of sulphur dioxide. Symbiosis 33:1–21

Nimis PL, Scheidegger C, Wolseley P (2002) Monitoring with lichens—monitoring lichens. Kluwer Academics, Dordrecht

Parmesan C, Yohe G (2003) A globally coherent fingerprint of climate change impacts across natural systems. Nature 421(6918):37–42. http://www.nature.com/nature/journal/v421/n6918/suppinfo/nature01286_S1.html

Perlmutter GB (2010) Bioassessing air pollution effects with epiphytic lichens in Raleigh, North Carolina, U.S.A. Bryologist 113(1):39–50. doi:10.1639/0007-2745-113.1.39

Pesch R, Schroeder W (2006) Mosses as bioindicators for metal accumulation: Statistical aggregation of measurement data to exposure indices. Ecol Ind 6(1):137–152. doi:10.1016/j.ecolind.2005.08.018

Purvis OW, Chimonides PJ, Jones GC, Mikhailova IN, Spiro B, Weiss DJ, Williamson BJ (2004) Lichen biomonitoring near Karabash Smelter Town, Ural Mountains, Russia, one of the most polluted areas in the world. Proc R Soc Lond Ser B: Biol Sci 271(1536):221–226. doi:10.1098/rspb.2003.2616

Riddell J, Jovan S, Padgett PE, Sweat K (2011) Tracking lichen community composition changes due to declining air quality over the last century: the Nash legacy in Southern California. Bibl Lichenologica 106:227–263

Rogers PC, Ryel RJ (2008) Lichen community change in response to succession in aspen forests of the southern Rocky Mountains. Forest Ecol Manage 256(10):1760–1770. doi:10.1016/j.foreco.2008.05.043

Root HT, Geiser LH, Fenn ME, Jovan S, Hutten MA, Ahuja S, Dillman K, Schirokauer D, Berryman S, McMurray JA (2013) A simple tool for estimating throughfall nitrogen deposition in forests of western North America using lichens. Forest Ecol Manage 306:1–8. doi:10.1016/j.foreco.2013.06.028

Rose CI, Hawksworth DL (1981) Lichen recolonization in London's cleaner air. Nature 289:289–292. doi:10.1038/289289a0

Rusu AM, Jones GC, Chimonides PDJ, Purvis OW (2006) Biomonitoring using the lichen *Hypogymnia physodes* and bark samples near Zlatna, Romania immediately following closure of a copper ore-processing plant. Environ Pollut 143(1):81–88. doi:10.1016/j.envpol.2005.11.002

Saeki M, Kunii K, Seki T, Sugiyama K, Suzuki T, Shishido S (1977) Metal burden of urban lichens. Environ Res 13(2):256–266. doi:10.1016/0013-9351(77)90102-5

Seaward MRD (1993) Lichens and sulphur dioxide air pollution: field studies. Environ Rev 1:73–91

Seaward MRD, Letrouit-Galinou MA (1991) Lichen recolonization of trees in the Jardin du Luxembourg, Paris. Lichenologist 23(02):181–186. doi:10.1017/S0024282991000324

Smith SD, Huxman TE, Zitzer SF, Charlet TN, Housman DC, Coleman JS, Fenstermaker LK, Seemann JR, Nowak RS (2000) Elevated CO_2 increases productivity and invasive species success in an arid ecosystem. Nature 408(6808):79–82. doi:10.1038/35040544

St. Clair LL (1999) A color guidebook to common rocky mountain lichens. M.L. Bean Life Science Museum of Brigham Young University Publisher, Provo

St. Clair LL, Johansen JR, Rushforth SR (1993) Lichens of soil crust communities in the Intermountain Area of the western United States. West North Am Nat 53(1):5–12

St. Clair SB, St. Clair LL, Mangelson NF, Weber DJ (2002a) Influence of growth form on the accumulation of airborne copper by lichens. Atmos Environ 36(36–37):5637–5644. doi:10.1016/s1352-2310(02)00688-x

St. Clair SB, St. Clair LL, Weber DJ, Mangelson NF, Eggett DL (2002b) Element accumulation patterns in foliose and fruticose lichens from rock and bark substrates in Arizona. Bryologist 105(3):415–421. doi:10.1639/0007-2745(2002)105[0415:EAPIFA]2.0.CO;2

St. Clair L, Johansen J, St. Clair S, Knight K (2007) The influence of grazing and other environmental factors on lichen community structure along an alpine tundra ridge in the Uinta Mountains, Utah, U.S.A. Arctic, Antarctic. Alp Res 39(4):603–613. doi:10.1657/1523-0430(06-071)[STCLAIR]2.0.CO;2

St. Clair LL, Leavitt SD (2013) The lichen air quality bio-monitoring program South Unit of the Duchesne Ranger District, Ashely National Forest and detailed elemental analysis of the Cart Creek Bridge and Selected sites along US highway 191. General technical report, Ashley National Forest Supervisor's Office, Vernal, UT, US Department of Agriculture, Forest Service

St. Clair LL, St. Clair SB, Leavitt SD (2013) Review of the lichen air quality bio-monitoring program and baseline for selected sites in the Anaconda-Pintler, Cabinet Mountains, and Selway Bitterroot wilderness areas Montana and Idaho (originally established in July–August 1992, 1993, and 1994); and Establishment and review of lichen air quality bio-monitoring programs and baselines in the Gates of the Mountains, Mission Mountains, and Bob Marshall Wilderness Areas (originally established in 2000, 2001, 2002, and 2003); and Establishment of lichen bio-monitoring air quality programs and baselines for the Gospel Hump and Scapegoat Wilderness Areas, Idaho and Montana Region 1. General technical report, US Northern Regional Office, Missoula, MT, US Department of Agriculture, Forest Service

Stolte KW (2001) Forest health monitoring and forest inventory analysis programs monitor climate change effects in forest ecosystems. Human Ecol Risk Assess: Int J 7(5):1297–1316. doi:10.1080/20018091095014

Sykes MT, Prentice IC, Laarif F (1999) Quantifying the impact of global climate change on potential natural vegetation. Clim Change 41(1):37–52. doi:10.1023/A:1005435831549

Thorpe A, Harrison RM (2008) Sources and properties of non-exhaust particulate matter from road traffic: a review. Sci Total Environ 400(1–3):270–282. doi:10.1016/j.scitotenv.2008.06.007

Ustin SL, Valko PG, Kefauver SC, Santos MJ, Zimpfer JF, Smith SD (2009) Remote sensing of biological soil crust under simulated climate change manipulations in the Mojave Desert. Remote Sens Environ 113(2):317–328. doi:10.1016/j.rse.2008.09.013

Will-Wolf S (2010) Analyzing lichen indicator data in the forest inventory and analysis program U.S. Department of Agriculture, Forest Service, Pacific Northwest Research Station, Portland, OR

Will-Wolf S, Geiser LH, Neitlich P, Reis AH (2006) Forest lichen communities and environment—how consistent are relationships across scales? J Veg Sci 17(2):171–184. doi:10.1111/j.1654-1103.2006.tb02436.x

Zhao Y, Xu M, Belnap J (2010) Potential nitrogen fixation activity of different aged biological soil crusts from rehabilitated grasslands of the hilly Loess Plateau, China. J Arid Environ 74(10):1186–1191. doi:10.1016/j.jaridenv.2010.04.006

Zhu X, Zhang J (2009) Quartered neighbor method: a new distance method for density estimation. Frontiers Biol Chin 4(4):574–578. doi:10.1007/s11515-009-0039-0

Community Metrics Under Air Pollution Stress Condition

6

Pulak Das, Jayashree Rout and Santosh Joshi

Contents

Abstract

The chapter deals with the pattern of changes in the diversity components of lichen communities around a paper mill in northeastern Indian region. Four indices (Shannon index, Pielou index, Margalef index, and Simpson index) are calculated for the seventeen sites situated within a radius of 25 km from the mill. The indices are studied vis-a-vis the distance (from the mill) and elevation. Positive correlation is found between number of species and distance, while a negative correlation is found between numbers of species and the elevation. Changes in the pattern of indices are seen along three distance categories; <10, 10–20, and >20 km. As expected, the dominance index increased (54 %) and the diversity, richness, and evenness index decreased near the mill; maximum decrease being for Shannon index (>14 %) of diversity. It is observed that neither of the indices showed significant correlation with the altitude or distance. The Shannon, Margalef, and Simpson index showed significant correlation with the number of species. The effect of stress on biodiversity is depicted with the help of changes in the shapes of Odum's model of 'concave' curve. It is observed from the graphical representation of the 'number of species' and 'number of individual per species' that the nearness to the mill converted the 'concave' hollow curve into a flattened one and ultimately into a 'convex' curve.

P. Das (✉)
Department of Environmental Science, Indira Gandhi National Tribal University (IGNTU), Amarkantak 484886, Madhya Pradesh, India
e-mail: pulakdas.ecology@gmail.com

J. Rout
Department of Ecology and Environmental Science, Assam University, Silchar 788011, Assam, India

S. Joshi
Lichenology Laboratory, Plant Biodiversity and Conservation Biology Division, National Botanical Research Institute, CSIR, Lucknow 226001, Uttar Pradesh, India

D.K. Upreti et al. (eds.), *Recent Advances in Lichenology*,
DOI 10.1007/978-81-322-2181-4_6

Keywords
Lichen • Diversity indices • Paper mill • Odum's concave curve

6.1 Introduction

An index is usually a count statistic that is obtained in the field and carries information about a population (Wilson et al. 1996a, b). The index is used when individuals of the species in question are difficult to observe and count, or capture and tag, or when a formal abundance estimate is too expensive or time-consuming (Gaines et al. 1999). In ecology, diversity in a sample, or an area sampled, is typically measured as either the number of species or an index of their relative abundance (Magurran 1988). Various indices and models have been developed to measure diversity within a community (Magurran 1988). In general, three main categories of measures are used to assess species diversity: (1) species richness indices, which measure the number of species in a sampling unit, (2) species abundance models, which have been developed to describe the distribution of species abundances, and (3) indices that are based on the proportional abundances of species such as the Shannon and Simpson indices (Magurran 1988). Numerous biodiversity indices (such as Shannon–Wiener and Margalef indices) based on the proportional abundance of species (Wilson et al. 1996a, b) characterize the biological community within an area as a whole (i.e., in terms of averages over the area).

Diversity indices are one of the favorite tools of ecologist world over to see, either the continuity of the naturalness of any environment or the perturbations occurred to it due to any internal or external factor. From last couple of decades, researchers have used special indices of species diversity as a means of indicating perturbations in almost all types of habitats and environments using different organisms (Adams and Barrett 1976; Bulan and Barrett 1971; Karr 1958; Wilham and Dorris 1968). A number of examples of the application of indices to assess diversity of species within a community are mentioned in the literature. Thiollay (1992) used species richness, the Shannon index, evenness index, and the Simpson index to assess the influence of selective logging on bird species diversity. Cushing and Gaines (1989) used the species richness, the Shannon index, and an equitability or evenness index to compare diversity of aquatic insects in three cold desert streams and to evaluate the effects of winter spates on diversity. Walker et al. (2006) in a study used species richness, Pielou evenness index, Shannon index, etc., of plant communities including lichens across the Tundra Biome to study the trend of climate warming.

Lichens, the unique and peculiar organism, are one of the most reliable and used worldwide for monitoring studies. High or low lichen diversity can result from various factors such as certain types of air pollution, changes in forest management or stand structure, diversity of plant substrates available for colonization, climate favorability, periodicity of fire, and so on (Jovan 2008). Lichens are sensitive to changes in atmosphere and micro-climate conditions, including changes in forest structure, and have long been used as environmental bioindicators for air pollution and environmental changes in tropical regions (Wolseley and Aguirre-Hudson 1997).

The present research work as a case study is an attempt to understand the species structure of the lichen communities by studying the different mathematical indices around a paper mill in northeastern region of India. It is presumed that the study around a potential pollution source is more likely to produce a differentiated pattern of the community structure. A comparative assessment of the indices was made to evaluate the effectiveness of the tool.

6.2 Materials and Methods

6.2.1 The Study Area

The study was conducted around a paper mill situated in Barak Valley (consists of three districts of Cachar, Hailakandi, and Karimganj, Fig. 6.1) in northeastern region of India. The paper mill (Latitude 24° 52′N and Longitude 92° 36′E) which was set up in the year 1973–1974 has a manufacturing capacity of 100,000 mega tones/year of writing and printing paper. The bamboo resource which is abundant in this region is used as a raw material. The Barak Valley is a swampy plain interspersed with low hills. The valley experiences subtropical climate with high humidity during the rainy seasons. The temperature drops to 6°–8° during the winter and reaches a maximum of 35°–38° during summer. The wind generally blows from the northeast in the morning and from the southeast in the afternoon. The yearly rainfall (mm) ranges from around 3,000 to as high as over 4,000.

6.2.2 Sampling Sites and Strategy

The sampling sites (seventeen) selected for the study are radially distributed (Fig. 6.1) around the mill up to a maximum distance of 24.1 km and minimum distance of 2.4 km. Out of all the sites, seven sites (Site nos.: 12, 11, 13, 8, 6, 14, and 4) are situated within 10 km distance from the mill, six sites (Site nos.: 7, 9, 3, 2, 1, and 5) are situated at a distance between 10 and 20 km from the mill, and four sites (Site nos.: 10, 17, 15, and 16) are situated beyond 20 km. The air

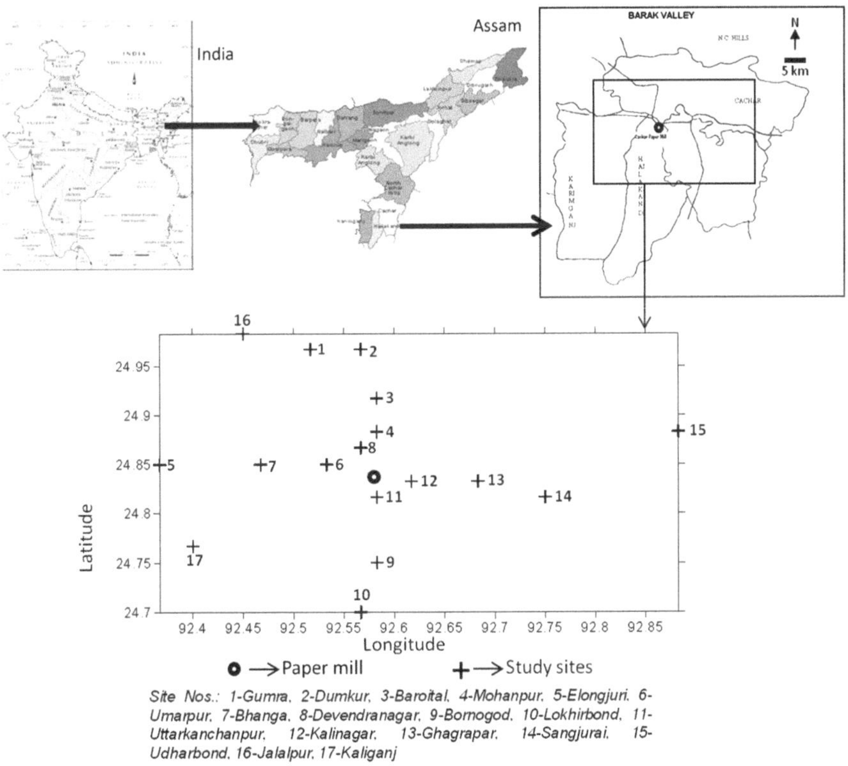

Fig. 6.1 Map of Barak Valley (consisting three districts of Cachar, Hailakandi, and Karimganj), showing the paper mill and seventeen sampling sites spread around it

quality measured during the year 2002–2005 inside the paper mill revealed that the concentration (μg/m^3) of SPM (Gradient: 0.011) and SO_2 (Gradient: 0.001) showed an increasing trend, while the concentration of NO (Gradient: −0.038) showed a decreasing trend. Site nos. 5 and 14 are situated adjacent to urban area (Karimganj and Silchar town, respectively) with moderate to heavy vehicular activities. Although the former is more close to the town, Site No. 15 is situated in an area amidst a dozen of stone crushers. Site No. 16 is comparatively an undisturbed area away from urban influence and vehicular activities. All the other sites have more or less similar characteristics. A single host tree species (*Artocarpus heterophyllus*) was selected to avoid the differences in species structure pattern due to the differences in bark characteristics. At each site, five numbers of host trees are selected with more or less similar height and girth measurements, and conditions of light, temperature, and humidity. Lichens are collected during the year 2005, up to a height of two meters from the base. All the other required information for the calculation of the indices was collected.

6.2.3 Calculation of Indices of Species Structure

The indices of species structure (Odum 1996) or single-number indices (Costello et al. 2001) used are as follows:

Shannon Diversity Index (*H*)

$$H = -\sum(\mathrm{ni}/N)\log(\mathrm{ni}/N)$$
or
$$-\sum(\mathrm{Pi})\log(\mathrm{Pi})$$

where

ni importance value for each species
N total of importance values
Pi importance probability for each species.

Margalef index of species richness (*d*)

$$d = (S-1)/\log N$$

where

S number of species
N number of individuals.

Pielou evenness index (*e*)

$$e = H/\log S$$

where

H Shannon index
S number of species.

Simpson index of dominance (*c*)

$$c = \sum(\mathrm{ni}/N)^2$$

where

ni importance value for each species
N total of importance value.

The Shannon–Wiener index gives equal weighting to rare and common species, while Simpson's dominance index gives more weight to common species and less weight to rare species (Magurran 1988). Thus, using both the Shannon–Wiener index and Simpson's dominance index provides measures of species heterogeneity. Heterogeneity is greater for communities with more species that are equally abundant compared to communities with fewer species that are not equally abundant. O'Neill et al. (1988) and Turner (1990) proposed indices of dominance (along with some other indices) as standard measures of landscape pattern. According to Gaines et al. (1999), dominance is an index of vegetation-type composition and equitability. Evenness was calculated to provide an estimate of the equitability of species abundances within each community.

6.2.4 Graphical Representation of the Species and Species Abundance

The general relationship between the organisms and the habitat conditions is demonstrated using the shape of the curves (Odum 1996) plotted against number of species and number of individual per species in *y*-axis and *x*-axis, respectively. According to Odum (1996), the hollowness or the 'concavity' of the curve depicts the trend of natural environment, but the stress factor tends to flatten the curve.

6.3 Result and Discussion

The seventeen sampling sites (Fig. 6.1) located around the paper mill are grouped into three categories; <10, 10–20, and >20 km. The numbers of sites in each category are seven, six, and four, respectively (Table 6.1). The altitude of sampling sites ranged from 17.1 m at site number 5 to 27.1 m at site number 15. The study revealed 53 species distributed among 13 families and 23 genera. The dominant family is observed to be the Graphidaceae consisting of 22 species followed by Pyrenulaceae with 17 species and Arthoniaceae with 11 species. Out of 23 genera, *Graphis* dominates with 10 species followed by *Pyrenula* and *Arthonia* with 5 species each. Out of the two growth forms (crustose and foliose), crustose consists of around 91 % of the lichen flora, whereas remaining is contributed by the foliose growth form. The foliose growth form is contributed by three families; Physciaceae (60 %), Parmeliaceae, and Biatoraceae (20 % each) consisting of five different genera (*Dirinaria*, *Parmotrema*, *Physcia*, *Phyllopsora*, and *Pyxine*).

Table 6.1 Indices of diversity and dominance (Shannon index of general diversity *H*, Margalef index of species richness *d*, Pielou index of evenness *e*, and Simpson's index of dominance) along with the number of species, distance, and altitude at each site around the paper mill

Site No.	Distance (km)	No. of species (S)	Altitude (m)	*H*	*d*	*e*	c
12	2.4	5	17.4	0.559	2.11	0.800	0.325
11	4.8	11	19.2	0.779	4.31	0.748	0.247
13	5.2	10	20.7	0.711	3.89	0.711	0.308
8	5.6	10	22.6	0.884	5.20	0.884	0.160
6	7.2	10	20.7	0.832	4.39	0.832	0.190
14	8.4	24	19.5	1.033	8.61	0.749	0.177
4	8.8	18	21.6	1.065	7.42	0.848	0.128
7	11.3	12	25.3	0.991	5.12	0.918	0.121
9	12.1	18	22	1.053	6.71	0.839	0.109
3	13.7	20	18.3	1.199	9.61	0.922	0.078
2	16.1	13	19.5	0.977	5.82	0.877	0.127
1	16.9	12	21	0.861	5.54	0.798	0.196
5	19.3	11	17.1	0.799	4.50	0.768	0.234
10	20.1	15	25.6	1.063	6.43	0.904	0.106
17	20.1	16	21.3	1.051	5.97	0.873	0.115
15	23.3	9	27.1	0.779	4.18	0.816	0.233
16	24.1	17	19.2	1.037	5.42	0.842	0.114

Table 6.2 Mean values and standard deviation of the indices of species structure in three distance categories

Distance category (km)	*N*	*H*	*d*	*e*	c	*S*
<10	7	0.838 ± 0.178	5.13 ± 2.21	0.796 ± 0.063	0.219 ± 0.076	12.57 ± 6.32
10–20	6	0.980 ± 0.142	6.22 ± 1.82	0.853 ± 0.063	0.853 ± 0.063	14.33 ± 3.72
>20	4	0.982 ± 0.137	5.50 ± 0.97	0.859 ± 0.038	0.142 ± 0.061	14.25 ± 3.59
From <10 to >20	17	**0.922 ± 0.164**	**5.60 ± 1.81**	**0.831 ± 0.063**	**0.175 ± 0.073**	**13.58 ± 4.74**

Bold indicates mean values and standard deviation of indices of species structure for the entire study area (<10 to >20)

The number of epiphytic lichens ranged (Tables 6.1 and 6.2) from 5 at Site nos. 12–24 at Site No. 14 with an average of 13.6. The numbers ranged from 5 to 24 within first 10 km from the mill with an average of 12.6, 11–20 within 10–20 km category with an average of 14.3, and from 9 to 17 beyond 20 km with an average of 14.3. The number of species/site showed (Fig. 6.2) an increasing trend along with the increase in distance from the mill. The number of species is showing a decreasing trend along with the increase in altitude.

The Shannon index of diversity (*H*) in the study area ranged from 0.559 to 1.199 with a mean of 0.922. Within different distance categories, its value ranged from 0.559 to 1.065 (Mean: 0.838) within 10 km category, from 0.799 to 1.199 (Mean: 0.980) (10–20 km category), and from 0.779 to 1.063 (Mean: 0.982) (>20 km category).

Index of species richness (*d*) of lichen community in the study area ranged from 2.108 to 9.607 with a mean of 5.60. The *d* value ranged from 2.108 to 8.607 (Mean: 5.13) in <10 km category, from 5.118 to 9.607 (Mean: 6.22) in 10–20 km category, and from 4.180 to 6.434 (Mean: 5.50) in >20 km category.

The mean value of evenness index (*e*) in the study area was observed to be 0.831 and it ranged from 0.711 to 0.922. Within the three different distance categories, the *e* value ranged from 0.711 to 0.884 (Mean: 0.796) in <10 km category, from 0.768 to 0.922 (Mean: 0.853) in 10–20 km category, and from 0.816 to 0.904 (Mean: 0.859) in >20 km category. The Simpson dominance index (*c*) ranged from 0.078 to 0.325

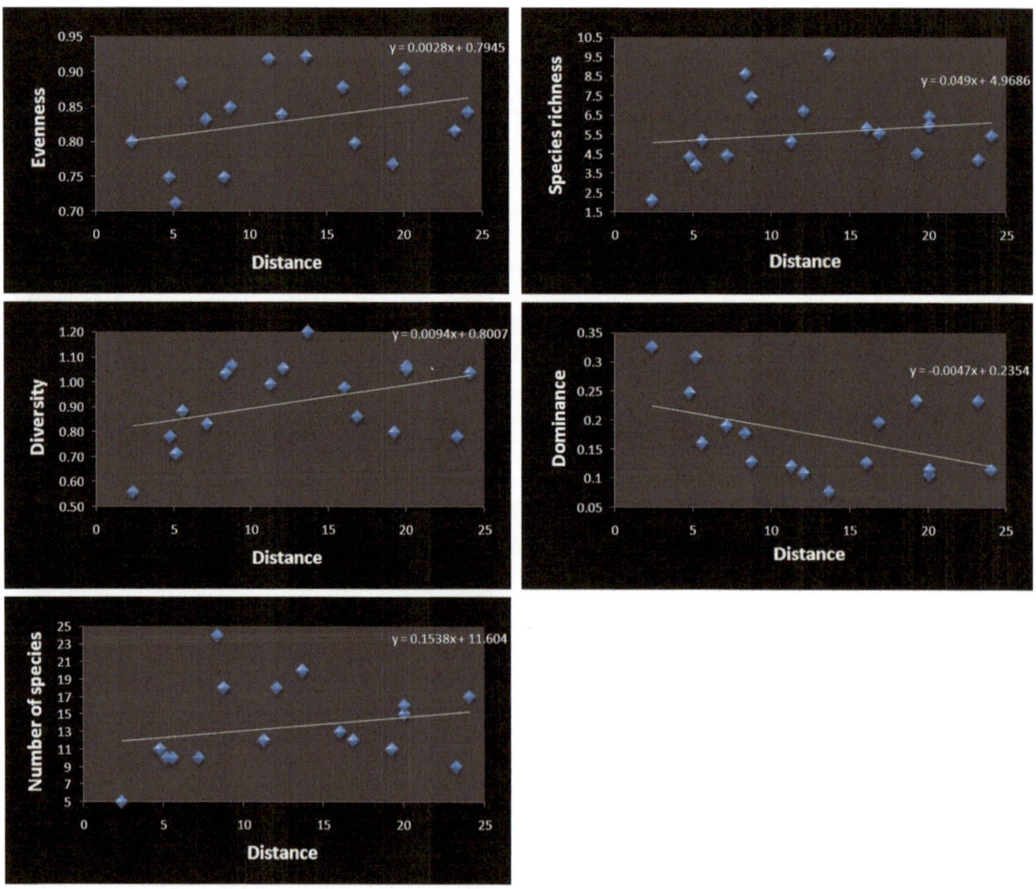

Fig. 6.2 Trend lines of different indices along the distance from the paper mill

with a mean value of 0.175 around the study area. The '*c*' value ranged from 0.128 to 0.325 (Mean: 0.219), from 0.078 to 0.234 (Mean: 0.144), and from 0.106 to 0.233 (Mean: 0.142) at three different distance categories of >10, 10–20, and >20 km category, respectively.

The three indices, diversity (*H*), evenness (*e*), and richness (*d*), showed (Table 6.3) a positive increasing trend with the increase in distance. With the increase in elevation, '*H*' and '*e*' showed an increasing trend, but the '*d*' value showed a decreasing trend. The number of lichen species showed an increasing trend with the distance and a decreasing trend along with the altitude (Table 6.3).

Relationship between the number of species and number of individuals/species around the paper mill is shown by using scatter plot diagram (Fig. 6.3a) as described by Odum (1996). Three 'curves' are obtained for three different distance categories. The curves are parabolic in nature with the axis of parabola being the *y*-axis (number of species).The thicker solid curve ($y = 0.009x^2 - 0.2576x + 7.2636$) is the trend line for category >20 km category. The thinner solid curve ($y = -0.019x^2 - 0.4214x + 4.2323$) and the thinner dotted one ($y = -0.0051x^2 - 0.1691x + 3.9163$) are the curves for 10–20 and <10 km category, respectively (Fig. 6.3a). Both the curves are convex in nature.

Diversity, richness, evenness, and abundance also called as community metrices (Therriault 2002) or single-number indices or heterogeneity indices or univariate indices can be the useful tools for summarizing patterns in biodiversity. These biodiversity indices, which are the functions of the number of species and the number of individuals in each species, reflect the effect of habitat conditions on them. Both spatial and temporal changes are of significance to understand the local variability in the habitat conditions.

In the present study, four indices, Shannon, Pielou, Simpson, and Margalef indices, are studied with respect to the changes in distance from the paper mill. The categorization of the distance into three types helped in more detailed analysis of the above-mentioned parameters (Table 6.2). The Shannon, Pielou, and Simpson indices revealed a continuous gradient with increasing distance. The Shannon diversity index increased from 0.838 at <10 km distance to 0.980 at 10–20 km and to 0.982 at >20 km distance with an overall increase of about 17 %. The evenness index increased about 7 %. The Simpson index decreased by about 34 % from <10 km distance category to >20 km category. The Margalef index increased from near to the remote distance. A similar correlation was noted at a coal-fired power plant where variation in distances (11–24 km) from the pollution source resulted in a decrease in the number of species, Margalef richness, and Shannon diversity by 33, 27, and 1 %, respectively, and an increase in evenness index and dominance index by 22 and 20 %, respectively, in Southeastern Ohio and West Virginia (Stefan and Rudolph 1979). Compared to this, in the present study, change in distance from >20 to <10 km reduced the number of species, species richness, Shannon diversity

Table 6.3 Correlations (*r*)

	H	*e*	*d*	*c*	*S*	*D*	*E*
H	1	**0.630**	**0.896**	**−0.951**	**0.858**	0.396	0.161
e	**0.630**	1	0.381	**−0.791**	0.194	0.313	0.375
d	**0.896**	0.381	1	**−0.751**	**0.922**	0.187	−0.006
c	**−0.951**	**−0.791**	**−0.751**	1	**−0.688**	−0.444	−0.251
S	**0.858**	0.194	**0.922**	**−0.688**	1	0.224	−0.077
D	0.396	0.313	0.187	−0.444	0.224	1	0.283
E	0.161	0.375	−0.006	−0.251	−0.077	0.283	1

Bold values are significant at $P < 0.05$

H Shannon index, *e* Pielou index, *d* Margalef index, *c* Simpson index, *S* number of species, *D* distance from paper mill, *E* elevation index, and evenness index by 12, 7, 15, and 7 %, respectively, and an increase in dominance index by 54 %. The trends of the four indices and the number of species at different distances from the mill are quite interesting (Fig. 6.2). Except the Simpson index, all the other showed an increasing trend away from the paper mill. The highest gradient was observed for the number of species (0.1538), followed by Margalef index (0.049), Shannon index (0.0094), and Pielou index (0.0028). Simpson index has a gradient of −0.0047.

Researchers have studied the changes in such parameters considering different organisms and in different stress conditions. Barrette (1969) found a decrease in species richness while an increase in evenness of arthropod population nearer to the environmental stress (acute insecticidal effect). A decrease in the number of species and Shannon value was observed for the nematodes due to increase in the Pb content in their habitat (Zullini and Peretti 1986). Wilhm (1967) demonstrated the changes in the Shannon value of the benthos downstream of a pollution outfall. Junshum et al. (2008) have shown the changes in the Shannon index value of algal diversity in water bodies in power plant. Wilhm (1975) has even classified the condition of water on the basis of Shannon index. Washburn and Cully (2006) found significant differences in the indices between various urban and non-urban sites in Ohio. Although some workers found little correlation between the H value and the extent of perturbation (Stefan and Rudolph 1979; Boyle et al. 1990), Therriault (2002) found only temporal differences within the community metrices and not due to differences in habitat conditions. Godfrey (1978) believes that the large evenness component within a community may mislead the interpretation of H value.

The H, d, and c value showed highly significant correlation ($P < 0.05$) with the number of species (Table 6.3). Evenness (e) does not exhibit a significant correlation with the number of species. Giordano et al. (2004) in a study found that the biodiversity is highly correlated with the total number of species ($r = 0.88$, $P < 0.05$). Most univariate indices and mathematical models, however, are designed to show greater diversity when abundance is more evenly distributed among the species sampled and when the number of species in a sample increases (Costello et al. 2001). In the present study, no significant correlation was found between evenness and richness (r = 0.381) although correlation between diversity and evenness (r = 0.63) and diversity and richness (r = 0.896) was statistically significant. Richness and evenness are not necessarily correlated always (Edgar 1983; Costello and Myers 1987; Mackey and Currie 2001), but richness and species diversity are often positively correlated with some exceptions (Hurlbert 1971).

The number of species showed negative correlation ($r = -0.077$, $P > 0.05$) with the elevation, while the Shannon index exhibit positive correlation ($r = 0.161$, $P > 0.05$) with the elevation. Earlier workers have found both positive (Insarov et al. 1999) and negative (Jovan and McCune 2006) correlations of number of species and diversity with the elevation. Unimodel relationship of Shannon diversity with the altitude has been studied in a lichen biodiversity study in northeast India (Pinokiyo et al. 2008). A positive correlation of number of species ($r = 0.224$, $P > 0.05$) and Shannon diversity ($r = 0.396$, $P > 0.05$) with the distance is observed in the present study. This, however, may not be similar for everywhere (Stefan and Rudolph 1979).

Present study encompassing the region around the paper mill with a maximum distance of over 20 km revealed a variation of different indices (Table 6.1). The Shannon index showed a small range from 0.559 to 1.199, Pielou index ranged from 0.711 to 0.922, the Margalef index exhibited a high range from 2.108 to 9.607, Simpson index ranged from 0.078 to 0.325, and the number of species ranged from 5 to 24. In a study conducted on epiphytic lichens of some European countries, Stofer (2006) found a varied range and differences of the indices within countries. For instance, Shannon index ranged from 1.49 to 3.1 in Switzerland, from 0.28 to 2.4 in Germany, and from 0.99 to 3.26 in Italy. Simpson index ranged from 0.06 to 0.33 in Switzerland, from 0.14 to 1 in Germany, and from 0.05 to 0.43 in Italy. Pielou index ranged from 0.57 to 0.88 in Switzerland, from 0.2 to 0.89 in Germany, and

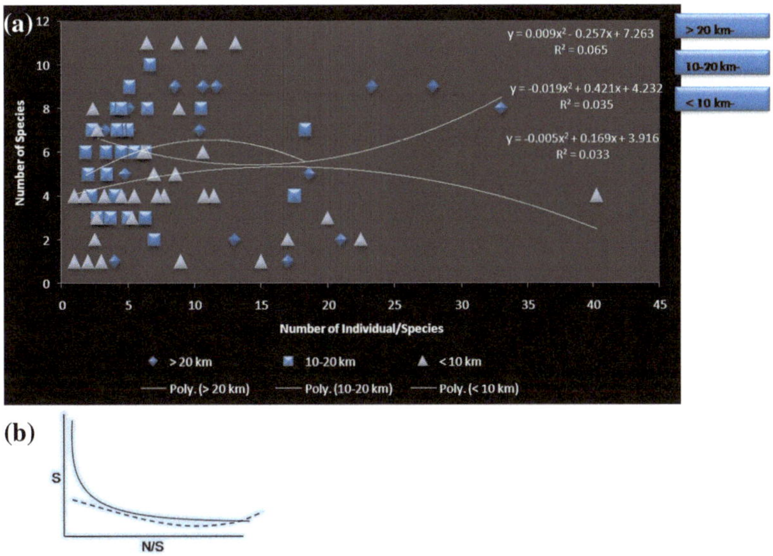

Fig. 6.3 Relationship between species (S) and number of individual per species (N/S) of lichens at three distance categories around the paper mill (**a**), from Odum (1996) (**b**)

from 0.64 to 0.87 in Italy. The number of species per plot (12 trees) ranged from 9 to 40 in Switzerland, from 1 to 24 in Germany, and from 4 to 47 in Italy.

Nearness to the pollution, source (paper mill) decreases the evenness and increases the dominance (Table 6.3; Fig. 6.2). Low evenness index value reveals dominance of opportunistic species or a species that can adapt better to the present ecological condition (Ramaharitra 2006) which also indicates a change in the dominance structure of the plant communities where fewer species produce a larger proportion of the cover. This response may represent a first step toward local species extinction (Walker et al. 2006).

Most natural communities contain few species with a large number of individuals (dominant species), and many species represented by a few individuals (rare species) (Odum 1996). The relationship can be pictured as a 'concave' curve (solid line, Fig. 6.3b). Rigorous physical environment (e.g., pollution) tends to flatten the curve as shown by the dotted line. To understand the concept more clearly, the distance from the pollution source in the study area has been divided into three categories (<10, 10–20, and >20 km) and the relationship between the number of species and number of individuals per species has been plotted (Fig. 6.3a). The curves do not reveal the characteristics of the natural communities. The curve for the third category (>20 km) is flattened (similar to the dotted curve of Fig. 6.3b), as the rare species (with less number of individuals) declines, indicating stress. The curves for the first two categories (<10 and 10–20 km) are quite interesting; it has gone beyond flattening assuming a 'convex' form. To be more specific, all the three curves are parabolic in nature having quadratic equation of the form '$ax^2 + bx + c$.' The 'a' value (indicating direction of the parabola) is positive (0.009) for the >20 km category, while it is negative and for 10–20 km category (−0.019) and for <10 km category (−0.005). In comparison with >20 km category, further declination of the rare species is observed in the remaining two categories. This decline is more pronounced in <10 km category as compared to the 10–20 km category.

The pattern of the community exhibits few number of both common/abundance and rare individuals species possessing large (common/abundant) and less (rare), while the species with moderate number of individuals are many. The stress (air pollution) has reduced the rare species and also the abundant species in both 'convex' curves in comparison with the '>20 km category' curve.

Graphis inamoena, *G. persulcata*, *G. chloroalba*, *Sarcographa subtricosa*, *Phaeographis dendritica*, *P. albolabiata*, and *Pyrenula fusco-olivaceae* are the abundant lichen species in

>20 km category. Some of the rare species in the same category are *Sarcographa leprieurii, Graphis assamensis, Arthothelium nigrodiscum, A. abnorme, Graphis subassimilis, G. scripta, G. persulcata, Arthonia arctata, Arthopyrenia indusiata, Chiodecton leptosporum, Bacidia inundata, Pyrenula costaricensis, Trypethelium albopruinosum, Dirinaria aegialita,* and *Thecaria austroindica*. In 10–20 km category, the abundant species are *S. subtricosa, P. fuscoolivaceae, D. aegialita, Cryptothecia lunulata, G. chloroalba*, etc., and the rare species are *A. indusiata, C. lunulata, G. inamoena, Arthothelium chiodectoides, Lecanora achroa, Graphis capillaceae, Platygramme pudica, Graphis nigroglauca, Porina subhibernica,* etc.

In the distance category nearest to the mill (<10 km), the abundant species are *D. aegialita, G. capillaceae*, and *Graphis subasahinae* and the rare species are *Trypethelium macrosporum, T. albopruinosum, P. subhibernica, Parmotrema saccatilobum, Arthonia catenatula, A. recedens, P. dendritica, A. chiodectoides,* and *Platygramme wattiana*.

It is interesting to note that in comparison with the species of >20 km category, the general abundance is greater (although not of the same species) in <10 km category. The species which contributed in the lowering of right arm (abundant species) of the curve from >20 km category to <10 km category are *A. indusiata, G. inamoena, P. albolabiata, P. dendritica, Pyrenula brunneae, Pyxine cocoes, and S. subtricosa* which are highly abundant in the former category, but its number reduced drastically in the latter.

Some of the rare species which are present in the >20 km category but are not found in <10 km category are *A. abnorme, A. nigrodiscum, B. inundata, C. leptosporum, G. assamensis, G. scripta, G. subassimilis, Physcia dilatata, Pyrenula oxysporiza, S. leprieurii, T. austroindica*, and *Trypethelium endosulphurium*. The rare species can be regarded to be locally extinct in the vicinity of the paper mill. Faith and Norris (1989) suggested that rare and common species respond differently to different environmental variables with rare species being more sensitive to changes in habitat condition. The species sensitivity depends on physiology, while abundant species may have greater tolerance for environmental change compared to rare species, a condition that allows them to become habitat generalists by dominating communities.

6.4 Conclusion

The diversity indices can be used as an effective tool to monitor changes due to the effect of pollution on the living organism around any pollution source. The result can be correlated to the result found by Stefan and Rudolph (1979), Barrette (1969), Zullini and Peretti (1986), Wilhm (1967), and Junshum et al. (2008). In the present study, it was found that Shannon diversity, Pielou evenness, and Margalef richness showed a reduction, while Simpson dominance index exhibits an increase as one approaches near the paper mill. Further, it is observed that the graph between number of species and number of individual per species (Odum 1996) precisely describes the effect of stressed environment on the richness and dominance factor of lichen community through the changes in the shape of the graph. The study also describes the lichen diversity pattern around a paper mill using four commonly used diversity indices. Application of Odum's graphical way of depicting diversity helps in understanding the effect of stress on the lichens and changes in lichens preferences to different levels of stress.

Acknowledgments The authors are grateful to the Director, National Botanical Research Institute (NBRI) and Head, Department of Ecology and Environmental Science, Assam University, Silchar for providing necessary laboratory and other facilities.

References

Adams DL, Barrett GW (1976) Stress effects on bird-species diversity within mature forest. Am Midl Nat 96:179–194

Barrett GW (1969) The effects of an acute insecticide stress on a semienclosed grassland ecosystem. Ecology 49:1019–1035

Boyle TC, Smillie GM, Anderson JC, Beeson JR (1990) A sensitivity analysis of nine diversity and seven similarity indices. J Water Pollut Control Fed 62:749–762

Bulan CA, Barrett GW (1971) The effects of two acute stresses on the arthropod component of an experimental grassland ecosystem. Ecology 52:597–605

Costello MJ, Myers AA (1987) Amphipod fauna of the sponges *Halichondria panicea* and *Hymeniacidon perleve* in Lough Hyne, Ireland. Mar Ecol Prog Ser 41:115–121

Costello MJ, Pohle G, Martin A (2001) Evaluation biodiversity in marine environmental assessments. Canadian Environmental Assessment Agency's research and development program, research and development monograph series

Cushing CE, Gaines WL (1989) Thoughts on the recolonization of endorheic cold desert spring-streams. J North Am Benthological Soc 8(3):277–287

Edgar GA (1983) The ecology of south-east Tasmanian phytal communities. J Exp Mar Biol Ecol 70:181–203

Faith DP and Norris RH (1989) Correlation of environmental variables with patterns of distribution and abundance of common and rare freshwater macroinvertebrates. Biol Conserv 50:77–98.

Gaines WL, Harrod RJ, Lehmkuhl JF (1999). Monitoring biodiversity: quantification and interpretation. General technical report. PNW-GTR-443. US Department of Agriculture, Forest Service, Pacific Northwest Research Station, Portland, OR

Giordano S, Sorbo S, Adamo P, Basile A, Spagnuolo V, Cobianchi CR (2004) Biodiversity and trace element content of epiphytic bryophytes in urban and extraurban sites of southern Italy. Plant Ecol 170:1–14

Godfrey PJ (1978) Diversity as a measure of benthic macroinvertebrates community response to water pollution. Hydrobiologia 57:111–122

Hurlbert SH (1971) The nonconcept of species diversity: a critique and alternative parameters. Ecology 52 (4):577–586

Insarov GE, Semenov SM, Insarova ID (1999) A system to monitor climate change with epilithic lichens. Environ Monit Assess 55:279–298

Jovan S (2008) Lichen bioindication of biodiversity, air quality, and climate: baseline results from monitoring in Washington, Oregon, and California. General technical report. PNW-GTR-737. US Department of Agriculture, Forest Service, Pacific Northwest Research Station, Portland, OR

Jovan S, McCune B (2006) Using epiphytic macrolichen communities for biomonitoring ammonia in forests of the greater Sierra Nevada, California. Water Air Soil Pollut 170:69–93

Junshum P, Somporn C, Traichaiyaporn S (2008) Biological indices for classification of water quality around Mae Moh power plant, Thailand, Maejo. Int J Sci Technol 2(01):24–36

Karr JR (1958) Habitat and avian diversity on strip mined land in east central Illinois. Condor 70:348–357

Mackey RL, Currie DJ (2001) The diversity-disturbance relationship: is it generally strong and peaked? Ecology 82(12):3479–3492

Magurran AE (1988) Ecological diversity and its measurement. Croom Helm, London

O'Neill RV, Krummel JR, Gardner RH et al (1988) Indices of landscape pattern. Landscape Ecol 1:153–166

Odum EP (1996) Fundamentals of ecology, 1st edn. Natraj Publishers, Dehradun

Pinokiyo A, Singh KP, Singh JS (2008) Diversity and distribution of lichens in relation to altitude within a protected biodiversity hot spot, north-east India. Lichenologist 40(1):47–62

Ramaharitra T (2006) The effects of anthropogenic disturbances on the structure and composition of rain forest vegetation. Trop Resour Bull 25:32–37

Stefan MB, Rudolph ED (1979) Terrestrial bryophytes as indicators of air quality in southeastern Ohio and adjacent West Virginia. Ohio Acad Sci Ohio Journal of Science 79(5):204

Stofer S (2006) Epiphytic Lichen Monitoring, Swiss Federal Research Institute WSL, CH-8903 Birmensdorf, Switzerland working report forest BIOTA

Therriault TW (2002) Temporal patterns of diversity, abundance and evenness for invertebrate communities from coastal freshwater and brackish water rock pools. Aquat Ecol 36:529–540

Thiollay Jean-Marc (1992) Influence of selective logging on bird species diversity in a Guianan rain forest. Conserv Biol 6(1):47–63

Turner MG (1990) Spatial and temporal analysis of landscape pattern. Landscape Ecol 4:21–30

Walker MD, Wahren CH, Hollister RD, Henry GHR, Ahlquist LE, Alatalo JM, Bret-Harte MS, Calef MP, Callaghan TV, Carroll AB, Epstein HE, Jónsdóttir IS, Klein JA, Magnússon B, Molau U, Oberbauer SF, Rewa SP, Robinson CH, Shaver GR, Suding KN, Thompson CC, Tolvanen A, Totland Ø, Turner PL, Tweedie CE, Webber PJ, Wookey PA (2006) Plant community responses to experimental warming across the tundra biome. In: Proceedings of the national academy of sciences USA, vol 103(5), pp 1342–1346, 31 Jan 2006

Washburn SJ, Culley TM (2006) Epiphytic macrolichens of the greater Cincinnati metropolitan area-part II: distribution, diversity and urban ecology. Bryologist 109:516–526

Wilhm JL (1967) Comparison of some diversity indices applied to populations of benthic macro invertebrates in a stream receiving organic wastes. J Water Pollut 39:1673–1683

Wilhm JL (1975) Biology indicators of pollution. In: Whitton BA (ed) Studies in ecology. River ecology, vol 2. Black Well Scientific Publications, London, pp 375–402

Wilhm JL, Dorris TC (1968) Biological parameters for water quality criteria. Bioscience 18:477–481

Wilson JB, Wells TCE, Trueman IC, Jones G, Atkinson MD, Crawley MJ, Dodd ME, Silvertown J (1996a) Are there assembly rules for plant species abundance? An investigation in relation to soil resources and successional trends. J Ecol 84:527–538

Wilson DE, Cole FR, Nichols JD et al (1996b) Measuring and monitoring biological diversity: standard methods for mammals. Smithsonian Institution Press, Washington, DC

Wolseley PA, Aguirre-Hudson B (1997) The ecology and distribution of lichens in tropical deciduous and evergreen forests of northern Thailand. J Biogeogr 24:327–343

Zullini A, Peretti E (1986) Lead Pollution and Moss-inhabiting Nematodes of an Industrial area. Water Air Soil Pollut 27:403–410

Remote Sensing and GIS for Biodiversity Conservation

7

Nupoor Prasad, Manoj Semwal and P.S. Roy

Contents

N. Prasad (✉)
GIS Consultant, JPS Consultants Pvt. Ltd., R-16, Hauz Khas Enclave, New Delhi 110016, India
e-mail: nupoor.prasad@gmail.com

M. Semwal
ICT Department, CSIR—Central Institute of Medicinal and Aromatic Plants, Lucknow, India
e-mail: m.semwal@cimap.res.in

P.S. Roy
Center for Earth and Space Sciences, University of Hyderabad, Hyderabad, India
e-mail: psroy13@gmail.com

D.K. Upreti et al. (eds.), *Recent Advances in Lichenology*, DOI 10.1007/978-81-322-2181-4_7

Abstract

This chapter focuses on the application of space-borne remote sensing and GIS for biodiversity conservation in the context of the state-of-the-art technology which has enhanced the classical approach. It reviews currently available instruments, i.e. space-borne or satellite sensors providing data which can be used without analysis or interpretation for studying individual organisms, species assemblages or ecological communities on ground. Subsequently, the image processing and GIS techniques developed to derive information from the captured satellite data are reviewed, and finally, this chapter concludes by reviewing the use of remote sensing and GIS techniques for mapping, monitoring and modelling lichens and their habitats.

Keywords

Remote sensing · GIS · Space-borne sensors · Biodiversity mapping · Lichen species richness

7.1 Introduction

Biogeographers are continuously researching methods to map species distributions and diversity that can have significant applications for conservation planning (Foody 2003; Whittaker et al. 2005). Exhaustive ground surveys and inventories of species in field are the precise ways to accumulate biogeographical data on species distributions. However, such ground-based methods require skilled individuals and significant amount of time in the field, making them highly expensive. Also, even for relatively well-studied areas, different field data sources can lead to dissimilar or biased results of species distributions and diversity (Graham and Hijmans 2006; Moerman and Estabrook 2006; Pautasso and McKinney 2007); in areas such as the tropics, species occurrence and distribution data are relatively coarse and not well collected (Phillips et al. 2003; Schulman et al. 2007), and in areas with high biodiversity but inaccessible terrain, surveying for mapping vegetation and other forest cover using conventional techniques is complex and time-consuming (Lele and Joshi 2009). Moreover, the traditional field ecological data do not translate readily to regional or global extents, and models derived purely from such local data are unlikely to predict the global consequences of human activities. Thus, ecologists and conservation biologists are turning to remote sensing techniques and data sources for mapping species diversity and for preparing scientific responses to environmental change such as management techniques and various decision-making (Kerr and Ostrovsky 2003).

Given the difficulties associated with field-based data collection, use of remote sensing for estimating environmental heterogeneity and species diversity is appropriate since it provides a synoptic view of an area with a high temporal resolution (Loarie et al. 2008), offering an inexpensive way of deriving complete spatial coverage of environmental information for large areas in a consistent manner that may be updated regularly (Muldavin et al. 2001; Duro et al. 2007). Because of the potential capacity for systematic observations at various scales, remote sensing technology extends possible data archives from present time to over several decades back (Xie et al. 2008). Also, it is not as labour-intensive and time-consuming as the field-observation-based studies. Remote sensing has considerable potential as a source of information on biodiversity at landscape, regional, continental and global spatial scales (Nagendra 2001; Willis and Whittaker 2002; Turner et al. 2003).

Satellite remote sensing's process of obtaining information about earth from space and then effectively processing it (collection, storage, analysis and dissemination) has swept the fields of biodiversity conservation and its management in recent years (Table 7.1). The latest advances in RS technologies strengthen its use in capturing/measurement of biodiversity on earth and in addressing in situ and ex situ biodiversity

Table 7.1 Advantages and disadvantages of remote sensing as compared to traditional in situ methods

Advantages	Disadvantages
Provides a continuous, repetitive, large-scale synoptic view relative to traditional point-based field measurements	Instruments used in satellite remote sensing are expensive to build and operate
Practical way to obtain data from dangerous or inaccessible areas	Remote sensing data are not direct samples of a phenomenon and so must be calibrated against reality
Relatively cheap and fast method of information acquisition over large geographical areas	Remote sensing data need to be geometrically corrected and georeferenced in order to be useful as maps, which is an intricate process
Easy to analyse with the computer and combine with other geographical data	Remote sensing data interpretation needs subject experts having knowledge of the phenomena being studied
Data acquisition from present time to over several decades back	Most sensors can monitor only features that can be viewed from above; characteristics of the understory must be inferred rather than directly observed

conservation efforts by adding value to the detection of species, ecological communities and patterns of species richness.

This chapter focuses on the space-borne remote sensing in the context of state-of-the-art technology, including instruments and techniques. The following section of this chapter reviews currently available instruments, i.e. space-borne or satellite sensors providing data that can be used without analysis or interpretation for studying individual organisms, species assemblages or ecological communities on ground. Subsequently, the image processing and GIS techniques developed to derive information from the captured satellite data are reviewed, and finally, this chapter reviews the use of remote sensing and GIS techniques for mapping, monitoring and modelling lichens and their habitats.

7.2 Sensors (Passive or Active) Used to Assess and Model Biodiversity

There has been a remarkable increase in earth observation satellites and sensors over the last few years that are being used to measure and model biodiversity from space (Table 7.2) (Gillespie et al. 2008). Advances in the resolutions of space-borne sensors available for ecologists are making it increasingly possible to distinguish species assemblages or identification of species of individual trees (Turner et al. 2003).

Passive sensors, which record reflected (visible and infrared wavelengths) and emitted energy (thermal wavelengths), are most frequently used in biodiversity studies. The highest spatial resolution data come from commercial satellites such as WorldView, QuickBird and IKONOS which contain visible and infrared bands used in species mapping. The NASA Landsat series is the most widely used sensor for biodiversity studies due to the ease in which the data can be obtained, long time series and low cost. MODIS and AVHRR sensors have provided extremely useful data for regional, continental and global studies of land-cover classification and diversity models. These sensors also provide data on temperature, precipitation and fire that have been incorporated into biodiversity studies (Gillespie et al. 2008). Other satellites and sensors such as IRS, SPOT and ASTER are also becoming common.

Active sensors, unlike passive sensors, can penetrate cloud cover, providing imagery of both day and night, regardless of weather conditions. Light detection and ranging (LIDAR) offers new and improved capabilities for vertical and structural characterization of objects, such as plant height and different growth stages/strata. The stand height and structural information derived from the LIDAR imagery contribute greatly to the characterization of biodiversity through vertical stratification (Dees et al. 2012). The Shuttle Radar Topography Mission (SRTM) provides 30–90-m resolution data on elevation and topography that has been used in species and diversity models. Radar backscatter from RADARSAT-1 has been used in land-cover classification and diversity models. ALOS PALSAR sensors have been shown to have very high level I accuracy when classifying several different forested regions (Kellndorfer et al. 2008). LIDAR pulses from the orbiting ICESat GLAS satellite have provided a wealth of data for forest biomass mapping and validation (Phlugmacher et al. 2008).

Based on the current state of the art in remote sensing instruments and their existing applications in the literature, we review four types of instruments: hyperspatial, hyperspectral, thermal infrared and LIDAR sensors with their applications in biodiversity conservation.

7.2.1 Hyperspatial Remote Sensing

Recently, many fine spatial resolution systems have been launched which allow ecosystems to be characterized over a range of scales and consider queries that were previously impractical to study from space or on the ground to be addressed (Table 7.3).

High spatial resolution imagery greatly increases the accuracy of identification and characterization of small objects at spatial scales which were previously only available from airborne

Table 7.2 Popular satellites and sensors used to measure and model biodiversity from space

Sensor	Launch date	Source	Spectrum	Details and global imagery (GI) extent
Passive				
Landsat 5 TM	1984	USA	V-TIR, 7 bands	Operating beyond expected lifetime. Continuous GI to 1985, spotty 1985 to present
IRS-1C/1D LISS-III, WIFS, PAN	1995, 1997	India	V-SWIR, 4 bands, 2 bands, V-NIR, 1 band	Constellation twin satellites, each hosts two moderate-resolution sensors. Pointable, panchromatic camera
IKONOS	2000 Commercial	USA	V-NIR, 5 bands	Pointable, stereo capability. Spotty GI, developing continuous GI
QuickBird	2001 Commercial	USA	V-NIR, 5 bands	
SPOT-5 HRVIR	2002 Commercial	France	V-SWIR, 5 bands	Pointable. Indonesia and Thailand use with Landsat data
RESOURCESAT-1 LISS-4	2003	India	V-NIR, 4 bands	Pointable and mounted with two moderate-resolution sensors
GeoEye-1	2008 Commercial	USA	V-NIR, 5 bands	Pointable, pan imagery released at 0.5 m. Spotty GI, developing continuous GI
WorldView-2	2009 Commercial	USA	V-NIR, 9 bands	Eight bands (+PAN) in VNIR, pointable
Active				
AVHRR	1978–2005	USA	V-TIR, 4–6 bands	A long-term satellite constellation, most recent is AVHRR/3. Continuous GI
MODIS Terra & Aqua	2000, 2002	USA	V-TIR, 36 bands	Mounted on two satellites. Vegetation bands are 250 m in resolution. Thermal bands used to detect fires. Continuous GI
SRTM	2000	USA	InSAR. X band, C band (Quadpolar)	Result of a 10-day shuttle mission
ENVISAT ASAR	2002	EU	SAR. C band. Quad-polar	Pointable. Continuous GI, with temporal interferometry in select areas
ICESAT GLAS	2004	USA	LIDAR	Tracks currently spaced at 15 km (equator) and 2.5 km (80° latitude)
RADARSAT-2	2006	Canada	SAR. C band. Quad-polar	Pointable. Continuous GI, with temporal interferometry in select areas
ALOS PALSAR	2006	Japan	SAR, L band. Quad-polar	Resolution and swath vary depending on mode. Continuous GI
TerraSAR-X	2007 Commercial	Germany	SAR, X band. Quad-polar	Commercial, variable resolution, and first in a satellite constellation

Adapted from Fagan and DeFries (2009)

platforms (Turner et al. 2003; Gillespie et al. 2008). With the availability of high-resolution imagery, the focus on mapping specific land-cover classes such as tree species rather than broad land-cover classes has gained significant attention (Boyd and Foody 2011). The derived information from mapping specific land-cover classes can aid both biodiversity assessment and conservation (Landenberger et al. 2003; Wilson et al. 2004). The availability of data from high spatial resolution sensors has opened up new opportunities for the development of operational mapping and monitoring of small features such as individual tree crowns (Hurtt et al. 2003; Levin et al. 2007; Rocchini 2007); mapping of fine-scale habitat elements, e.g. trees, at a landscape scale (Pitkanen 2001; Culvenor 2002; Hirschmugl et al. 2007; Rocchini 2007); mapping of invasive species; and

Table 7.3 Main space-borne hyperspatial sensors currently available for biodiversity studies

Sensor	Spatial grain (pixel size)	Spatial extent (swath width)	Spectral grain (no. of bands)	Launch date
IKONOS	1 m PAN, 4 m MSS	11.3 km	5	September 1999
QuickBird	0.65 m PAN, 2.6 m MSS	16.5 km	4	October 2001
SPOT-5 HRVIR	2.5–5 m PAN, 10 m MSS, 20 m SWIR	60 km	5	May 2002
RESOURCESAT-1 LISS-4	5.8 m	70 km PAN, 23.9 km MSS	3	October 2003
WorldView-1	0.5 m PAN, 2 m MSS	17.6 km	1	September 2007
GeoEye-1	0.5 m PAN, 2 m MSS	15.2 km	4	September 2008
WorldView-2	46 cm PAN, 1.84 cm MSS	16.4 km	8	October 2009
SPOT-6	1.5 m PAN, 8 m MSS	60 km	5	September 2012

development of species-level distribution maps along with structural information on dominance, canopy diameters and age–class distribution (Singh et al. 2010). Turner et al. (2003) have pointed out the applicability and feasibility of direct identification of certain species and species assemblages ; Gillespie et al. (2008) provide several examples of accurately identifying plant species based on the high spatial resolution imagery.

IKONOS, QuickBird, GeoEye-1 and Satellite Pour l'Observation de la Terre-5 (SPOT-5) are the commonly used systems. IKONOS imagery with a spatial resolution of 1 m (panchromatic) and 4 m (multispectral) enables the study of local-scale features from space (Read et al. 2003); QuickBird system from Digital Globe with multispectral imagery at resolutions of 2.4–2.8 m and panchromatic imagery 0.6–0.8 m makes direct identification of certain species and species assemblages feasible; WorldView-2 (WV-2) imagery combines high spatial resolution (0.5 m), sufficient spectral resolution (8 bands) and a Red-Edge detector (705–745 nm) for conducting vegetative analyses to reveal plant type, age, health and diversity in unprecedented detail. It is anticipated that the joint use of these data sets can be used for a variety of natural resource management applications.

Recent studies (Omar 2010; Kanniah 2011) have reported that the spectral bands of WV-2 ranging from 400 to 1,040 nm are suitable for discriminating tree species. IKONOS imagery has been used in conjunction with Landsat to map the expansion of an invasive plant species (Gillespie et al. 2008); to study forested environments at the scale of individual tree crowns (Hurtt et al. 2003; Clark et al. 2004); and for quantification and evaluation of the spatial structure of critical habitats and its effect on endemic species (Perotto-Baldivieso et al. 2009). Recently, Eckert (2012) explored the potential of WV-2 data for biomass and carbon estimation of tropical humid rainforests; the results on linking biomass and carbon inventory data for tropical humid rainforests with very high-resolution WV-2 satellite data are promising.

High spatial resolution imagery can be employed to assess the accuracy of remote sensing precuts derived from moderate or coarse spatial resolution imagery. For instance, Wabnitz et al. (2008) assessed the accuracy of LANDSAT-based large-scale seagrass mapping against patterns detectable with very high-resolution IKONOS images.

High-resolution data are also being increasingly used for ex situ biodiversity conservation; for example, scientists at the Royal Botanical Garden, Kew, discovered some previously unknown species in a remote highland forest in Mozambique, identified using very high-resolution (VHR) imagery from Google Earth (Timberlake et al. 2007). High-resolution images are being keyed to tabular data for providing additional dimensions of access to specimens (Bisby 2000; Edwards et al. 2000; Oliver et al. 2000),

Table 7.4 Main space-borne hyperspectral sensors currently available for biodiversity studies

Sensor	Satellite	Spatial grain (pixel size)	Spatial extent (swath width) (km)	Spectral grain (no. of bands)	Spectral grain (bandwidth)	Spectral resolution (nm)
Compact high-resolution imaging spectrometer (CHRIS)	Project for on-board autonomy (PROBA)	19–36 m	14	Up to 62, programmable	1–12 nm	Up to 410–1,050
Hyperion EO-1	Earth observing-1	10 m PAN, 30 m all other bands	7.7	220	10 nm	356–2,578
Global imager (GLI)	Advanced earth observing satellite-II (ADEOS-II)	250 m—6 bands corresponding to Landsat, 1 km—all other bands	1,600	36	12–2,985 nm	380–12,000
Medium-resolution imaging spectrometer (MERIS)	ENVISAT	250 m–1 km	1,150	15	2.5–12.5 nm, programmable	390–1,040
Moderate-resolution imaging spectrometer (MODIS)	TERRA	250 m VNIR, 500 m VNIR-SWIR, 1 km TIR	2,330	36	10–15 nm VNIR-SWIR	405–14,385

and these data sets are being made available through the Internet, e.g. New York Botanical Garden, Museum of Vertebrate Zoology, University of California at Berkeley, Missouri Botanical Gardens and Instituto Nacional de Biodiversidad, Costa Rica.

Although high spatial resolution satellite remote sensing is a very useful source of data, Nagendra and Rocchini (2008) have rightfully pointed out that it is the most potentially powerful yet underutilized source for tropical research on biodiversity and stimulating discussion on its possible applications should be the first step in promoting a more extensive use of such data.

7.2.2 Hyperspectral Remote Sensing

Hyperspectral remote sensors acquire images across many narrow contiguous spectral bands throughout visible, near-infrared and mid-infrared portions of electromagnetic spectrum and measure the reflected spectrum at wavelengths between 350 and 2,500 nm using 150–300 contiguous bands of 5–10 nm bandwidths (Table 7.4). According to Shippert (2004), several recently launched hyperspectral sensors are acquiring imagery from space, including Hyperion sensor on NASA's EO-1 satellite, CHRIS sensor on European Space Agency's PROBA satellite and FTHSI sensor on US Air Force Research Lab's MightySat II satellite. Many airborne hyperspectral sensors, including NASA's AVIRIS sensor, are also available to collect data. Of these sensors, the first-civilian and most commonly used data are derived from the Hyperion, which is operated by the Earth Resources Observation and Science (EROS) at a relatively low cost to the general public.

Hyperspectral data can discriminate fine-scale, species-specific land cover (Turner et al. 2003), such as vegetation categories or soil types which make remarkable contribution to any study regarding biodiversity patterns. These data are well suited for vegetation studies since reflectance/absorption spectral signatures from individual species as well as more complex mixed-pixel communities can be better differentiated from the much wider spectral bands of hyperspectral imagery (Varshney and Arora 2004). Nagendra and Rocchini (2008) summarize that hyperspectral data have been successfully applied in recording information regarding critical plant properties (e.g. leaf pigment, water content and chemical composition), in discriminating tree species in landscapes and in fairly accurate distinction between different species. Xie et al. (2008) have emphasized how vegetation extraction from hyperspectral imagery has been increasing; recent studies imply that this imagery is capable of separating plant species, which may be difficult using multispectral images.

Recent applications of Hyperion hyperspectral imagery mainly include ecology and biodiversity in forest, grassland (Guerschman et al. 2009), agriculture (Bannari et al. 2008), vegetation (Walsh et al. 2008), fragmented ecosystem and ecosystem succession and coastal environment (Lee et al. 2007). Vegetation types and densities have also been classified using Hyperion data to support of wildfire management (Keramitsoglou et al. 2008). Foster et al. (2008) state that hyperspectral imagery from EO-1 Hyperion is capable of mapping low-lying woody lianas, which are critical to tropical forest dynamics; Pignatti et al. (2009) successfully analysed the capability of Hyperion data for discriminating land cover according to the structure of current European standard classification system (CORINE Land Cover 2000). Besides the application of land-cover classification, the relationships between LAI and spectral reflectance were studied (Twele et al. 2008) using narrowband (EO-1 Hyperion) and broadband Landsat-enhanced thematic mapper plus (ETM+) remotely sensed data. Integration of Hyperion and IKONOS imagery has also been tested successfully for differentiation of subtle spectral differences of land-use/land-cover types on household farms with an emphasis on secondary and successional forests (Walsh et al. 2008). The MODIS vegetation continuous fields product pictures heterogeneous vegetation more realistically and improves the detection of changes in complex landscapes (Pfeifer et al. 2011).

There are many drawbacks associated with hyperspectral remote sensing as its high cost of acquisition data and complex technical aspects of processing make the whole process outside the expertise of most ecologists, and also the huge volume of data require a large data storage capacity and can be time intensive to process. Nagendra and Rocchini have preliminarily discussed strengths and drawbacks of hyperspatial (i.e. high spatial resolution) and hyperspectral data (Nagendra and Rocchini 2008).

7.2.3 Thermal Remote Sensing

Thermal remote sensing detects the energy emitted from Earth's surface in thermal infrared (TIR, 3–15 μm), which can be radiated by all bodies above absolute zero. Theoretically, TIR sensors measure surface temperature and thermal properties of targets (Canada Centre for Remote Sensing 2007), which are essential for developing a better understanding and more robust models of land surface energy balance interactions. TIR remote sensing is capable of uncovering the principles of ecological patterns of structure and function due to the development of ecological thermodynamics (Quattrochi and Luvall 2009).

The well-known sensors with TIR bands include the advanced VHR radiometer (AVHRR) on board the Polar Orbiting Environmental Satellites (POES), Landsat Thematic Mapper (TM) and ETM+, the advanced space-borne thermal emission and reflection radiometer (ASTER) on Terra Earth observing satellite platform (Quattrochi and Luvall 2009).

There have been several successful attempts in using TIR remote sensing for biodiversity conservation; for example, biophysical variables have been derived from thermal and multispectral remote sensing data and coupled with a Soil-Vegetation-Atmosphere-Transfer (SVAT) model (Gillies and Temesgen 2004); Duro et al. (2007) have illustrated that TIR region is important to study environmental disturbance because of negative relationship between vegetation density and land surface temperatures (LST), and Mildrexler et al. (2007) have successfully applied disturbance detection index using Moderate Resolution Imaging Spectroradiometer (MODIS) 16-day Enhanced Vegetation Index (EVI) and 8-day LST to detect continental-scale disturbance events such as wildfire, irrigated vegetation, precipitation variability and incremental process of recovery of disturbed landscapes. Another good use of TIR remote sensing data is to measure evapotranspiration, evaporation and soil moisture; for example, Crow and Zhan have analysed the continental-scale performance of surface soil moisture retrieval algorithms depending on satellite passive microwave, scatterometer and thermal remote sensing observations (Crow and Zhan 2007), and Petropoulos et al. (2009) have reviewed surface temperature/vegetation index remote sensing-based methods for the retrieval of land surface energy fluxes and soil surface moisture.

7.2.4 LIDAR Remote Sensing

LIDAR, also called laser altimetry, is an active remote sensing technology that utilizes a laser to illuminate a target object and a photodiode to register the backscatter radiation (Lim et al. 2008; Hyyppa et al. 2009). The longer wavelength pulses of radars can penetrate clouds, and the longest radar wavelengths (i.e. L band and beyond) penetrate tree canopies or, in cases of bare and loamy soil, the surface of the earth to depths of a metre or more (Ulaby et al. 1982). LIDAR sensors use the return signals to detect the height of the canopy top, ground elevation and the positions of leaves and branches in between, giving it the ability to penetrate forest canopies and making it a potential tool for measuring biomass and determining vegetation structure (Turner et al. 2003). LIDAR is particularly useful for measuring height (Van der Meer et al. 2002), which may then be incorporated into further ecological analysis.

Besides airborne LIDAR with limitations of large data volumes, footprint size and high costs (Duncanson et al. 2010), space-borne LIDAR has come through with launch of Ice, Cloud, and land Elevation Satellite/Geoscience Laser Altimeter System (ICESat/GLAS), which is the first laser-ranging instrument for continuous global observations. Space-borne LIDAR focuses on the forest vertical structure, especially forest canopy height and aboveground biomass estimation. Lefsky et al. (2007) have estimated forest canopy height with an RMSE of 5 m in varied forest types including evergreen needle leaf, deciduous broadleaf and mixed and tropical evergreen broadleaf stands. The height and aboveground biomass of mangrove forests have also been successfully measured and mapped based on SRTM elevation data, GLAS waveforms and field data (Simard et al. 2008). Pflugmacher et al. (2008) have compared GLAS height and biomass estimates with reference data from the Forest Inventory and Analysis (FIA) program of the US Forest Service at a regional scale to obtain promising results. Helmer et al. (2009) have combined Landsat time series and GLAS to estimate the biomass accumulation of Amazonian secondary forest to have results in agreement with ground-based studies. Duncanson et al. (2010) successfully tested simulated GLAS data for areas with dense forests, high relief or heterogeneous vegetation cover to demonstrate the capability of GLAS waveforms as supplemental model input to improve estimates of canopy height. The application of LIDAR technology to marine biodiversity conservation also shows promise for detecting habitats (Turner et al. 2003). Large-footprint LIDAR information is fused with MODIS data to generate forest height maps (Lefsky 2010), and the P band of the synthetic aperture radar (SAR) shows good agreement with boreal forest biomass.

In many recent studies, the detection of individual vegetation objects has been facilitated by increasing spatial resolutions of remotely sensed imagery, LIDAR and use of object-based approaches to data classification (Falkowski et al. 2006; Koch et al. 2006; Zhang et al. 2010).

In ex situ conservation efforts also, LIDAR seems to have found a use; for example, professionals at Montgomery Botanical Center, Florida, were able to improve the assessment of their property with deeper evaluation of the collections and natural resources using remote sensing imagery and data. By adding LIDAR imagery to maps and employing techniques normally used at larger regional scales, new information was discovered about the garden and its collections. (http://arnoldia.arboretum.harvard.edu/pdf/articles/2011-69-1-remote-sensing-as-a-botanic-garden-tool.pdf).

7.3 Satellite Image Processing and GIS Techniques Used to Measure and Model Biodiversity

Several researchers have pursued cutting-edge studies on biodiversity conservation using remote sensing and GIS approaches, some of which have been presented in Table 7.5.

Existing studies of biodiversity conservation using remote sensing and GIS can be grouped into three categories:

- Biodiversity mapping (individual species, species assemblages and habitats)
- Biodiversity monitoring
- Biodiversity modelling.

7.3.1 Biodiversity Mapping for Individual Species, Species Assemblage and Habitats

The improved spatial, temporal and spectral resolution of satellite data has made it possible to generate information at diverse scales on major climatic, physiognomic vegetation types as well as region-specific vegetation types depicting local gregarious formations and distinctive habitats. The higher-resolution imagery shows great potential in identification of tree species and canopy attributes, with several studies undertaking targeted mapping to estimate the variety, type and extent of specific species; for example, economically and medicinally important species in remote and inaccessible areas have been mapped by several researchers (Behera et al. 2000; Roy et al. 2001; Porwal et al. 2003; Joshi et al. 2005) using Indian Remote Sensing (IRS) satellite data; QuickBird data have been used to map giant reed (Arundo donax) in southern Texas with 86–100 % accuracy (Everitt et al. 2006); high-resolution data have been used to identify mangrove species (Dahdouh-Guebas et al. 2006; Wang et al. 2004). Mapping of species assemblages has been attempted using satellite data; for example, evergreen forests/vegetation of Mouling National Park, Arunachal Pradesh, India, was mapped by integrating a digital elevation model (DEM) with Linear Imaging Self-scanning Sensor (LISS) III multi-band data (Singh et al. 2005); seven species of trees were classified with an overall accuracy of 86 % in temperate forests (Carleer and Wolff 2004).

Remote sensing-based habitat maps and information on species habitat associations are generally being used to derive information on the distribution of species. Satellite data can be combined with field-based habitat data, landscape structure and species abundance information to identify the habitat boundaries and biophysical characteristics of species (Kerr et al. 2001; Scribner et al. 2001). Satellite data classification has long been used to link species distributions with vegetation types and associated habitat preference (Nagendra 2001; Gottschalk et al. 2005; Leyequien et al. 2007). Way back in 1994, Franklin et al. carried out an analysis using satellite imagery to differentiate compositionally distinct vegetation communities, and in a later study, the preliminary observations in a tropical forest in northern India suggested that supervised classification of satellite data permits the discrimination of vegetation types with distinct species composition (Ravan et al. 1995). In a detailed analysis in tropical forests of Western Ghats of India, Nagendra and Gadgil (1999) mapped a landscape into seven habitat types ranging from secondary evergreen forests to paddy fields, using supervised and unsupervised classification of IRS satellite imagery. Vegetation maps as a surrogate for habitat preference have provided insights into the distributions of birds (Peterson et al. 2006), herpetofauna (Raxworthy et al. 2003) and insects (Luoto et al. 2002).

Table 7.5 Some important biodiversity conservation studies, techniques employed and data sources

Issue addressed	Technique used	RS data used	References
Biodiversity mapping	Individual species mapping, species assemblage mapping	IRS LISS-III, QuickBird MSS	Franklin et al. (1994), Behera et al. (2000), Roy et al. (2001), Porwal et al. (2003), Carleer and Wolff (2004), Dahdouh-Guebas et al. (2006), Wang et al. (2004), Joshi et al. (2005), Singh et al. (2005), Everitt et al. (2006), Kandwal et al. (2009), Kimothi et al. (2010)
	Habitat mapping	IRS LISS-III, Landsat ETM+	Ravan et al. (1995), Nagendra and Gadgil (1999), Peterson et al. (2006), Raxworthy et al. (2003), Luoto et al. (2002), Giriraj et al. (2008, 2009), Roy and Tomar (2001), Amarnath et al. (2003), Roy et al. (2012)
Biodiversity monitoring and assessment	Monitoring habitat modification/ degradation	IRS LISS-II, LISS-III, Landsat TM/ETM/MSS, SPOT MSS	Nagendra and Gadgil (1999), Singh et al. (2004), Joshi et al. (2005), Jha et al. (2005), Reddy et al. (2007), Sharma et al. (2002), Giriraj et al. (2009), Behera and Roy (2010), Roy (2011), Joseph et al. (2011), Roy et al. (2013)
	Monitoring human-induced habitat threat	LISS-III, Landsat TM/ ETM/MSS, IKONOS, DMSP-OLS, IRS P6-AWIFS	Roy and Tomar (2001), Roy (2003), Gupta et al. (2004), Talukdar (2004), Kiran Chand et al. (2006), Kandwal et al. (2009), Kimothi et al. (2010), Nagendra et al. (2010), Thakur et al. (2011), Sudheesh and Reddy (2013), Krishna and Reddy (2012), Badrinath et al. (2011), Reddy et al. (2012), Saranya et al. (2014)
	Monitoring vegetation versus climate change	IRS LISS-III, LISS-IV, Landsat TM/MSS	Panigrahy et al. (2010), Dudley et al. 2010, Bharti et al. (2012), Volante et al. 2012, Caride et al. 2012
Biodiversity modelling	Species distribution modelling	NA	Engler et al. (2004), Araujo et al. (2005), Akcakaya et al. (2006), Hamann and Wang (2006), Austin (2007), Botkin et al. (2007), Carvalho et al. (2010), Luoto et al. (2007), Williams et al. (2009), Adhikari et al. (2012), Yang et al. (2013), Giriraj et al. (2008), Kumar and Stohlgren (2009), Irfan ulah et al. (2006), Saran et al. (2010a, b), Nativi et al. (2009), Richardson et al. (2010), Barik and Adhikari (2011), Lorenzen et al. (2011)
	Species diversity modelling	MODIS, QSCAT, SRTM	Gould (2000), Oindo and Skidmore (2002), Foody and Cutler (2006), Tuomisto et al. (2003), Fairbanks and McGwire (2004), Waser et al. (2007), Rocchini (2007), Stickler and Southworth (2008), Lucas and Carter (2008), He and Zhang (2009), Rocchini et al. (2009), Oldeland et al. (2010)

Giriraj et al. (2009) applied data generated from RS and GIS to categorize habitats and determined the relationship between habitat categorizations and species distribution patterns in tropical rainforests of southern Western Ghats, India.

In order to spatially delineate habitats, remote sensing-derived habitats were analysed in conjunction with landscape metrics, species assemblages, microhabitats such as slope, topography, species endemism and proportion of core and edge species (Giriraj et al. 2008). Under a national-level project on biodiversity characterization at landscape-level, vegetation-type maps were analysed in conjunction with climate and topography using geographical information system (GIS) to identify habitats a priori and determine the relationship between remotely sensed habitat categories and species distribution patterns (Roy and Tomar 2000; Amarnath et al. 2003; Roy et al. 2012). Giriraj et al. (2009) have used data generated using remote sensing and geographical information systems to categorize habitats and determine the relationship between habitat categorizations and species distribution patterns. Nandy and Kushwaha (2010) have attempted habitat classification of mangroves in Sunderban Biosphere Reserve (SBR) in West Bengal province of India using IRS satellite data using different classification approaches to delineate four distinct mangroves classes. Habitats with clear boundaries (e.g. grassland and agriculture) can generally be mapped with greater accuracy (Bock et al. 2005; Lucas et al. 2007; Forster and Kleinschmit 2008). SAC (2001) attempted grassland mapping in Gujarat using remote sensing and GIS techniques. Recently, Rahman et al. (2013) used Landsat ETM+ data for mapping and inventory of mangrove forest in Sundarbans using different classification methods.

New mapping approaches such as object-oriented classification (Collingwood et al. 2009), decision tree, support vector machine, multilayer perception and radial basis function neutral networks significantly improve classification accuracy (Dalponte et al. 2009; Mukhopadhyay and Maulik 2009; Kumar et al. 2007; Foody and Mathur 2004; Sanchez-Hernandez et al. 2007a; Boyd et al. 2006). Remote sensing may be used to monitor a habitat of interest with a one-class classification approach adopted to focus effort and resources on the class of interest (Boyd et al. 2006; Sanchez-Hernandez et al. 2007b), and there is a considerable scope for different types of classification methods such as soft or fuzzy classifications which allow the study of environmental gradients, transition zones and subpixel land cover (Jensen et al. 2009; Lu and Weng 2007; Plourde et al. 2007; Rocchini and Ricotta 2007). Integration of radar data improves classification accuracy (Saatchi et al. 2001; Boyd and Danson 2005; Li and Chen 2005), and LIDAR data demonstrate the potential for mapping emergent tree species and subcanopy layers that are important indicators of stratification for forest bird species. Further research is underway on information extraction techniques such as image classifier development for derivation of thematic maps.

More advanced techniques have examined the variability of spectral signals in satellite imagery which has been demonstrated to have an intrinsic power in evaluating species diversity (Palmer et al. 2002) since it is expected that the higher the spectral variability is, the higher the habitat and species variability will be (Carlson et al. 2007; Rocchini et al. 2007).

7.3.2 Biodiversity Monitoring

Effective monitoring is critical to evaluate and improve biodiversity conservation practice and a well-conceived, designed and implemented biodiversity monitoring agenda should (i) deliver information on trends in key aspects of biodiversity; (ii) provide early warning of problems that might otherwise be difficult or expensive to reverse; (iii) generate quantifiable evidence of conservation successes and failures; (iv) highlight ways to make management more effective; and (v) provide information for return on conservation investment (Lindenmayer et al. 2012).

Ecological studies increasingly require biophysical and habitat data through time and over significant areas, and also background rate and direction of change in ecological systems are essential to detect the signature of anthropogenic impacts. These prerequisites are generating a lot of interest in methods to aid tracking and interpretation of biodiversity changes through time to produce monitoring data sets. Technologies such as remote sensing and GIS can facilitate rapid collection of huge amount of large-scale data (Youngentob et al. 2011), establish baselines of the extent and condition of habitats and associated species diversity as well as quantify losses, degradation or recovery associated with specific events or processes more effectively as compared to more traditional, field-based methods (Nagendra et al. 2013; Kerr and Ostrovsky 2003).

In current times, when the satellite record is around four decades old, it can provide potential means to study impacts of environmental change. Remote sensing can be applied to habitat inventory and evaluation, assessment of habitat attributes and identification of suitable sites for protected areas (Kamat 1986; Panwar 1986). Mostly, the use of high-to-moderate spatial resolution data, such as provided by the Landsat and the Indian Remote Sensing (IRS) satellite, may be sufficient to capture the broad extent and spatial patterns of habitats (Lucas et al. 2007, 2011). Satellite data sets from IRS, Landsat and Système Pour l'Observation de la Terre (SPOT) have been used effectively in monitoring and analysing biodiversity (Nagendra and Gadgil 1999; Joshi et al. 2005; Behera and Roy 2010; Roy 2011). In recent years, the advent of VHR satellites has provided opportunities for more detailed mapping and studies of changes in habitat coverage, landscape fragmentation and human pressure.

Habitat modification and degradation are prevalent even in intact landscapes, so developing methods to quantify and monitor changes in habitats is critical. Many researchers have been working in this direction; for example, Singh et al. (2004) highlighted the application of satellite remote sensing in assessment and monitoring of mangrove habitats along Goa state coastline, India, by analysing the changes in mangrove forest cover using IRS data; Reddy et al. (2007) performed periodic assessment and monitoring of mangroves of Bhitarkanika Wildlife Sanctuary, Orissa, India, using temporal multisensor satellite data to find changes in mangrove and other land-cover categories during the last 30 years. Disruption of habitat connectivity due to landscape fragmentation can impact species dispersion, habitat colonization, gene flows, population diversity, species mortality and reproduction. Quantitative analyses of changes in landscape structure can provide early warnings of habitat degradation. Sharma et al. (2002) analysed the spatial patterns of different attributes and explored the extent and patterns of forest fragmentation in a Himalayan landscape to find that out of total landscape in study area, 41 % was fragmented. Fragmentation led vegetation changes in Vindhya hills over a decade, and the resultant impact on local biodiversity was studied using remote sensing data (Jha et al. 2005). Giriraj et al. (2009) studied remnant intact patches of evergreen forest using multitemporal satellite data and explored the likelihood of their sustenance in coming decades with respect to vegetation composition, changes in patch characteristics and regeneration potential in Kalakad Mundanthurai Tiger Reserve, Tamil Nadu (India). Joseph et al. (2011) reviewed the state of remote sensing technology in characterization of tropical forest degradation by describing the factors responsible and its likely impacts. The authors conclude by requesting an additional momentum in research to answer numerous unresolved questions of tropical forest degradation. A recent exemplary work by Roy et al. (2013) used a moving window approach to identify potential areas of forest fragmentation in Indian landscape to assess the impact of anthropogenic pressures and cultural practices on forest fragmentation that provides critical inputs for prioritization and conservation of forests and associated biodiversity.

There can be many types of threats to habitats depending on the landscape, context and time period of focus, and more common types include urbanization, road construction, mining, logging, agriculture, fire, invasion by alien species, hunting, grazing and drought (DeFries et al. 2005; Nagendra 2008). Remote sensing data sets of medium-to-fine spatial resolution can provide important information on signature of human pressure related to land use, management and other disturbances (Fuller et al. 2007). Satellite sensors such as AWiFS, LISS-III, ETM+, MODIS, OCM, AVHRR and MODIS provide synergistic data sets that have potential in forest fire detection and monitoring (Sudheesh and Reddy 2013). The forest fire detection capabilities of Indian Defense Meteorological Satellite Program Operational Linescan System (DMSP-OLS) satellite data for effective monitoring of forest fires have been evaluated with positive outcomes (Kiran Chand et al. 2006; Badrinath et al. 2011). The state of forests, methodology, models and case studies of forest fire risk and degradation assessment in the context of Indian forests have been discussed by Roy (2003), emphasizing the utility of geospatial techniques as powerful tools to assess the forest fire risk and degradation assessment. The critical spatial information for determination of forest burnt areas has been achieved using IRS P6 AWIFS satellite data which emphasize the need of remote sensing-based time series analysis in forest fire management (Reddy et al. 2012; Krishna and Reddy 2012). Sudheesh and Reddy (2013) have highlighted the utility of multitemporal satellite data in effective planning and conservation of forest resources by analysing the fire-prone areas in Nagarjunasagar–Srisailam Tiger Reserve (NSTR), Andhra Pradesh, India. Saranya et al. (2014) have accentuated that spatial databases offer excellent opportunity to understand the ecological impact of fires on biodiversity and are helpful in formulating conservation action plans by locating and estimating the spatial extent of forest burnt areas and fire frequency covering decadal fire events in Similipal Biosphere Reserve.

Satellite measurements of broadscale trends in vegetation also provide direct estimates of habitat loss, increasing the power of applied ecological studies to detect changes in species distributions or to model extinction rates. Roy and Tomar (2001) analysed the modification of natural landscape due to anthropogenic activities using temporal IRS and Landsat Multi-spectral Scanner (MSS) data by studying the land-cover dynamics pattern of biologically rich landscapes of Meghalaya, India. The linkages between socio-economic drivers and consequent forest loss in Nicobar group of islands have been investigated on the basis of indicators derived from LULC mapping of satellite remote sensing data (Gupta et al. 2004). Geostatistical analysis of forest cover changes in Meghalaya, India, has been attempted using LULC trends for 1980, 1989 and 1995, revealing the incessant impact of shifting cultivation and mining by generation of the likely landscape pattern for year 2025 (Talukdar 2004). The detection, mapping and patch information extraction of invasive weeds using high-resolution satellite data have also been attempted revealing the suitable seasons for weed detection and developing of a new technique by analysis of 29 vegetation indices (VIs) for discrimination of invasive weeds (Kandwal et al. 2009; Kimothi et al. 2010). Thakur et al. (2011) analysed cumulative impacts of human-induced processes such as livestock grazing and fuel wood extraction on forest cover using satellite data and field observations and revealed a significant loss of forest cover around the fringes of Kedarnath Wildlife Sanctuary in the last 29 years. Nagendra et al. (2010) used Landsat TM and ETM+ imagery to find a clear signal of forest fragmentation and deforestation at the periphery of an Indian tiger park because of extraction by local residents of villages outside the boundary.

Remote sensing data can provide insights into the impacts of climatic variability through analysis of changes in the extent and condition of vegetation (e.g. phenological shifts and species range shifts). Species in transition zones are especially vulnerable to climate change, as they have limited scope to move further (ICIMOD 2010) like in Indian Himalayan region, a unique habitat for distinct biological assemblages, and native and endemic, floral and faunal species, economically important species (Dhar 2000); it is speculated that under continuing climate change, tree species will be affected in different ways and ranges will adjust at different rates and by different processes (Subedi 2009). Many researchers are focussing their efforts in this direction; for example, Panigrahy et al. (2010) reported an upward shift of timberline vegetation by 300 m and a considerable reduction in snow cover in the Nanda Devi Biosphere Reserve (NDBR), Garhwal Himalaya, India. More recently, Bharti et al. (2012) studied temporal changes in the timberline ecotone vegetation of Nanda Devi National Park (NDNP), Indian Himalaya, on the basis of remote sensing analysis of Landsat MSS and TM images, over a period of 30 years. Satellite-derived data along with field data can be effectively used both to estimate and to monitor the level of provision of ecosystem services over large areas (e.g. He et al. 2005; Nosetto et al. 2005). Remotely sensed data and modelling techniques can provide estimates of carbon-related services for forest regional quantification of carbon gains in protected areas to support their protection given that they help address ecological problems related to climate change issues (Volante et al. 2012; Caride et al. 2012; Dudley et al. 2010). Though numbers of such studies are increasing, the full potential of remote sensing is yet to be explored in terms of monitoring forest health and monitoring range shift.

7.3.3 Biodiversity Modelling

Techniques that can greatly aid ecological modelling activities in support of conservation efforts are becoming increasingly available to the ecological community (Santos et al. 2006). Remote sensing and GIS have been used to predict species distributions and risks to biodiversity (Spens et al. 2007), to aid visualization, exploration and modelling of data on species distributions (Lopez-Lopez et al. 2006; Vogiatzakis et al. 2006; Zhang et al. 2007) and to study the effect of major variables such as disturbance events (Pennington 2007).

7.3.4 Species Distribution Modelling, Ecological Niche Modelling or Habitat Suitability Modelling

Species distribution modelling, ecological niche modelling or habitat suitability modelling has been growing at a striking rate in the recent years. Species distribution or environmental (or ecological) niche models (ENMs) provide a tool for addressing lack of species distribution data and can be used to fill knowledge gaps by projecting habitat suitability in areas with few or no occurrence records. Species distribution modelling (SDM) can combine georeferenced occurrence data points (latitude, longitude) with different data sets that characterize the environment where the focal species occur. These sets are combined and analysed aiming to build a representation of ecological requirements of the focal species, i.e. a representation of their ecological niche. The final result can be projected in geographical space, indicating areas suitable for focal species that can be potentially occupied by it (Giannini et al. 2011).

Usually, the data sets used are comprised by abiotic features such as temperature, precipitation and altitude, which describe the environment of species occurrence. But SDM can also include data about occurrences of interacting species (biotic features) responsible for shaping geographical distribution and species dispersal capacities, to estimate their capability of occupying new environments. Intermingled with these environmental (abiotic) and interaction (biotic) features, geographical distribution is also related to species evolutionary trends, determining patterns of genetic diversity and trait variation across space. These models utilize presence, absence or abundance data from museum vouchers or field surveys and environmental predictors to create probability models of species distributions within landscapes, regions and continents (Guisan and Thuiller 2005). These models can forecast potential future impacts of environmental changes on species distributions, allowing practitioners to assess alternative policies and actions to plan for the change (Nielson et al. 2008). It also estimates the relative suitability of habitat known to be occupied by the species, relative suitability of habitat in geographical areas not known to be occupied by the species, changes in the suitability of habitat over time given a specific scenario for environmental change and the species niche (Warren and Seifert 2011).

After years of ongoing development and testing, species distribution models are now offering valuable contribution to facilitation in site selection for reintroduction of critically rare and endangered species (Williams et al. 2009); design of their management plans (Adhikari et al. 2012); modelling the distribution and abundance of medicinally important species (Yang et al. 2013); and economically important species (Giriraj et al. 2008) or threatened and endangered species (Kumar and Stohlgren 2009; Irfan-ullah et al. 2006; Saran et al. 2010a). These models are extremely helpful in studying impacts of climate change on spatial distribution of biodiversity to facilitate future conservation activities (Skov and Svenning 2004; Nativi et al. 2009; Richardson et al. 2010; Araujo et al. 2005; Akcakaya et al. 2006; Hamann and Wang 2006; Austin 2007; Botkin et al. 2007) using bioclimate envelope models (Berry et al. 2002; de Garzon et al. 2007). The outcomes may be used to project future distribution of species under a set of climate change scenarios (Pearson and Dawson 2003; Luoto et al. 2007) which can provide a valuable initial assessment of likely climate change impacts. A variety of other applications also use these models such as invasive species management and prediction of their geographical distribution (Barik and Adhikari 2011); elucidation of paleodistributions (Lorenzen et al. 2011); design of field surveys (Engler et al. 2004); or design and selection of reserves (Carvalho et al. 2010).

New approaches are necessary to analyse the importance of these complex features. Recently, Pavoine et al. (2011) suggested a framework based on a mathematical method of ordination to analyse phylogeny, traits, abiotic variables and space in a plant community. Another example can be found in Diniz-Filho et al. (2009) proposing an integrated framework to study spatial patterns in genetic diversity within local populations, coupling genetic data, SDM and landscape genetics. The new trends on SDM, regarding the impacts of global changes on species diversity, are niche evolution and phylogeographic and phylogenetic research (Zimmermann et al. 2010).

Many taxonomic groups, however, are highly specious with little information on distribution or other attributes of every component species, and more new species are still being discovered. In such situations, SDM is limited in its ability to predict occurrences for all species in a taxon and hence in predicting patterns for biodiversity as a whole (Mokany and Ferrier 2011). Community-level modelling has the capacity to complement species-level approaches by predicting spatial patterns in biodiversity for highly diverse, poorly studied taxa. It may confer significant benefits for applications involving very large numbers of species, particularly where a sizeable proportion of these species is rarely recorded in the data set. Unlike species-level modelling, for which species with too little data are usually excluded from further analysis (for statistical reasons), many community-level modelling strategies make use of all available data across all species, regardless of the number of records per species. Among the broad range of community-level modelling approaches (Ferrier and Guisen 2006), species richness prediction of communities (alpha-diversity) and dissimilarity in community composition between pairs of sites (beta-diversity) are the most commonly employed ones. Alpha-diversity accounts for local species richness or abundance within each sampling unit, whereas beta-diversity is related to species compositional turnover among sampling units.

The prediction of diversity has substantially relied on simple univariate regression or multiple regression models appropriately scaling sensor imagery to field data on vascular plants (Fairbanks and McGwire 2004; Rocchini 2007; Stickler and Southworth 2008), lichens (Waser et al. 2007) and mammals (Oindo and Skidmore 2002). While these approaches provide a basic understanding of patterns and can be used to create predictive diversity maps for a landscape, region or continent, increasingly more sophisticated statistical and spatial techniques are being examined and developed to model patterns of diversity (Foody 2004, 2005) such as use of hyperspectral imagery for predicting species richness and abundance using univariate statistics (Lucas and Carter 2008; Oldeland et al. 2010); explicit mapping of uncertainty in species diversity prediction by spectral variability (Gould 2000; Oindo and Skidmore 2002); use of neural networks for predicting species diversity (Foody and Cutler 2006); or use of spectral distances between sampling units for estimating turnover in species composition (Tuomisto et al. 2003; He and Zhang 2009; Rocchini et al. 2009). Spatial statistics such as geographically weighted regression analyses have also resulted in improved models of diversity (Foody 2005). Furthermore, increased accuracy of predictions can be obtained using more complex approaches such as neural networks (Foody and Cutler 2006). There has also been an increasing interest in the combination of passive and active sensors to improve species diversity models, e.g. using a combination of passive sensors (MODIS) and active sensors (QSCAT, SRTM) to model tree diversity for the entire Amazon Basin (Saatchi et al. 2008).

7.4 Limitations of Using Remote Sensing and GIS for Biodiversity Conservation

Using remote sensing data for biodiversity mapping, monitoring and modelling requires a trade-off between data availability, spatial resolution and coverage, spectral resolution, timing of image acquisition, practicality of ground validation combined with overall cost of imagery and analytical effort. Any of these criteria can potentially limit the use of RS data for biodiversity conservation (Biodiversity report, 2012); hence, there are several limitations to using satellite images for biodiversity conservation, including

i. Possible end of key satellite programs; that is, the uncertainty in long-term continuity of remote sensing data restrains individual and organizations to invest in biodiversity monitoring projects utilizing remote sensing data.
ii. Except LIDAR and RADAR sensors, other sensors can monitor only features viewed from above, and characteristics of understory are unable to monitor/sense.
iii. Restricted data dissemination policy and low investment by public organizations lead to assignment of satellite development to private sector, making remote sensing data expensive and unaffordable. Many space agencies and countries are now offering free and open data access to their satellite data (e.g. Landsat, MODIS), but most data products especially high spatial resolution imagery have expensive and restricted access, making financial cost a major challenge to most biodiversity researchers and conservation practitioners (Leidner et al. 2012) (Table 7.6).
iv. Low temporal resolution leading to long temporal repeat cycle and short time series for trend analysis limits the sensitivity of remote sensing products to detect surface changes. The time-composited satellite products are insensitive to some natural phenomena which occur on finer timescales such as phenological changes (Cleland et al. 2007).
v. Technical expertise required to handle satellite imagery and other data products remains a challenge for ecologists and conservation biologists. Application of ecological models that can convert remote sensing data products into actual knowledge of species distributions and richness also requires additional software and analytical and technical skills.

7.5 Prospects

Key areas of research identified by Task Force on Mountain Ecosystems (Planning Commission of India 2006) were to understand the relationship between climatic patterns and species response to predict future distribution of biota, especially in subalpine and alpine habitats, to understand the

Table 7.6 Current cost of most common and popular satellite data

Satellite (sensor)	Pixel size (m)	Minimum order area (km^2)	Approximate cost ($)
NOAA (AVHRR)	1,100	Free	No cost
EOS (MODIS)	250, 500, 1,000	Free	No cost
SPOT-VGT	1,000	Free	No cost
Landsat	15, 30, 60, 100, 120	Free	No cost
ENVISAT (MERIS)	300	Free	No cost
ENVISAT (ASAR)	150	Free	No cost
SRTM (DEM)	90	Free	No cost
EO-1 (Hyperion)	30	Free	No cost
EOS (ASTER)	15, 30, 90	3,600	100
SPOT-4	10, 20	3,600	1,600–2,500
SPOT-5	2.5, 5, 10	400	1,300–4,000
SPOT-6	1.5, 6.0	500	1,000–3,000
RapidEye	5	500	700
IKONOS	1, 4	100	1,000–2,000
QuickBird	0.6, 2.4	100	2,500
GeoEye	0.25, 1.65	100	2,000–4,000
WorldView	0.5, 2, 4	100	2,600–7,400

Source IKONOS, QuickBird, GeoEye, WorldView and RapidEye: Landinfo. SPOT-4 and SPOT-5: Astrium EADS. Aster: GeoVAR. SRTM DEM, Landsat, Hyperion, MERIS, ASAR, AVHRR, SPOT-VGT and MODIS: NASA, ESA and Land Cover Facility

current distribution of sensitive species, to identify species having better carbon dioxide sequestration potential for cold deserts and to understand the impact of retreating glaciers at the species and ecosystem levels.

There is an acute paucity of information on the fringe habitats in high-altitude regions, such as timberline and snowline zones, which are under direct influence of climate change (Planning Commission of India 2006). Some key/gap areas for further ecological research as identified by Singh et al. (2010) include macroecological studies to understand environment and species richness, habitat/species transitions and losses, landscape-level solutions to adaptation and mitigation strategies to climate change (Singh et al. 2010).

The ongoing and persistent loss of biodiversity is the main challenge of today, which can be addressed and mitigated by targeted approach in conservation. The site, amount and kind of biodiversity to be conserved should be the focus of research, which relates very much to the contemporary developments of remote sensing. The recently launched satellites and many in various stages of development can provide us with tools and methods to address these challenges. To make progress, ecologists, evolutionary biologists and conservation biologists must bring their data sets on species distributions, levels of species richness, areas of endemism and so on to the table and combine them with the global, regional and local data sets of, for example, primary productivity and climate, which have been generated by remote sensing researchers.

7.6 Exploring Lichen Species Richness Using Remote Sensing and GIS

Various possible applications of remote sensing and GIS have barely been explored; here, we discuss their potential use in detection, mapping and modelling for lichens. Niche modelling approaches are gaining acceptance for lichen habitat mapping but needs comprehensive and robust database on environmental parameters.

Lichens are fixed components of almost all known ecosystems, but despite the diversity of their morphological and ecological forms as well as their capability of colonizing extreme habitants, they remain generally little known. They are natural sensors of our changing environment: the sensitivity of particular lichen species and assemblages to a very broad spectrum of environmental conditions, both natural and unnatural, is widely appreciated. Lichens disappear or die out as a result of the environmental changes, which directly or indirectly are anthropogenic in nature. Lichen-dominated landscapes are among the most sensitive to mechanical damage, and they are affected by various forms of anthropogenic disturbance such as agricultural and forest management (Scheidegger and Goward 2002), atmospheric pollution and climate change (Nash 1996; Nimis et al. 2002). Lichens are therefore used increasingly in evaluating threatened habitats, in environmental impact assessments and in monitoring environmental perturbations, particularly those resulting from a disturbingly large and growing number of chemical pollutants.

In the year 1866, a study was published on the use of epiphytic lichens as bioindicators (Nylander 1866), and since then, many studies have stressed the possibility of using lichens as bioindicators or biomonitors of air quality in view of their sensitivity to various environmental factors, which can provoke changes in some of their components and/or specific parameters (Gilbert 1973; Nimis 1990; Loppi 1996). Remote sensing technology, owing to its synoptic coverage and repeatability, can be used in mapping of lichen species and characterization of lichen land cover (Richardson 1991; Seaward 1993). Also, the changes in reflectance and in ecophysiological responses, such as chlorophyll levels, gaseous exchange and water absorbance, caused by anthropogenic disturbances to lichen-dominated communities, can be detected in remotely sensed images (Petzold and Goward 1988; O'Neill 1994; Karnieli et al. 2001).

Way back in 1976, researchers used lichens as indicator of pollution extent, their presence and abundance being related to pollution level (Kuliyev 1977, 1979; Kuliyev and Lobanov 1978). In a similar assessment later, Virtanen et al. (2002) attempted evaluation of lichen bioaccumulation to resolve principal impact zones, where satellite images provided evidence of long-term cumulative pollution impacts by revealing dramatic changes in plant species composition, most notably reduction of lichens. Nordberg and Allard (2002) used change in lichen cover as a tool for monitoring disturbed ecosystem; Waser et al. (2007) attempted ecological modelling for predicting species richness of lichens by combining remote sensing data and regression models; Olthof et al. (2008) studied 20 years of vegetation-specific responses to northern climate warming, revealing that lichen-dominated communities exhibit lower trends than those dominated by vascular plants; Gilichinsky et al. (2011) utilized NFI data as training data in mapping lichen classes and demonstrated high classification accuracy of SPOT imagery for classification of lichen-abundant and lichen-poor areas.

Lichen habitat mapping using remote sensing imagery was initiated for understanding of antelope population dynamics and spatial behaviour for development of adequate wildlife management and conservation plans (Crittenden 2000). Several such studies were conducted successfully in the late 1990s and early 2000s; for example, Nordberg (1998) developed a Normalized Difference Lichen Index (NDLI), derived from Landsat TM spectral bands. Later, Nordberg and Allard (2002) found Normalized Difference Vegetation Index (NDVI), to be a better predictor of lichen cover than NDLI. Dahlberg (2001) argued that NDVI might be better representative of land-cover classes than lichen biomass and recommended topography or other ancillary data to be used together with NDLI or NDVI to achieve better estimates of lichen biomass. Normalized Difference Moisture Index (NDMI) first introduced by Wilson and Sader (2002) also proved to have a potential for lichen biomass detection (Rees et al. 2004).

Measurement of lichen ground cover from satellite images largely relies on a variety of supervised and hybrid supervised classification methods to distinguish among a few rough classes of abundance, yielding a good classification accuracy for most lichen-dominated vegetation classes (Colpaert et al. 2003; Gilichinsky et al. 2011; Nordberg and Allard 2002; Tømmervik et al. 2003). Theau et al. (2005) used Landsat TM imagery for mapping lichen and evaluated results from enhancement–classification method (ECM) and spectral mixture analysis (SMA) for their suitability to characterize lichen land cover. In more recent works, Nelson et al. (2013) attempted modelling of lichen abundance by capitalizing on unique spectral characteristics of specific lichens using Landsat 7 ETM+ imagery; Falldorf et al. (2014) developed Lichen Volume Estimator and used it for obtaining a continuous prediction of lichen volume based on two previously developed indices derived from Landsat TM: NDLI and NDMI with promising results.

Like remote sensing, the number of ecological publications using GIS has grown very rapidly, and it has been used to predict species distributions and risks to biodiversity (Spens et al. 2007), to aid the visualization, exploration and modelling of data on species distributions (Lopez-Lopez et al. 2006; Vogiatzakis et al. 2006; Zhang et al. 2007) and to study the effect of major variables such as disturbance events (Pennington 2007). GIS data and its derivatives are being increasingly used now for lichen-related studies owing to their capability to capture and analyse the spatial patterns of lichen cover.

Berryman and McCune (2006) estimated biomass of epiphytic macrolichen by developing regression models based on GIS-derived topographic variables, stand structure and lichen community data; Holt et al. (2008) used GIS-based categorical variables to explore scaling effects on vegetation using lichens and identified potentially useful stratifying variables such as remote sensing-derived land-cover types for future studies; Lattman et al. (2014) used GIS for species classification while surveying a selection of lichen species on trees to show the effect of urbanization on species number and cover of lichens.

Relationship between lichen diversity and environmental factors can be even better characterized by analysing them at different scales using geostatistical methods which use preliminary GIS data. Geostatistics allows development of spatial models of variables used in interpolation with known uncertainty (Store and Jokimaki 2003) along with providing generalization and scaling-up methods (Burrough 2001). Geostatistical modelling theory is being used in basic and applied ecology (Perry et al. 2002; Liebhold and Gurevitch 2002) to provide models for studying spatial pattern of biodiversity variables and is being increasingly seen as important tools for biomonitoring studies.

Several authors have related lichen variables to neighbourhood land cover considering land-cover types as potential pollutant sources (Tømmervik et al. 1998; Aarrestad and Aamlid 1999; Augusto et al. 2004); other authors have considered distance to roads and traffic intensity (Gombert et al. 2003) and distance to farms and cattle grazing (Ruoss 1999; van Herk 1999, 2001; Wolseley et al. 2006) and associated NO_2 and NH_3 (Frati et al. 2006). Most popular and studied types of neighbourhood land cover that may induce changes in lichen diversity are urban and industrial as observed by Kapusta et al. (2004), who analysed the spatial pattern of lichen species richness in a forest ecosystem impacted for 50 years by industrial emissions using geostatistical tools to identify factors influencing it. Pinho et al. (2008) used geostatistical modelling to describe spatial relations between biodiversity data, land-cover categories and atmospheric pollutant concentrations. Ribeiro et al. (2013) applied a multivariate geostatistical method to describe relationships between abundance of a lichen species used as an ecological indicator and environmental factors at multiple spatial scales.

Intensive lichen monitoring is a necessary component of any programme aimed at effective long-term observation of environmental disturbances, both natural and man-made. The proper use of lichens as indicators and samplers of ambient conditions is a valuable resource for the environmentalist for appraisals and impact studies (Seaward 2004). Studies of changes in lichen zonation through detailed ground truthing coupled with remote sensing and GIS techniques for comprehensive mapping can form the basis for large-scale monitoring of quantitative and qualitative changes in environmental regimes, ranging from air pollution to climatic disturbances/global warming (e.g. van Herk et al. 2002; Belnap and Lange 2005). In this review, we have extensively examined biodiversity conservation studies using GIS and remote sensing and we recommend that future biodiversity conservation research prioritizes macroecological studies such as lichen cover monitoring using GIS and remote sensing.

Acknowledgments Author Mr. Manoj Semwal is thankful to Director, CSIR-CIMAP, Lucknow, for providing the facilities and encouragement.

References

Aarrestad PA, Aamlid D (1999) Vegetation monitoring in South-Varanger, Norway-species composition of ground vegetation and its relation to environmental variables and pollution impact. Environ Monit Assess 58(1):1–21

Adhikari D, Barik SK, Upadhaya K (2012) Habitat distribution modelling for reintroduction of *Ilex khasiana* Purk., a critically endangered tree species of northeastern India. Ecol Eng 40:37–43

Akcakaya HR, Butchart SHM, Mace GM, Stuart SN, Hilton-Taylor C (2006) Use and misuse of the IUCN red list criteria in projecting climate change impacts on biodiversity. Glob Change Biol 12:2037–2043

Amarnath G, Murthy MSR, Britto SJ, Rajashekar G, Dutt CBS (2003) Diagnostic analysis of conservation zones using remote sensing and GIS techniques in wet evergreen forests of the Western Ghats—an ecological hotspot, Tamil Nadu, India. Biodivers Conserv 12:2331–2359

Araujo MB, Whittaker RJ, Ladle RJ, Erhard M (2005) Reducing uncertainty in projections of extinction risk from climate change. Glob Ecol Biogeogr 14:529–538

Augusto S, Pinho P, Branquinho C, Pereira MJ, Soares A, Catarino F (2004) Atmospheric dioxin and furan deposition in relation to land-use and other pollutants: a survey with lichens. J Atmos Chem 49:53–65

Austin MP (2007) Species distribution models and ecological theory: a critical assessment of some possible new approaches. Ecol Model 200:1–19

Badarinath KVS, Sharma AR, Kharol SK (2011) Forest fire monitoring and burnt area mapping using satellite data: a study over the forest region of Kerala State, India. Int J Remote Sens 32(1):85–102

Bannari A, Khurshid KS, Staenz K, Schwarz J (2008) Potential of Hyperion EO-1 hyperspectral data for wheat crop chlorophyll content estimation. Can J Remote Sens 34:S139–S157

Barik SK, Adhikari D (2011) Predicting geographic distribution of an invasive species *Chromolaena odorata* L. (King) and H.E. Robins. In: Bhatt JR, Singh JS, Tripathi RS, Singh SP, Kohli RK (eds) Invasive alien plants—an ecological appraisal for the Indian sub-continent. CABI, Oxfordshire, pp 77–88

Behera MD, Roy PS (2010) Assessment and validation of biological richness at landscape level in part of the Himalayas and Indo-Burma hotspots using geospatial modeling approach. J Indian Soc Remote Sens 38:415–429

Behera MD, Srivastava S, Kushwaha SPS, Roy PS (2000) Stratification and mapping of *Taxus baccata* L. bearing forests in Talle Valley using remote sensing and GIS. Curr Sci 78(8):1008–1013

Belnap J, Lange OL (2005) Biological soil crusts and global changes: what does the future hold? In: Dighton J, White JF, Oudemans P (eds) The fungal community: its organization and role in the ecosystem, 3rd edn. CRC Press, Boca Raton, pp 697–712

Berry PM, Dawson TP, Berry PA, Harrison PA (2002) SPECIES: a spatial evaluation of climate impacts on the envelope of species. Ecol Model 154:289–300

Berryman S, McCune B (2006) Estimating epiphytic macrolichen biomass from topography, stand structure and lichen community data. J Veg Sci 17(2):157–170

Bharti RR, Adhikari BS, Rawat GS (2012) Assessing vegetation changes in timberline ecotone of Nanda Devi National Park, Uttarakhand. Int J Appl Earth Obs Geoinf 18:472–479

Bisby FA (2000) The quiet revolution: biodiversity informatics and the internet. Science 289 (5488):2309–2312

Bock M, Xofis P, Mitchley J, Rossner G, Wissen M (2005) Object-oriented methods for habitat mapping at multiple scales—case studies from Northern Germany and Wye Downs, UK. J Nat Conserv 13:75–89

Botkin DB, Saxe H, Araujo MB, Betts R, Bradshaw RHW, Cedhagen T, Chesson P, Dawson TP, Etterson JR, Faith DP, Ferrier S, Guisan A, Hansen AS, Hilbert DW, Loehle C, Margules C, New M, Sobel MJ, Stockwell DRB (2007) Forecasting the effects of global warming on biodiversity. Bioscience 57:227–236

Boyd DS, Danson FM (2005) Satellite remote sensing of forest resources: three decades of research development. Prog Phys Geogr 29:1–26

Boyd DS, Foody GM (2011) An overview of recent remote sensing and GIS based research in ecological informatics. Ecol Inform 6:25–36

Boyd DS, Sanchez-Hernandez C, Foody GM (2006) Mapping a specific class for priority habitats monitoring from satellite sensor data. Int J Remote Sens 27:2631–2644

Burrough PA (2001) GIS and geo-statistics: essential partners for spatial analysis. Environ Ecol Stat 8:361–377

Canada Centre for Remote Sensing (2007) Fundamentals of remote sensing: a remote sensing tutorial. http://ccrs.nrcan.gc.ca/resource/tutor/fundam/pdf/fundamentals_e.pdf

Caride C, Pineiro G, Paruelo JM (2012) How does agricultural management modify ecosystem services in the argentine Pampas? The effects on soil C dynamics. Agric Ecosyst Environ 154:23–33

Carleer A, Wolff E (2004) Exploitation of very high resolution satellite data for tree species identification. Photogram Eng Remote Sens 70:135–140

Carlson KM, Asner GP, Hughes RF, Ostertag R, Martin RE (2007) Hyperspectral remote sensing of canopy biodiversity in Hawaiian lowland rainforests. Ecosystems 10:536–549

Carvalho SB, Brito JC, Pressey RL, Crespo E, Possingham HP (2010) Simulating the effects of using different types of species distribution data in reserve selection. Biol Conserv 143(2):426–438

Clark DB, Read JM, Clark ML, Cruz AM, Dotti MF, Clark DA (2004) Application of 1-M and 4-M resolution satellite data to ecological studies of tropical rain forests. Ecol Appl 14:61–74

Cleland EE, Chuine I, Menzel A, Mooney HA, Schwartz MD (2007) Shifting plant phenology in response to global change. Trends Ecol Evol 22(7):357–365

Collingwood A, Franklin SE, Guo X, Stenhouse G (2009) A medium-resolution remote sensing classification of agricultural areas in Alberta grizzly bear habitat. Can J Remote Sens 35:23–36

Colpaert A, Kumpula J, Nieminen M (2003) Reindeer pasture biomass assessment using satellite remote sensing. Arctic 56:147–158

Crittenden PD (2000) Aspects of the ecology of mat-forming lichens. Rangifer 20(2):127–139

Crow WT, Zhan X (2007) Continental-scale evaluation of remotely sensed soil moisture products. IEEE Geosci Remote Sens Lett 4:451–455

Culvenor DS (2002) TIDA: an algorithm for delineation of tree crowns in high spatial resolution remotely sensed imagery. Comput Geosci 28:33–44

Dahdouh-Guebas F, Collin S, Lo SD, Ronnback P, Depommier D, Ravishankar T, Koedam N (2006) Analysing ethnobotanical and fishery related importance of mangroves of the East-Godavari Delta (Andhra Pradesh, India) for conservation and management purposes. J Ethnobiol Ethnomed 2(1):1–22

Dahlberg U (2001) Quantification and classification of Scandinavian mountain vegetation based on field data and optical satellite images. Licentiate thesis, Report 12. Umeå: Swedish University of Agricultural Sciences, Department of Forest Resource Management and Geomatics, 82 pp

Dalponte M, Bruzzone L, Vescovo L, Gianelle D (2009) The role of spectral resolution and classifier complexity in the analysis of hyperspectral images of forest areas. Remote Sens Environ 113:2345–2355

de Garzon MB, Dios RS, Ollero HS (2007) Predictive modelling of tree species distributions on the Iberian Peninsula during the last glacial maximum and mid-Holocene. Ecography 30:120–134

Dees M, Straub C, Koch B (2012) Can biodiversity study benefit from information on the vertical structure of forests? Utility of LiDAR remote sensing. Curr Sci 102(8):25

DeFries R, Hansen A, Newton AC, Hansen MC (2005) Increasing isolation of protected areas in tropical forests over the past twenty years. Ecol Appl 15:19–26

Dhar U (2000) Prioritization of conservation sites in the timberline zone of West Himalaya. In: Singh S, Sastry ARK, Mehta R, Uppal V (eds) Setting biodiversity conservation priorities for India. BCPP Process Document WWF India, New Delhi, pp 193–211

Diniz-Filho JAF, Nabout JC, Bini LM, Soares TN, Campos Telles MP, Marco P, Collevatti RG (2009) Niche modelling and landscape genetics of *Caryocar brasiliense* (Pequi tree, Caryocaraceae) in Brazilian Cerrado. Tree Genet Genomes 5(4):617–627. ISSN 1614-2942

Dudley N, Stolton S, Belokurov A, Krueger L, Lopoukhine N, MacKinnon K, Sandwith T, Sekhran N (2010) Natural solutions: protected areas helping people cope with climate change. IUCNWCPA, TNC, UNDP, WCS, The World Bank and WWF, Gland, Washington DC, p 130

Duncanson LI, Niemann KO, Wulder MA (2010) Estimating forest canopy height and terrain relief from GLAS waveform metrics. Remote Sens Environ 114:138–145

Duro D, Coops NC, Wulder MA, Han T (2007) Development of a large area biodiversity monitoring system driven by remote sensing. Prog Phys Geogr 31:235–260

Eckert S (2012) Improved forest biomass and carbon estimations using texture measures from WorldView-2 satellite data. Remote Sens 4:810–829

Edwards JL, Lane MA, Nielsen ES (2000) Interoperability of biodiversity databases: biodiversity information on every desktop. Science 289(5488):2312–2314

Engler R, Guisan A, Rechsteiner L (2004) An improved approach for predicting the distribution of rare and endangered species from occurrence and pseudo-absence data. J Appl Ecol 41:263–274

Everitt JH, Yang C, Deloach CJ Jr (2006) Remote sensing of giant reed with QuickBird satellite imagery. J Aquat Plant Manage 43:81–85

Fagan M, DeFries RS (2009) Measurement and monitoring of the world's forests: a review and summary of remote sensing technical capability, 2009–2015. Resources for the Future

Fairbanks DHK, McGwire KC (2004) Patterns of floristic richness in vegetation communities of California: regional scale analysis with multi-temporal NDVI. Glob Ecol Biogeogr 13:221–235

Falkowski MJ, Smith AMS, Hudak AT, Gessler PE, Vierling LA, Crookston NL (2006) Automated estimation of individual conifer tree height and crown diameter via two-dimensional spatial wavelet analysis of lidar data. Can J Remote Sens 32:153–161

Falldorf T, Strand O, Panzacchi M, Tømmervik H (2014) Estimating lichen volume and reindeer winter pasture quality from Landsat imagery. Remote Sens Environ 140:573–579

Ferrier S, Guisan A (2006) Spatial modelling of biodiversity at the community level. J Appl Ecol 43:393–404
Foody GM (2003) Remote sensing of tropical forest environments: towards the monitoring of environmental resources for sustainable development. Int J Remote Sens 24:4035–4046
Foody GM (2004) Supervised image classification by MLP and RBF neural networks with and without an exhaustively defined set of classes. Int J Remote Sens 25:3091–3104
Foody GM (2005) Mapping the richness and composition of British breeding birds from coarse spatial resolution satellite sensor imagery. Int J Remote Sens 26:3943–3956
Foody GM, Cutler ME (2006) Mapping the species richness and composition of tropical forests from remotely sensed data with neural networks. Ecol Model 195(1):37–42
Foody GM, Mathur A (2004) A relative evaluation of multiclass image classification by support vector machines. IEEE Trans Geosci Remote Sens 42 (6):1335–1343
Forster M, Kleinschmit B (2008) Object-based classification of Quickbird data using ancillary information for the detection of forest types and NATURA 2000 habitats. In: Blaschke T, Lang S, Hay GJ (eds) Object-based image analysis: spatial concepts for knowledge-driven remote sensing applications. Lecture notes in geoinformation and cartography, Sect. 3. Springer, Berlin, pp 275–290
Foster JR, Townsend PA, Zganjar CE (2008) Spatial and temporal patterns of gap dominance by low-canopy lianas detected using EO-1 Hyperion and Landsat thematic mapper. Remote Sens Environ 112 (5):2104–2117
Franklin SE (1994) Discrimination of subapline forest species and canopy density sing digital CASI, SPOT PLA, and Landsat TM data. Photogram Eng Remote Sens 60:1223–1241
Frati L, Caprasecca E, Santoni S, Gaggi C, Guttova A, Gaudino S, Pati A, Rosamilia S, Pirintsos SA, Loppi S (2006) Effects of NO_2 and NH_2 from road traffic on epiphytic lichens. Environ Pollut 142:58–64
Fuller RM, Devereux BJ, Gillings S, Hill RA, Amable GS (2007) Bird distributions relative to remotely sensed habitats in Great Britain: towards a framework for national modeling. J Environ Manage 84:586–605
Giannini TC, Lira-Saade R, Ayala R, Saraiva AM, Alves dos Santos I (2011) Ecological niche similarities of *Peponapis* bees and non-domesticated *Cucurbita* species. Ecol Model 222(12):2011–2018, 0304-3800
Gilbert OL (1973) Lichens and air pollution. In: Ahmadjian V, Hale ME (eds) The Lichens. Academic Press, New York, pp 443–472
Gilichinsky M, Sandstrom P, Reese H, Kivinen S, Nilsson M (2011) Mapping ground lichens using forest inventory and optical satellite data. Int J Remote Sens 32:455–472
Gillespie TW, Foody GM, Rocchini D, Giorgi AP, Saatchi S (2008) Measuring and modeling biodiversity from space. Prog Phys Geogr 32:203–221
Gillies RR, Temesgen B (2004) Coupling thermal infrared and visible satellite measurements to infer biophysical variables at the land surface. Thermal Remote Sens Land Surf Process 160–184
Giriraj A, Irfan-Ullah M, Ramesh BR, Karunakaran PV, Jentsch A, Murthy MSR (2008) Mapping the potential distribution of Rhododendron arboreum Sm. ssp. nilagiricum (Zenker) Tagg (Ericaceae), an endemic plant using ecological niche modelling. Curr Sci 94:12
Giriraj A, Murthy MSR, Ramesh BR, Dutt CBS (2009) A method for assessing evergreen habitats using phytodiversity and geospatial techniques in tropical rain forests of Southern Western Ghats (India). Ecol Res 24:749–760
Gombert S, Asta J, Seaward MRD (2003) Correlation between the nitrogen concentration of two epiphytic lichens and the traffic density in an urban area. Environ Pollut 123:281–290
Gottschalk TK, Huettmann F, Ehler M (2005) Thirty years of analyzing and modeling avian habitat relationships, using satellite imagery: a review. Int J Remote Sens 26:2631–2656
Gould W (2000) Remote sensing of vegetation, plant species richness, and regional biodiversity hotspots. Ecol Appl 10(186):187
Graham CH, Hijmans RJ (2006) A comparison of methods for mapping species ranges and species richness. Glob Ecol Biogeogr 15:578–587
Guerschman JP, Hill MJ, Renzullo LJ, Barrett DJ, Marks AS, Botha EJ (2009) Estimating fractional cover of photosynthetic vegetation, non-photosynthetic vegetation and bare soil in the Australian tropical savanna region upscaling the EO-1 Hyperion and MODIS sensors. Remote Sens Environ 113:928–945
Guisan A, Thuiller W (2005) Predicting species distribution: offering more than simple habitat models. Ecol Lett 8:993–1009
Gupta S, Porwal MC, Roy PS (2004) Human modification of the tropical rain forest of Nicobar Islands: indicators from land use land cover mapping. J Hum Ecol 16 (3):163–171
Hamann ML, Wang TL (2006) Potential effects of climate change on ecosystem and tree species distribution in British Columbia. Ecology 87:2773–2786
He KS, Zhang J (2009) Testing the correlation between beta diversity and differences in productivity among global ecoregions, biomes, and biogeographical realms. Ecol Inform 4:93–98
He H, Yang M, Pan Y, Zhu W (2005) Measurement of terrestrial ecosystem service value in China based on remote sensing. In: International geoscience and remote sensing symposium (IGARSS), pp 2965–2968
Helmer EH, Lefsky MA, Roberts DA (2009) Biomass accumulation rates of Amazonian secondary forest and biomass of old-growth forests from Landsat time series and the Geoscience Laser Altimeter System. J Appl Remote Sens 3:033505
Hirschmugl M, Ofner M, Raggam J, Schardt M (2007) Single tree detection in very high resolution remote sensing data. Remote Sens Environ 110:533–544

Holt EA, McCune B, Neitlich P (2008) Spatial scale of GIS derived categorical variables affects their ability to separate sites by community composition. Appl Veg Sci 11(3):421–430

Hurtt G, Xiao XM, Keller M, Palace M, Asner GP, Braswell R, Brondizio ES, Cardoso M, Carvalho CJR, Fearon MG, Guild L, Hagen S, Hetrick S, Moore B, Nobre C, Read JM, Sa T, Schloss A, Vourlitis G, Wickel AJ (2003) IKONOS imagery for the large scale biosphere-atmosphere experiment in Amazonia (LBA). Remote Sens Environ 88:111–127

Hyyppa J, Wagner W, Hollaus M, Hyyppa H (2009) Airborne laser scanning. In: Warner TA, Nellis MD, Foody GM (eds) The SAGE handbook of remote sensing. SAGE Publications Ltd., London, pp 199–211

ICIMOD (International Centre for Integrated Mountain Development) (2010) Climate change impact and vulnerability in the Eastern Himalayas-synthesis report. Kathmandu, Nepal

Irfan-Ullah M, Amarnath G, Murthy MSR, Peterson AT (2006) Mapping the geographic distribution of Aglaia bourdillonii Gamble (Meliaceae), an endemic and threatened plant using ecological niche modeling. Biodivers Conserv 16:1917–1925

Jensen JR, Im J, Hardin P, Jensen RR (2009) Image classification. In: Warner TA, Nellis MD, Foody GM (eds) The SAGE handbook of remote sensing. SAGE Publications Ltd., London, pp 269–281

Jha CS, Goparaju L, Tripathi A, Gharai B, Raghubanshi AS, Singh JS (2005) Forest fragmentation and its impact on species diversity: an analysis using remote sensing and GIS. Biodivers Conserv 14:1681–1698

Joseph S, Murthy MSR, Thomas AP (2011) The progress on remote sensing technology in identifying tropical forest degradation: a synthesis of the present knowledge and future perspectives. Environ Earth Sci 64 (3):731–741

Joshi PK, Rawat GS, Padaliya H, Roy PS (2005) Land use/land cover identification in an alpine and arid region (Nubra Valley, Ladakh) using satellite remote sensing. J Indian Soc Remote Sens 22:371–380

Kamat DS (1986) An integrated approach to remote sensing studies for wildlife habitat evaluation. In: Proceedings of seminar-cum-workshop on wildlife habitat evaluation using remote sensing techniques, Dehradun, pp 165–182

Kandwal R, Jeganathan C, Tolpekin V, Kushwaha SPS (2009) Discriminating the invasive species, lantana using vegetation indices. J Indian Soc Remote Sens 37:275–290

Kanniah KD (2011) Worldview-2 remote sensing data for tropical mangrove species classification. Research report submitted to DigitalGlobe incorporated, USA for the WorldView-2, 8-Bands Research Challenge. Longmont, CO, DigitalGlobe

Kapusta P, Szarek-Lukaszewska G, Kiszka J (2004) Spatial analysis of lichen species richness in a disturbed ecosystem (Niepolomice Forest, S Poland). Lichenologist 36(3–4):249–260

Karnieli A, Kokaly R, West NE, Clark RN (2001) Remote sensing of biological soil crusts. In: Belnap J, Lange OL (eds) Biological soil crusts: structure, function and management. Springer, Berlin, pp 431–455

Kellndorfer J, Walker W, Nepstad D, Stickler C, Brando P, Lefebvre P, Rosenqvist A, Shimada M (2008) Implementing REDD: the potential of ALOS/PALSAR for forest mapping and monitoring. Paper presented at the second GEOSS Asia-Pacific symposium, Tokyo, Japan

Keramitsoglou I, Kontoes C, Sykioti O, Sifakis N, Xofis P (2008) Reliable, accurate and timely forest mapping for wildfire management using ASTER and Hyperion satellite imagery. For Ecol Manage 255:3556–3562

Kerr JT, Ostrovsky M (2003) From space to species: ecological applications for remote sensing. Trends Ecol Evol 18(6):299–305

Kerr JT, Southwood TRE, Cihlar J (2001) Remotely sensed habitat diversity predicts butterfly species richness and community similarity in Canada. Proc Natl Acad Sci USA 98:11365–11370

Kimothi MM, Anitha D, Vasistha HB, Soni P, Chandola SK (2010) Remote sensing to map the invasive weed, Lantana camara in forests. Trop Ecol 51:67–74

Kiran Chand TR, Badarinath KVS, Krishna Prasad V, Murthy MSR, Elvidge CD, Tuttle BT (2006) Monitoring forest fires over the Indian region using defense meteorological satellite program-operational linescan system nighttime satellite data. Remote Sens Environ 103(2):165–178

Koch B, Heyder U, Weinacker H (2006) Detection of individual tree crowns in airborne lidar data. Photogram Eng Remote Sens 72:357–363

Krishna PH, Reddy CS (2012) Assessment of increasing threat of forest fires in Rajasthan, India using multitemporal remote sensing data (2005–2010). Curr Sci 102(10) (00113891)

Kuliyev AN (1977) The effect of the dust pollution to tundra vegetation in Vorkuta outskirts (the eastern part of the Bolshezemelskaya Tundra). In: Syroyechkovskiy YeYe (ed) Scientific fundamentals of nature protection, Moscow, pp 22–28

Kuliyev AN (1979) A study of the dust pollution effects to tundra vegetation in Vorkuta outskirts. Syroyechkovskiy YeYe (ed) The influence of human activities on natural ecosystems, Moscow, pp 60–66

Kuliyev AN, Lobanov VA (1978) Distribution of dust pollution in Vorkuta outskirts. In: Syroyechkovskiy YeYe (ed) The influence of human activities on natural ecosystems, Moscow, pp 81–89

Kumar S, Stohlgren TJ (2009) Maxent modeling for predicting suitable habitat for threatened and endangered tree Canacomyrica monticola in New Caledonia. J Ecol Nat Environ 1:94–98

Kumar A, Ghosh SK, Dadhwal VK (2007) Full fuzzy land cover mapping using remote sensing data based on fuzzy c-means and density estimation. Can J Remote Sens 33(2):81–87

Landenberger RE, McGraw JB, Warner TA, Brandtberg T (2003) Potential of digital color imagery for censusing

Haleakala silverswords in Hawaii. Photogram Eng Remote Sens 69:915–923

Lattman H, Bergman KO, Rapp M, Talle M, Westerberg L, Milberg P (2014) Decline in lichen biodiversity on oak trunks due to urbanization. Nord J Bot 32 (4):518–528

Lee Z, Casey B, Arnone R, Weidemann A, Parsons R, Montes MJ, Gao BC, Goode W, Davis CO, Dye J (2007) Water and bottom properties of a coastal environment derived from Hyperion data measured from the EO-1 spacecraft platform. J Appl Remote Sens 1:011502

Lefsky MA (2010) A global forest canopy height map from the Moderate Resolution Imaging Spectroradiometer and the Geoscience Laser Altimeter System. Geophys Res Lett 37:L15401

Lefsky MA, Keller M, de Pang Y, Camargo PB, Hunter MO (2007) Revised method for forest canopy height estimation from geoscience laser altimeter system waveforms. J Appl Remote Sens 1:013537

Leidner AK, Turner W, Pettorelli N, Leimgruber P, Wegmann M (2012) Satellite remote sensing for biodiversity research and conservation applications: A Committee on Earth Observation Satellites (CEOS) workshop. http://remote-sensing-biodiversity.org/images/workshops/ceos/CEOS_SBA_Biodiversity_WorkshopReport_Oct2012_DLR_Munich.pdf

Lele N, Joshi PK (2009) Analyzing deforestation rates, spatial forest cover changes and identifying critical areas of forest cover changes in north-east India during 1972–1999. Environ Monit Assess 156:159–170

Levin N, Shimida A, Levanoni O, Tamari H, Kark S (2007) Predicting mountain plant richness and rarity from space using satellite-derived vegetation indices. Divers Distrib 13:1–12

Leyequien E, Verrelst J, Slot M, Schaepman-Strub G, Heitkonig IMA, Skidmore A (2007) Capturing the fugitive: applying remote sensing to terrestrial animal distribution and diversity. Int J Appl Earth Obs Geoinf 9:1–20

Li J, Chen W (2005) A rule-based method for mapping Canada's wetlands using optical, radar, and DEM data. Int J Remote Sens 26:5051–5069

Liebhold AM, Gurevitch J (2002) Integrating the statistical analysis of spatial data in ecology. Ecography 25:553–557

Lim K, Treitz P, Wulder M, St-Onge B, Flood M (2008) LIDAR remote sensing of forest structure. Prog Phys Geogr 27:88–106

Lindenmayer DB, Gibbons P, Bourke MAX, Burgman M, Dickman CR, Ferrier S, Zerger A (2012) Improving biodiversity monitoring. Aust Ecol 37(3):285–294

Loarie SR, Joppa LN, Pimm SL (2008) Satellites miss environmental priorities: response to Loveland et al. and Kark et al. Trends Ecol Evol 23(4):183–184

Lopez-Lopez P, Garcia-Ripolles C, Aguilar J, Garcia-Lopez F, Verdejo J (2006) Modelling breeding habitat preferences of Bonelli's eagle (Hieraaetus fasciatus) in relation to topography, disturbance, climate and land use at different spatial scales. J Ornithol 147:97–106

Loppi S (1996) Lichens as bioindicators of geothermal air pollution in central Italy. Bryologist 99(1):41–48

Lorenzen ED, Nogués-Bravo D, Orlando L, Weinstock J, Binladen J, Marske KA, Ugan A, Borregaard MK, Gilbert MTP, Nielsen R, Ho SYW, Goebel T, Graf KE, Byers D, Stenderup JT, Rasmussen M, Campos PF, Leonard JA, Koepfli K-P, Froese D, Zazula G, Stafford TW Jr, Aaris-Sørensen K, Batra P, Haywood AM, Singarayer JS, Valdes PJ, Boeskorov G, Burns JA, Davydov SP, Haile J, Jenkins DL, Kosintsev P, Kuznetsova T, Lai X, Martin LD, McDonald HG, Mol D, Meldgaard M, Munch K, Stephan E, Sablin M, Sommer RS, Sipko T, Scott E, Suchard MA, Tikhonov A, Willerslev R, Wayne RK, Cooper A, Hofreiter M, Sher A, Shapiro B, Rahbek C, Willerslev E (2011) Species-specific responses of late quaternary megafauna to climate and humans. Nature 479:359–364

Lu D, Weng Q (2007) A survey of image classification methods and techniques for improving classification performance. Int J Remote Sens 28:823–870

Lucas KL, Carter GA (2008) The use of hyperspectral remote sensing to assess vascular plant species richness on Horn Island, Mississippi. Remote Sens Environ 112:3908–3915

Lucas R, Rowlands A, Brown A, Keyworth S, Bunting P (2007) Rule-based classification of multi-temporal satellite imagery for habitat and agricultural land cover mapping. ISPRS J Photogram Remote Sens 62:165–185

Lucas R, Medcalf K, Brown A, Bunting P, Breyer J, Clewley D, Key-worth S, Blackmore P (2011) Updating the phase 1 habitat map of Wales UK, using satellite sensor data. ISPRS J Photogram Remote Sens 66:81–102

Luoto M, Kuussaari M, Toivonen T (2002) Modelling butterfly distribution based on remote sensing data. J Biogeogr 29:1027–1037

Luoto M, Virkkala R, Heikkinen RK (2007) The role of land cover in bioclimatic models depends on spatial resolution. Glob Ecol Biogeogr 16:34–42

Mildrexler DJ, Zhao MS, Heinsch FA, Running SW (2007) A new satellite-based methodology for continental-scale disturbance detection. Ecol Appl 17:235–250

Moerman DE, Estabrook GF (2006) The botanist effect: counties with maximal species richness tend to be home to universities and botanists. J Biogeogr 33:1969–1974

Mokany K, Ferrier S (2011) Predicting impacts of climate change on biodiversity: a role for semi-mechanistic community-level modelling. Divers Distrib 17:374–380

Mukhopadhyay A, Maulik U (2009) Unsupervised pixel classification in satellite imagery using multiobjective fuzzy clustering combined with SVM classifier. IEEE Trans Geosci Remote Sens 47(4):1132–1138

Muldavin EH, Neville P, Harper G (2001) Indices of grassland biodiversity in the Chihuahuan Desert ecoregion derived from remote sensing. Conserv Biol 15:844–855

Nagendra H (2001) Using remote sensing to assess biodiversity. Int J Remote Sens 22:2377–2400 (NRSA, 2007)

Nagendra H (2008) Do parks work? Impact of protected areas on land cover clearing. AMBIO J Human Environ 37(5):330–337

Nagendra H, Gadgil M (1999) Biodiversity assessment at multiple scales: linking remotely sensed data with field information. Proc Natl Acad Sci USA 96:9154–9158

Nagendra H, Rocchini D (2008) High resolution satellite imagery for tropical biodiversity studies: the devil is in the detail. Biodivers Conserv 17:3431–3442

Nagendra H, Rocchini D, Ghate R (2010) Beyond parks as monoliths: spatially differentiating park-people relationships in the Tadoba Andhari Tiger Reserve in India. Biol Conserv 143:2900–2908

Nagendra H, Lucas R, Honrado JP, Jongman RH, Tarantino C, Adamo M, Mairota P (2013) Remote sensing for conservation monitoring: assessing protected areas, habitat extent, habitat condition, species diversity, and threats. Ecol Ind 33:45–59

Nandy S, Kushwaha SPS (2010) Study on the utility of IRS 1D LISS-III data and the classification techniques for mapping of Sunderban mangroves. J Coast Conserv 15(1):123–137

Nash T (ed) (1996) Lichen biology. Cambridge University Press, Cambridge

Nativi S, Mazzettia P, Saarenma H, Kerr J, Tuama ÉÓ (2009) Biodiversity and climate change use scenarios framework for the GEOSS interoperability pilot process. Ecol Inf 4:23–33

Nelson PR, Roland C, Macander MJ, McCune B (2013) Detecting continuous lichen abundance for mapping winter caribou forage at landscape spatial scales. Remote Sens Environ 137:43–54

Nielson RM, McDonald LL, Sullivan JP, Burgess C, Johnson DS, Johnson DH, Bucholtz S, Hyberg S, Howlin S (2008) Estimating the response of Ring-necked Pheasants (*Phasianus colchicus*) to the conservation reserve program. Auk 125:434–444

Nimis PL (1990) Air quality indicators and indices: the use of plants as bioindicators for monitoring air pollution. EUR 13060:93–126

Nimis PL, Scheidegger C, Wolseley PA (2002) Monitoring with lichens-monitoring lichens. Springer, Netherlands, pp 1–4

Nordberg ML (1998) Vegetation and biomass changes in mountainous areas in Sweden using satellite and airborne imaging scanner data. ISRSE/NSC. In: Proceedings for the 27th international symposium on remote sensing of environment, Tromso, Norway, pp 431–435

Nordberg ML, Allard A (2002) A remote sensing methodology for monitoring cover. Can J Remote Sens 28:262–274

Nosetto MD, Jobbagy EG, Paruelo JM (2005) Land-use change and water losses: the case of grassland afforestation across a soil textural gradient in central Argentina. Glob Change Biol 11:1101–1117

Nylander W (1866) Les lichens du Jardin de Luxembourg (Lichens from the Luxembourg Gardens)

O'Neill AL (1994) Reflectance spectra of microphytic soil crusts in semi-arid Australia. Int J Remote Sens 15:675–681

Oindo BO, Skidmore AK (2002) Interannual variability of NDVI and species richness in Kenya. Int J Remote Sens 23:285–298

Oldeland J, Wesuls D, Rocchini D, Schmidt M, Jurgens N (2010) Does using species abundance data improve estimates of species diversity from remotely sensed spectral heterogeneity? Ecol Ind 10:390–396

Oliver I, Pik A, Britton D, Dangerfield JM, Colwell RK, Beattie AJ (2000) Virtual biodiversity assessment systems. Bioscience 50:441–449

Olthof I, Pouliot D, Latifovic R, Chen W (2008) Recent (1986–2006) vegetation-specific NDVI trends in northern Canada from satellite data. Arctic 61(4):381–394

Omar H (2010) Commercial timber tree species identification using multispectral Worldview-2 data. Research report submitted to DigitalGlobe Incorporated, USA for the WorldView-2, 8-Bands Research Challenge. DigitalGlobe, Longmont

Palmer MW, Earls P, Hoagland BW, White PS, Wohlgemuth T (2002) Quantitative tools for perfecting species lists. Environmetrics 13:121–137

Panigrahy S, Anitha D, Kimothi MM, Singh SP (2010) Timberline change detection using topographic map and satellite imagery. Trop Ecol 51:87–91

Panwar HS (1986) Forest cover mapping for planning tiger corridors between Kanha and Bandhavgarh—a proposed project. In: Proceedings of seminar-cum-workshop wildlife habitat evaluation using remote sensing techniques, IIRS Dehradun, pp 209–212

Pautasso M, McKinney ML (2007) The botanist effect revisited: plant species richness, county area, and human population size in the United States. Conserv Biol 21:1333–1340

Pavoine S, Vela E, Gachet S, Belair G, Bonsall MB (2011) Linking patterns in phylogeny, traits, abiotic variables and space: a novel approach to linking environmental filtering and plant community assembly. J Ecol 99(1):165–175. ISSN 1365-2745

Pearson RG, Dawson TP (2003) Predicting the impacts of climate change on the distribution of species: are bioclimate envelope models useful? Glob Ecol Biogeogr 12:361–371

Pennington DD (2007) Exploratory modelling of forest disturbance scenarios in central Oregon using computational experiments in GIS. Ecol Inf 2:387–403

Perotto-Baldivieso HL, Melendez-Ackerman E, Garcia MA, Leimgruber P, Cooper SM, Martinez A, Calle P, Ramos Gonzales OM, Quinones M, Christen CA, Pons G (2009) Spatial distribution, connectivity, and the influence of scale: habitat availability for the endangered Mona Island rock iguana. Biodivers Conserv 18:905–917

Perry JN, Liebhold AM, Rosenberg MS, Dungan J, Miriti M, Jakomulska A, Citron-Pousty S (2002) Illustrations and guidelines for selecting statistical methods for quantifying spatial pattern in ecological data. Ecography 25:578–600

Peterson AT, Sánchez-Cordero V, Martínez-Meyer E, Navarro-Sigüenza AG (2006) Tracking population extirpations via melding ecological niche modeling with land-cover information. Ecol Model 195:229–236

Petropoulos G, Carlson TN, Wooster MJ, Islam S (2009) A review of Ts/VI remote sensing based methods for the retrieval of land surface energy fluxes and soil surface moisture. Prog Phys Geogr 33:224–250

Petzold DE, Goward SN (1988) Reflectance spectra of subarctic lichens. Remote Sens Environ 24:481–492

Pfeifer M, Disney M, Quaife T, Marchant R (2011) Terrestrial ecosystems from space: a review of earth observation products for macroecology applications. Glob Ecol Biogeogr. doi:10.1111/j.1466-8238.2011.00712

Pflugmacher D, Cohen W, Kennedy R, Lefsky M (2008) Regional applicability of forest height and above-ground biomass models for the Geoscience Laser Altimeter System. For Sci 54:647–657

Phillips OL, Martínez RV, Vargas PN, Monteagudo AL, Zans MEC, Galiano WG, Pena Cruz A, Timana M, Yli-Halla M, Rose S (2003) Efficient plot-based floristic assessment of tropical forests. J Trop Ecol 19:629–45

Pignatti S, Cavalli RM, Cuomo V, Fusilli L, Pascucci S, Poscolieri M, Santini F (2009) Evaluating Hyperion capability for land cover mapping in a fragmented ecosystem: Pollino National Park, Italy. Remote Sens Environ 113:622–634

Pinho P, Augusto S, Martins-Loução M, Pereira MJ, Soares A, Máguas C, Branquinho C (2008) Causes of change in nitrophytic and oligotrophic lichen species in a Mediterranean climate: impact of land cover and atmospheric pollutants. Environ Pollut 154:380–389

Pitkanen J (2001) Individual tree detection in digital aerial images by combining locally adaptive binarization and local maxima methods. Can J For Res 31:832–844

Planning Commission, Government of India (2006) Report of task force on the mountain ecosystem: environmental and forest sector. Planning Commission India, New Delhi

Plourde LC, Ollinger SV, Smith ML, Martin ME (2007) Estimating species abundance in a northern temperate forest using spectral mixture analysis. Photogram Eng Remote Sens 73:829–840

Porwal MC, Sharma L, Roy PS (2003) Stratification and mapping of Ephedra Gerardiana Wall. in Poh (Lahul and Spiti) using remote sensing and GIS. Curr Sci 84:208212

Quattrochi DA, Luvall JC (2009) Thermal remote sensing in Earth science research. In: Warner TA, Nellis MD, Foody GM (eds) The SAGE handbook of remote sensing. SAGE Publications Ltd., London, pp 64–78

Rahman M, Ullah R, Lan M, Sri Sumantyo JT, Kuze H, Tateishi R (2013) Comparison of Landsat image classification methods for detecting mangrove forests in Sundarbans. Int J Remote Sens 34(4):1041–1056

Ravan SA, Roy PS, Sharma CM (1995) Space remote sensing for spatial vegetation characterisation. J Biosci 20:427–438

Raxworthy CJ, Martinez-Meyer E, Horning N, Nussbaum RA, Schneider GE, Ortega-Huerta MA, Peterson AT (2003) Predicting distributions of known and unknown reptile species in Madagascar. Science 426:837–841

Read JM, Clark DB, Venticinque EM, Moreira MP (2003) Application of merged 1-m and 4-m resolution satellite data to research and management in tropical forests. J Appl Ecol 40:592–600

Reddy CS, Pattanaik C, Murthy MSR (2007) Assessment and monitoring of mangroves of Bhitarkanika Wildlife Sanctuary, Orissa, India using remote sensing and GIS. Curr Sci 92(10):1409–1415

Reddy CS, Hari Krishna P, Anitha K, Joseph S (2012) Mapping and inventory of forest fires in Andhra Pradesh, India: current status and conservation needs. ISRN Forest 2012, Article ID 380412

Rees WG, Tutubalina OV, Golubeva EI (2004) Reflectance spectra of subarctic lichens between 400 and 2400 nm. Remote Sens Environ 90:281–292

Ribeiro MC, Pinho P, Llop E, Branquinho C, Sousa AJ, Pereira MJ (2013) Multivariate geostatistical methods for analysis of relationships between ecological indicators and environmental factors at multiple spatial scales. Ecol Ind 29:339–347

Richardson DHS (1991) Lichens as biological indicators - recent developments. In: Jeffrey DW, Madden B (eds) Bioindicators and environmental management. Academic Press, London, pp 263–272

Richardson DM, Iponga DM, Roura-Pascual N, Krug RM, Milton SJ, Hughes GO, Thuiller W (2010) Accommodating scenarios of climate change and management in modelling the distribution of the invasive tree Schinus molle in South Africa. Ecography 33:1049–1061

Rocchini D (2007) Distance decay in spectral space in analysing ecosystem beta-diversity. Int J Remote Sens 28:2635–2644

Rocchini D, Ricotta C (2007) Are landscapes as crisp as we may think? Ecol Model 204:535–539

Rocchini D, Ricotta C, Chiarucci A (2007) Using satellite imagery to assess plant species richness: the role of multispectral systems. Appl Veg Sci 10:325–331

Rocchini D, He KS, Zhang JT (2009) Is spectral distance a proxy of beta diversity at different taxonomic ranks? A test using quantile regression. Ecol Inf 4:254–259

Roy PS (2003) Forest fire and degradation assessment using satellite remote sensing and geographic information system. In: Proceedings of a training workshop satellite remote sensing and GIS applications in agricultural meteorology, pp 361–400

Roy PS (2011) Geospatial characterization of biodiversity: need and challenges. ISPRS Archives XXXVIII-8/W20. In: Workshop proceedings on earth observation for terrestrial ecosystem, Bhopal, Nov 8, pp 10–16

Roy PS, Tomar S (2000) Biodiversity characterization at landscape level using geospatial modelling technique. Biol Conserv 95(1):95–109

Roy PS, Tomar S (2001) Landscape cover dynamics pattern in Meghalaya. Int J Remote Sens 22 (18):3813–3825

Roy PS, Porwal MC, Sharma L (2001) Mapping of Hyppophae rhamnoides Linn. in the adjoining areas of Kaza in Lahul and Spiti using remote sensing and GIS. Curr Sci 80:1107–1111

Roy PS, Kushwaha SPS, Murthy MSR, Roy A, Kushwaha D, Reddy CS, Behera MD, Mathur VB, Padalia H, Saran S, Singh S, Jha CS, Porwal MC (2012) Biodiversity characterisation at landscape level: national assessment, Indian Institute of Remote Sensing, Dehradun, India, p 140. ISBN 81-901418-8-0

Roy PS, Murthy MSR, Roy A, Kushwaha SPS, Singh S, Jha CS, Behera MD, Joshi PK, Jagannathan C, Karnatak HC, Saran S, Reddy CS, Kushwaha D, Dutt CBS, Porwal MC, Sudhakar S, Srivastava VK, Padalia H, Nandy S, Gupta S (2013) Forest fragmentation in India. Curr Sci 105(6):774–780

Ruoss E (1999) How agriculture affects lichen vegetation in Central Switzerland. Lichenologist 31:63–73

Saatchi S, Agosti D, Alger K, Delabie J, Musinsky J (2001) Examining fragmentation and loss of primary forest in Southern Bahian Atlantic forest of Brazil with radar imagery. Conserv Biol 15:867–875

Saatchi S, Buermann W, ter Mori S, Steege H, Smith T (2008) Modeling distribution of Amazonian tree species and diversity using remote sensing measurements. Remote Sens Environ 112(5):2000–2017

SAC (2001) Report on Grassland mapping in Gujarat using remote sensing and GIS techniques. Space Application Centre (SAC) ISRO, Ahmedabad

Sanchez-Hernandez C, Boyd DS, Foody GM (2007a) Mapping specific habitats from remotely sensed imagery: support vector machine and support vector data description based classification of coastal saltmarsh habitats. Ecol Inf 2:83–88

Sanchez-Hernandez C, Boyd DS, Foody GM (2007b) One-class classification for mapping a specific land-cover class: SVDD classification of fenland. IEEE Trans Geosci Remote Sens 45:1061–1073

Santos X, Brito JC, Sillero N, Pleguezuelos JM, Llorente GA, Fahd S, Parellada X (2006) Inferring habitat-suitability areas with ecological modelling techniques and GIS: a contribution to assess the conservation status of Vipera latastei. Biol Conserv 130:416–425

Saran S, Joshi R, Sharma S, Padalia H, Dadhwal VK (2010) Geospatial modeling of brown oak (Quercus semecarpifolia Sm.) habitats in the Kumaun Himalaya under climate change scenario. J Indian Soc Remote Sens 38:534–547

Saranya KRL, Reddy CS, Rao PP, Jha CS (2014) Decadal time-scale monitoring of forest fires in Similipal biosphere reserve, India using remote sensing and GIS. Environ Monit Assess 1–14(186):3283–3296

Scheidegger C, Goward T (2002) Monitoring lichens for conservation: red lists and conservation action plans. In: Nimis PL, Scheidegger C, Wolseley PA (eds) Monitoring with lichens—monitoring lichens. IV. Earth and environmental science. Kluwer Academic, Dordrecht, pp 163–181

Schulman L, Toivonen T, Ruokolainen K (2007) Analysing botanical collecting efforts in Amazonia and correcting for it in species range estimation. J Biogeogr 34:1388–1399

Scribner KT, Arntzen JW, Cruddace N, Oldham RS, Burke T (2001) Environmental correlates of toad abundance and population genetic diversity. Biol Conserv 98(2):201–210

Seaward MRD (1993) Lichens and sulphur dioxide air pollution: field studies. Environ Rev 1:73–91

Seaward MRD (2004) The use of lichens for environmental impact assessment. Symbiosis 37:293–305

Sharma S, Palni LMS, Roy PS (2002) Vegetation fragmentation and anthropogenic disturbance in a Himalayan landscape. Asian J Geoinformatics 12 (4):62–69

Shippert P (2004) Why use hyperspectral imagery? Photogram Eng Remote Sens 70(4):377–396

Simard M, Rivera-Monroy VH, Mancera-Pineda JE, Castaneda-Moya E, Twilley RR (2008) A systematic method for 3D mapping of mangrove forests based on shuttle radar topography mission elevation data, ICEsat/GLAS waveforms and field data: application to Cienaga Grande de Santa Marta, Colombia. Remote Sens Environ 112:2131–2144

Singh IJ, Singh SK, Kushwaha SPS, Ashutosh S, Singh RK (2004) Assessment and monitoring of estuarine mangrove forests of Goa using satellite remote sensing. J Indian Soc Remote Sens 32(2):167–174

Singh S, Singh TP, Srivastava G (2005) Vegetation cover type mapping in Mouling National Park in Arunachal Pradesh, Eastern Himalayas—an integrated geospatial approach. J Indian Soc Remote Sens 33:547–563

Singh JS, Roy PS, Murthy MSR, Jha CS (2010) Application of landscape ecology and remote sensing for assessment, monitoring and conservation of biodiversity. J Indian Soc Remote Sens 38:365–385

Skov F, Svenning JC (2004) Potential impact of climatic change on the distribution of forest herbs in Europe. Ecography 27:366–380

Spens J, Englund G, Lundqvist H (2007) Network connectivity and dispersal barriers: using geographical information system (GIS) tools to predict landscape scale distribution of a key predator (Esox lucius) among lakes. J Appl Ecol 44(6):1127–1137

Stickler CM, Southworth J (2008) Application of a multi-scale spatial and spectral analysis to predict primate occurrence and habitat associations in Kibale National Park, Uganda. Remote Sens Environ 112:2170–2186

Store R, Jokimaki J (2003) A GIS-based multi-scale approach to habitat suitability modeling. Ecol Model 169:1–15

Subedi MR (2009) Climate change and its potential effects on tree line position: an introduction and analysis. Greenery J Environ Biodivers 7:17–21

Sudeesh S, Reddy CS (2013) Forest fire monitoring in Nagarjunasagar-Srisailam Tiger Reserve, Andhra Pradesh, India using geospatial techniques. Natl Acad Sci Lett 36(4):437–446

Talukdar G (2004) Geospatial modelling of shifting cultivation induced landscapes of Meghalaya. Doctoral dissertation, PhD thesis

Thakur AK, Singh G, Singh S, Rawat GS (2011) Impact of pastoral practices on forest cover and regeneration in the outer fringes of Kedarnath Wildlife Sanctuary, Western Himalaya. J Indian Soc Remote Sens 39:127–134

Theau J, Peddle DR, Duguay CR (2005) Mapping lichen in a caribou habitat of Northern Quebec, Canada, using an enhancement-classification method and spectral mixture analysis. Remote Sens Environ 94:232–243

Timberlake J, Bayliss J, Alves T, Baena S, Francisco J, Harris T, da Sousa C (2007) The biodiversity and conservation of Mount Chiperone, Mozambique. Report Produced under the Darwin Initiative Award, 15, 036

Tommervik H, Johansen ME, Pedersen J, Guneriussen PT (1998) Integration of remote sensed and in-situ data in an analysis of the air pollution effects on terrestrial ecosystems in the border areas between Norway and Russia. Environ Monit Assess 49:51–85

Tommervik H, Hogda KA, Solheim I (2003) Monitoring vegetation changes in Pasvik (Norway) and Pechenga in Kola Peninsula (Russia) using multi-temporal Landsat MSS/TM data. Remote Sens Environ 85:370–388

Tuomisto H, Poulsen AD, Ruokolainen K, Moran RC, Quintana C, Celi J, Canas G (2003) Linking floristic patterns with soil heterogeneity and satellite imagery in Ecuadorian Amazonia. Ecol Appl 13:352–371

Turner W, Spector S, Gardiner N, Fladeland M, Sterling E, Steininger M (2003) Remote sensing for biodiversity science and conservation. Trends Ecol Evol 18 (6):306–314

Twele A, Erasmi S, Kappas M (2008) Spatially explicit estimation of leaf area index using EO-1 Hyperion and Landsat ETM+ data: implications of spectral bandwidth and shortwave infrared data on prediction accuracy in a tropical montane environment. GIScience Remote Sens 45:229–248

Ulaby FT, Moore RK, Fung AK (1982) Microwave remote sensing, active and passive, II, radar remote sensing and surface scattering and emission theory. Artech House, Norwood

Van Der Meer F, Schmidt KS, Bakker W, Bijker W (2002) New environmental remote sensing systems. In: Skidmore A (ed) Environmental modelling with GIS and remote sensing. Taylor & Francis, London, pp 26–51

van Herk CM (1999) Mapping of ammonia pollution with epiphytic lichens in the Netherlands. Lichenologist 31:9–20

van Herk CM (2001) Bark pH and susceptibility to toxic air pollutants as independent causes of changes in epiphytic lichen composition in space and time. Lichenologist 33:419–441

van Herk CM, Aptroot A, van Dobben HF (2002) Long-term monitoring in the Netherlands suggests that lichens respond to global warming. Lichenologist 34:141–154

Varshney PK, Arora MK (2004) Advanced image processing techniques for remotely sensed hyperspectral data. Springer, Berlin

Virtanen T, Mikkola K, Patova E, Nikula A (2002) Satellite image analysis of human caused changes in the tundra vegetation around the city of Vorkuta, north-European Russia. Environ Pollut 120 (3):647–658

Vogiatzakis IN, Mannion AM, Griffiths GH (2006) Mediterranean ecosystems: problems and tools for conservation. Prog Phys Geogr 30:175–200

Volante JN, Alcaraz-Segura D, Mosciaro MJ, Viglizzo EF, Paruelo JM (2012) Ecosystem functional changes associated with land clearing in NW Argentina. Agric Ecosyst Environ 154:12–22

Wabnitz CC, Andrefouet S, Torres-Pulliza D, Muller-Karger FE, Kramer PA (2008) Regional-scale seagrass habitat mapping in the Wider Caribbean region using Landsat sensors: applications to conservation and ecology. Remote Sens Environ 112:3455–3467

Walsh SJ, Shao Y, Mena CF, McCleary AL (2008) Integration of Hyperion satellite data and a household social survey to characterize the causes and consequences of reforestation patterns in the northern Ecuadorian Amazon. Photogram Eng Remote Sens 74:725–735

Wang L, Sousab WP, Gong P, Biging GS (2004) Comparison of IKONOS and QuickBird images for mapping mangrove species on the Caribbean coast of Panama. Remote Sens Environ 91:432–440

Warren DL, Seifert SN (2011) Ecological niche modeling in Maxent: the importance of model complexity and the performance of model selection criteria. Ecol Appl 21(2):335–342

Waser LT, Keuchler M, Schwarz M, Ivits E, Stofer S, Scheidegger C (2007) Prediction of lichen diversity in an UNESCO biosphere reserve—correlation of high resolution remote sensing data with field samples. Environ Model Assess 12:315–328

Whittaker RJ, Araujo MB, Jepson P, Ladle RJ, Watson JEM, Willis KJ (2005) Conservation biogeography: assessment and prospect. Divers Distrib 11:3–23

Williams JN, Seo C, Thorne J, Nelson JK, Erwin S, O'Brien JM, Schwartz MW (2009) Using species distribution models to predict new occurrences for rare plants. Divers Distrib 15(4):565–576

Willis KJ, Whittaker RJ (2002) Species diversity-scale matters. Science 295:1245–1248

Wilson EH, Sader SA (2002) Detection of forest harvest type using multiple dates of Landsat TM imagery. Remote Sens Environ 80:385–396

Wilson RJ, Thomas CD, Fox R, Roy DB, Kunin WES (2004) Spatial patterns in species distributions reveal biodiversity change. Nature 432:393–396

Wolseley PA, James PW, Theobald MR, Sutton MA (2006) Detecting changes in epiphytic lichen communities at sites affected by atmospheric ammonia from agricultural sources. Lichenologist 38:161–176

Xie Y, Sha Z, Yu M (2008) Remote sensing imagery in vegetation mapping: a review. J Plant Ecol 1(1):9–23
Yang XQ, Kushwaha SPS, Saran S, Xu J, Roy PS (2013) Maxent modeling for predicting the potential distribution of medicinal plant, Justicia adhatoda L. in Lesser Himalayan foothills. Ecol Eng 51:83–87
Youngentob KN, Wallis IR, Lindenmayer DB, Wood JT, Pope ML, Foley WJ (2011) Foliage chemistry influences tree choice and landscape use of a gliding marsupial folivore. J Chem Ecol 37:71–84
Zhang JT, Pennington DD, Liu XH (2007) GBD-explorer: extending open source java GIS for exploring ecoregion-based biodiversity data. Ecol Inf 2:94–102
Zhang XY, Feng XZ, Jiang H (2010) Object-oriented method for urban vegetation mapping using IKONOS imagery. Int J Remote Sens 31:177–196
Zimmermann HE, Edwards TC Jr, Graham CH, Pearman PB, Svenning JC (2010) New trends in species distribution modelling. Ecography 33(6):985–989 (0906-7590)

8 Reflectance Spectra of High-altitude Lichens Based on In Situ Measurements

C.P. Singh and C. Prabakaran

Contents

Abstract

The unique chemistry together with typical physiology, morphology and anatomy enables lichens to survive in any harsh environment and same forms the basis for their unique spectral signatures. In this study, five lichen species found at Ravanshila, Tungnath in Garhwal Himalaya region were studied for their spectral characteristics. The reflectance measurements were taken using ASD Fieldspec 3 portable spectroradiometer at every 1 nm from 350 to 2,500 nm. The discrimination analysis was performed by band ratios and stepwise discriminant analysis (SDA). The band ratios of 400/685 and 773/685 nm distinctly provided the spectral characteristics of different colour and growth forms of lichen species. A set of 11 wavelengths were selected using SDA, and most of the variances in the spectra were observed in the NIR and SWIR regions. The maximum variability between lichen species was found at 950 nm ($F = 7{,}572.86$; $L = 0.00$). It was also observed that all lichens showed a more gradual increase through 400–1,300 nm with a maximum inflection point at about 680 nm (red edge). Spectra of all lichens were similar in shape beyond 700 nm, but the magnitude of reflectance varied among them. Between 1,500 and 1,800 nm, all lichens exhibited similar pattern, but with difference in the level of reflectance. Study on spectral properties of high-altitude lichens will be helpful in developing methods for classifying and monitoring lichens using hyperspectral remote sensing.

C.P. Singh (✉) · C. Prabakaran
EHD/BPSG/EPSA, Space Applications Centre (ISRO), Satellite Road, Ambawadi Vistar P.O., Ahmedabad, Gujarat, India
e-mail: cpsingh@sac.isro.gov.in

C. Prabakaran
e-mail: karanc@sac.isro.gov.in

D.K. Upreti et al. (eds.), *Recent Advances in Lichenology*,
DOI 10.1007/978-81-322-2181-4_8

Keywords
Lichens · Himalaya · Hyperspectral · Discriminant analysis

8.1 Introduction

Lichens are symbiotic organisms which consist of a fungal and an algal component. The fungal component provides a structure and algal component provides food. Lichens are distributed worldwide and represented by about 17,000 species (Hale 1983). The Indian Himalayan region exhibits luxuriant growth of different plant groups including lichens. The lichens form dominant terrestrial vegetation in higher alpine habitats in the Himalaya. The higher alpine habitats in the Himalaya also exhibit dominant terrestrial vegetation of lichens together with other herbaceous plants (Upreti 1998). The unique chemistry together with typical physiology, morphology and anatomy enables lichens to survive in any harsh environment. The adaptability of the lichens in the harsh environments resulted in different growth forms such as crustose, foliose, fruticose, leprose and squamulose forms. Lichens have an ability to colonize on any perennial substrate such as bark, rock, soil and many man-made artefacts. Lichens are slow growing organisms and need stable conditions for habitat establishment. Though lichens have great ability to survive in harsh environment but are most sensitive to microclimate changes, thus they act as a natural sensor and successfully utilized as biological indicator of air quality and climate change studies. To understand the changing environmental condition of an area the monitoring of lichen habitat cover, their distribution and change in distribution pattern provide useful data. Hyperspectral remote sensing methods play a major role in the study of lichen characterization. The Ager and Milton (1987), Petzold and Goward (1988), Nordberg and Allard (2002), Bechtel et al. (2002), Rees et al. (2004) and Zhang et al. (2005) studied the spectral properties of lichens worldwide. Most of the spectral properties of lichens from high latitude lichens, especially from arctic and subarctic regions, are available. Many researchers used hyperspectral images to retrieve the lichen spectra. However, lichens are growing in and around many substrates, the discriminating lichens from others become inevitable, and many methods have been used. Zhang et al. (2005) used a derivative spectral unmixing method by combining spectral mixture analysis (SMA) and derivate spectra to discriminate the lichens from the rocks. Bechtel et al. (2002) proposed a band ratio of 400/685 and 773/685 nm to discriminate lichen species and 2,132/2,198 and 2,232/2,198 nm to distinguish features between rocks with OH bearing minerals and lichens. Spectral discrimination of lichens was done using the technique of principle components analysis (PCA) by Rees et al. (2004). This method constructs a variance–covariance matrix for the spectra to identify correlations between different wavelengths which identify the set of wavelengths containing most of the information in the spectra.

The present study is an attempt to investigate the spectral properties of high-altitude lichens of Indian Himalaya and to explore the common features and variations among the lichen species.

8.2 Study Site

The study site, Ravanshila peak (30° 29′ 09.8″ N, 79° 12′ 38.5″ E) at Tungnath, is situated at 3,433 m above mean sea level in Rudraprayag district of Uttarakhand, India. The in situ spectral measurements of lichens were collected during post-monsoon season in the month of October 2013. The study site is completely covered by snow during winter period from late November to March. The area comes under the alpine habitats above the tree lines in the western Himalayan region, and the lichens are exposed to various climatic stress factors such as low atmospheric pressure, low precipitation, high UV

Table 8.1 Lichen species sampled for their spectral measurements

Lichen species	Growth form	Colour	Total observations
Umbilicaria indica Frey	Foliose	Greyish green	25
Porpidia macrocarpa (DC.) Hertel and Hertel and A. J. Schwab	Crustose	Greyish green	25
Cladonia furcata (Huds.) Schrad.	Fruticose	Greyish green	25
Stereocaulon foliolosum Nyl.	Fruticose	Greyish green	25
Ioplaca pindarensis (Räsänen) Poelt and Hinter.	Crustose	Yellowish orange	25

radiation and high wind speed with limited nutrient resource and less growing season. Five lichen species growing luxuriantly and having different growth form were selected for studying the reflectance spectra (Table 8.1).

8.3 Experimental Approach

The measurements were taken using ASD Fieldspec 3 portable spectroradiometer that operates in the spectral region between 350 and 2,500 nm, with a spectral resolution of 3 nm at 700 nm, 10 nm at 1,400 nm and 10 nm at 2,100 nm, and the sampling intervals for the spectral region of 350–1,000 nm and 1,000–2,500 nm are 1.4 and 2 nm, respectively (Analytical Spectral Devices 2010).

The spectroradiometer instrument was battery operated and connected with Lenovo notebook laptop through Ethernet cable. Spectral measurements were collected using a fibre optic (FOV 25°) sensor. Spectroradiometer was calibrated with white reference panel made up of polytetrafluoroethylene (PTFE) and cintered halon, before taking sample measurements. The sensor height was adjusted depending on the specimen size in order to preclude viewing a mixture of lichen and rock. Measurements were viewed and recorded taken by using RS3 software and stored in the laptop memory in 'asd' format. These measurements were recorded in reflectance mode, and 25 samples were taken for each lichen specimen to calculate an average spectrum. Later, 'asd' files were viewed and exported as reflectance in text format by ViewSpec Pro software (Analytical Spectral Devices 2008) for further analysis. The reflectance measurements were taken at every 1 nm from 350 to 2,500 nm. Each sample data was subsampled at 10-nm interval from 400 to 2,500 nm, thus making each sample by up to 211 values of the spectral reflectance. Water absorption wavelengths were excluded from the analysis to reduce the noise from the data. An average endmember spectrum was created by taking average of 25 samples to characterize the spectral properties of each lichen species (Table 8.1).

8.3.1 Discrimination of Lichen Spectra

The discrimination of lichens was performed by band ratio and stepwise discriminant analysis (SDA) methods using Wilks' lambda. Following the Bechtel et al. (2002) method, lichen species at Ravanshila peak of Tungnath, Garhwal Himalaya were studied for their spectral characteristics. Ratio of reflectance at 400/685 and 773/685 nm was adopted to discriminate among the lichen species and other materials such as rock from lichens. The ratio distinctly provided the spectral characteristics of different colour, types and species of lichens. In another method, discriminant analysis was performed using SPSS 16.0. Total of 125 samples (10-nm subsamples) were grouped into five according to species for the analysis.

8.4 Results

8.4.1 Spectral Features of Lichens

From the reflectance features of five lichen species (Fig. 8.1), *Porpidia macrocarpa* showed a reflectance between 17 and 22 % in the visible

Fig. 8.1 Reference lichen species selected for spectral signatures in Chopta–Tungnath region of Garhwal Himalaya. **a** *Cladonia furcata*, **b** *Ioplaca pindarensis*, **c** *Porpidia macrocarpa*, **d** *Stereocaulon foliolosum*, **e** Umbilicaria *indica* and **f** spectral measurement of lichen thallus being taken in field

region, while other lichens showed a reflectance between 4 and 16 % in the visible region. *Ioplaca pindarensis* exhibited 4–5 % of reflectance at 400–500 nm and quickly rose to approximately 6–16 % from 500 nm until the chlorophyll absorption at approximately 680 nm. *Cladonia furcata, I. pindarensis* and *Stereocaulon foliolosum* showed a strong absorption feature at 680 nm which is attributed to the presence of chlorophyll, and a very weak absorption feature

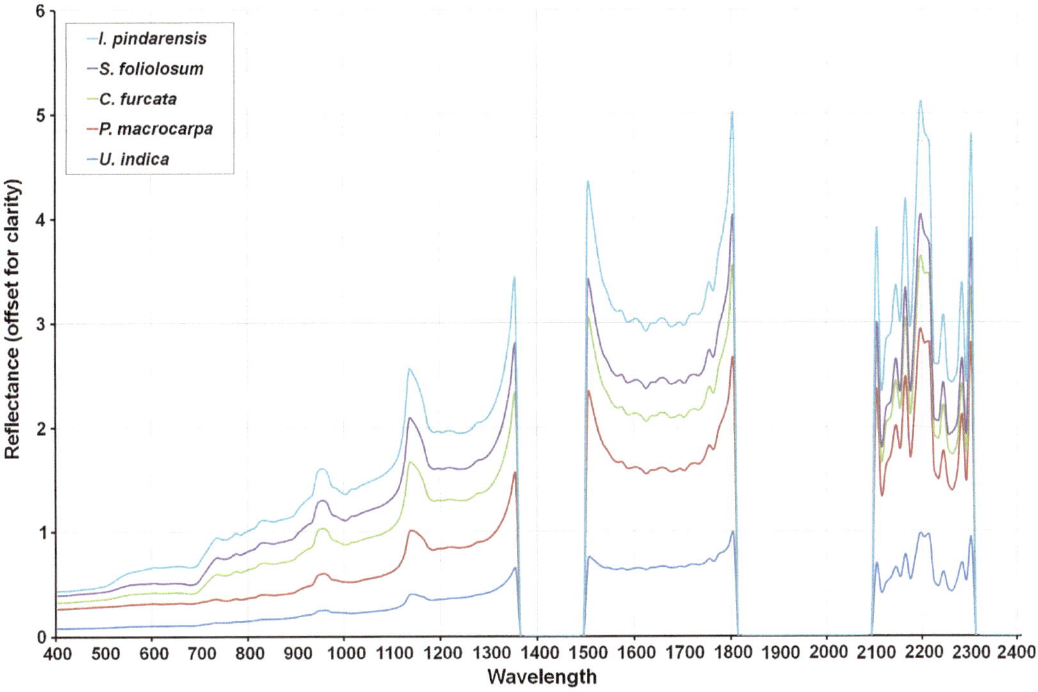

Fig. 8.2 Reflectance spectral profile of five lichen species in Tungnath area of Garhwal Himalaya (reflectance values in *y*-axis were offset for clarity. The value of second profile can be obtained by subtracting the first profile and so on)

observed in the spectra of *P. macrocarpa* and *Umbilicaria indica* at 680 nm. *C. furcata* reflected a greater proportion of the NIR. *U. indica* and *P. macrocarpa* exhibited a strong absorption in visible wavelengths and a gradual increase thereafter. All the lichen species showed a similar pattern of reflectance between 720 and 1,320 nm. A weak absorption feature was observed in the water absorption band at 950 nm in all lichen spectra, except that foliose lichen, *U. indica* which showed a moderate absorption feature.

All lichen species had a gradual increase in reflectance to 1,380 nm followed by water absorption feature centred near 1,430 nm. Reflectance features in the region from 1,500 to 1,800 nm were same for all the lichen species. A minimum absorption at 1,600–1,800 nm centred around 1,700 nm, attributed to absorption by cellulose, which was observed in all samples, though less absorption found in foliose and crustose lichens (Fig. 8.2).

8.4.2 Discrimination of Lichen Species Using Band Ratios

The discrimination of lichen spectra was performed by a plot of the ratio of reflectance at 400/685 and 773/685 nm (Fig. 8.3). The reflectance values at 400 nm were less than 6 % except *P. macrocarpa* and strong absorption at 680 nm due to the presence of chlorophyll pigments. All lichen species were showing high reflectance at NIR (773 nm), yet with low variations among species (Fig. 8.2). *I. pindarenesis*, a yellowish orange lichen, had very low value for both 400/685 and 773/685 nm ratios, thus forming a separate group than other lichens which were having a different colour (greyish green). Fruticose lichens, *C. furcata* and *S. foliolosum,* were having high values, while crustose lichens have less values for 773/685 nm ratio and moderate for 400/685 nm ratio. Foliose lichen was found to be somewhere between fruticose and crustose, for both ratios. The high and low reflectance values

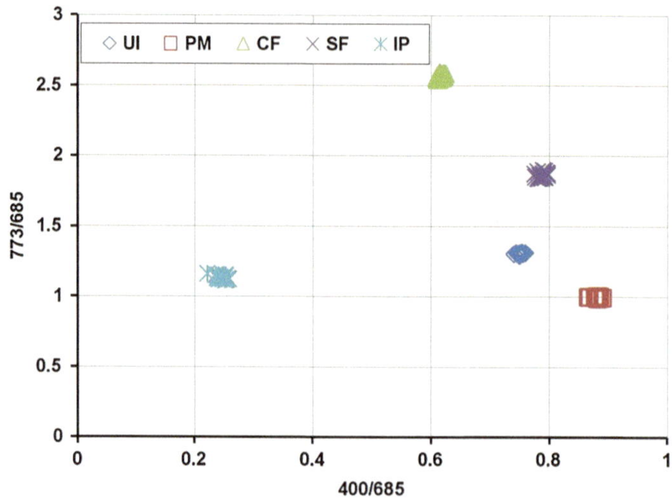

Fig. 8.3 The band ratios of 400/685 nm against 773/685 nm were showing the separation of different lichens species (UI: *Umbilicaria indica*, PM: *Porpidia macrocarpa*, CF: *Cladonia furcata*, SF: *Stereocaulon foliolosum*, IP: *Ioplaca pindarensis*)

at 773 nm may be explained by the structural nature of fruticose lichens and in crustose, respectively. Values for 400/685 nm ratio were mainly influenced by the amount of chlorophyll present in the lichens. Since reflectance values at 400 nm were low and less variation among lichen species, denominator value of 685 nm played a significant role in the discrimination. Among greyish green lichens, *P. macrocarpa* had high value for 400/685 nm ratio which stated the presence of more chlorophyll concentration and yellowish orange lichen *I. pindarensis* had less chlorophyll concentration (Fig. 8.3).

8.4.3 Discrimination of Lichens Using SDA Method

SDA was performed, and a set of 11 wavelengths were selected for discriminating lichens spectra. Most of the variances in the lichen spectra were accounted in the NIR and SWIR regions. The maximum variability between lichen species was found at 950 nm (F = 7,572.86; L = 0.00) (Table 8.2). Variability in green wavelength band at 520 nm showed because of the colour variation among them. Rest of the wavelength sets were having minimum difference in Wilks' lambda value.

Table 8.2 Stepwise discriminant analysis (SDA), combination of selected wavelengths, Wilks' lambda and F value in each steps of SDA

Step	Wavelengths	Wilks' lambda (L)	F value
1	950	1.2e-011	7,572.862
2	520, 950	1.3e-005	7,999.789
3	520, 950, 1,060	1.4e-011	8,460.630
4	520, 950, 1,060, 1,670	8.9e-012	9,709.012
5	520, 950, 1,060, 1,500, 1,670	1.0e-011	11,438.580
6	470, 520, 950, 1,060, 1,500, 1,670	2.1e-011	13,454.240
7	470, 520, 740, 950, 1,060, 1,500, 1,670	6.2e-008	13,739.930
8	470, 520, 740, 860, 950, 1,060, 1,500, 1,670	1.7e-011	14,242.570
9	470, 520, 710, 740, 860, 950, 1,060, 1,500, 1,670	2.7e-011	17,890.660
10	470, 520, 550, 710, 740, 860, 950, 1,060, 1,500, 1,670	1.2e-010	18,536.340
11	470, 520, 550, 710, 740, 860, 880, 950, 1,060, 1,500, 1,670	7.1e-010	21,988.300

8.5 Discussion

8.5.1 Spectral Features of Lichens

The spectral reflectances of lichens are differed from that of vascular plants in the visible region as a result of lichens' symbiotic character and consequential photosynthetic system (Petzold and Goward 1988). Lichens are having much lower values than that of green vegetations in NIR region (Nordberg and Allard 2002) because of anatomical differences. Lichen spectrum showed a little (in yellowish orange colour lichen) or no reflectance peak (in greyish green colour lichens) in green light (500–600 nm). Petzold and Goward (1988) observed a similar spectral reflectance in the medium dark brown lichen (*Cetraria ericetorum*) which exhibited strong absorption in visible region (400–700 nm). According to Nordberg and Allard (2002), the darker colour lichen (*Cetraria islandica*) exhibits much lower reflectance in visible region as compared to pale counterpart, which helps to discriminate clear difference between pale and darker lichen species spectral responses.

In the present study, it was observed that all lichens showed a more gradual increase through 400–1,300 nm with a maximum inflection point at about 680 nm (red edge). Similar to studies carried out by Nordberg and Allard (2002) on lichens under laboratorial condition, the three lichen species (*C. furcata*, *S. foliolosum*, *I. pindarensis*) of the study area exhibit highest reflectance in red band of visible region. Reflectance differences among lichens in the visible regions are mainly due to variable pigmentation of the lichens (Petzold and Goward 1988) and type of fungal components in the lichens which are responsible for the pigments colouration and secondary metabolic by-products (Hale 1983).

Similar to studies carried out by Rees et al. (2004) on subarctic lichens, *P. macrocarpa a* crustose lichen in the study area exhibit a more gradual increase in reflectance up to 1,300 nm, while *I. pindarensis* showed a week absorption feature at 950 nm, indicated that lichens were having lower intercellular water reserves in their thallus and measurements were performed during post-monsoon season a more or less dry winter period. Lichens have no root system and cannot receive water from soil, and they dry quickly after rainfall (Petzold and Goward 1988). Peak reflectance at 1,380 and 1,800 nm followed by water absorption features was reported in lichen species by Zhang et al. (2005), Rivard and Arvidson (1992) and Ager and Milton (1987). Cellulose presence is observed by the absorption at 1,700 nm, but Ager and Milton (1987) identified three cellulose absorption features near at 1,730, 2,100 and 2,300 nm. However, weak and strong cellulose absorption at 1,700 and 2,300 nm were observed, respectively, in the present study.

8.5.2 Discrimination of Lichen Species

The band ratio of 400/685 and 773/685 nm showed an interesting result in discriminating different coloured lichens. Yellowish orange lichen has low value than that of other lichens having greyish green colour and having higher concentration of chlorophyll. The band ratio placed fruticose lichens in upper level and crustose in lower level in the scatter plot (Fig. 8.2) due to high reflectance in NIR region which attributes to internal structures of thallus. The foliose lichens exhibited a moderate value between fruticose and crustose forms. Thus, the band ratio method can be used to discriminate various growth forms of lichens. However, Bechtel et al. (2002) reported that band ratio method cannot be helpful to distinguish all lichen species. According to Elvidge's (1990), lichens can be clearly distinguished from non-photosynthetic vegetation in the visible region. The present studies clearly indicate that the band ratio method can be useful to discriminate the lichens in terms of colour and growth forms, but not as individual species.

The SDA brought out a total of 11 wavelengths out of 211 wavelengths which can be used to distinguish the five lichen species. Most of the variation between the lichen spectra could be accounted between 700 and 1,100 nm region. Spectra of all lichens were similar in shape beyond 700 nm, but the magnitude of reflectance varied among them. A similar pattern was reported by Zhang et al. (2005). Two green wavelengths at 520 and 550 nm and NIR bands at 710 and 740 nm can be used to distinguish the colour differences and structural differences among lichens, respectively. Petzold and Goward (1988) observed that light coloured and dark coloured lichens were having spectral difference in NIR and MIR reflectance wavelengths (Nordberg and Allard 2002). Two MIR bands—1,500 and 1,670 nm—can be helpful in discriminating the wet and dry lichens. Between 1,500 and 1,800 nm, all lichens exhibited similar pattern, but with difference in the level of reflectance.

8.6 Conclusion

This present study was attempted to investigate the spectral properties of high-altitude lichens from Indian Himalaya. Spectral analyses of five lichen species were carried out. Greyish green and yellowish orange lichens exhibited different patterns in the visible region. All lichens had a similar reflectance beyond 700 nm. Colour and growth forms of lichen species can be distinguished using 400/685 and 773/685 nm band ratio. This method was applicable to colour and growth forms, but not to distinguish each species. The individual species of lichens can be used to discriminate by a set of 11 wavelengths from SDA. Study on spectral properties of high-altitude lichens will be helpful in developing methods for classifying and monitoring lichens using hyperspectral remote sensing.

Acknowledgments We are thankful to Dr. A.S. Kiran Kumar, Director, SAC (ISRO) for his support. We are also thankful to Dr. M.C. Nautiyal, Director, HAPPRC, HNB Garhwal University, Srinagar, Garhwal, Uttarakhand for all logistic support during fieldwork. Authors are also thankful to Dr. J.S. Parihar, DD, EPSA, SAC, Ahmedabad and Dr. R.P. Singh, Head, EHD, SAC, Ahmedabad for their constant encouragements. Thanks are also due to Dr. (Mrs.) S. Panigrahy for her encouragement in formulating this programme. We are also thankful to Dr. K.R. Manjunath for his valuable comments. Authors are thankful to Dr. D.K. Upreti and Dr. R. Bajpai, CSIR-NBRI, Lucknow for collection and identification of lichens. This work has been carried out under the PRACRITI-II project, 'Alpine ecosystem dynamics and impact of climate change in Indian Himalaya'.

References

Ager CM, Milton NM (1987) Spectral reflectance of lichens and their effects on the reflectance of rock substrates. Geophysics 52:898–906

Analytical Spectral Devices (2008) ViewSpec Pro™ User manual. ASD Document 600555 Rev. A. www.asdi.com. Boulder

Analytical Spectral Devices (2010) FieldSpec® 3 User manual. ASD Document 600540 Rev. I. www.asdi.com. Boulder

Bechtel R, Rivard B, Sánchez-Azofeifa A (2002) Spectral properties of foliose and crustose lichens based on laboratory experiments. Remote Sens Environ 82:389–396

Elvidge CD (1990) Visible and near infrared reflectance characteristics of dry plant materials. Int J Remote Sens 11:1775–1795

Hale ME (1983) The biology of lichens. Edward Arnold, London

Nordberg ML, Allard A (2002) A remote sensing methodology for monitoring lichen cover. Can J Remote Sens 28:262–274

Petzold DE, Goward SN (1988) Reflectance spectra of subarctic lichens. Remote Sens Environ 24:481–492

Rees WG, Tutubalina OV, Golubeva EI (2004) Reflectance spectra of subarctic lichens between 400 and 2400 nm. Remote Sens Environ 90:281–292

Rivard B, Arvidson RE (1992) Utility of imaging spectrometry for lithologic mapping in Greenland. Photogram Eng Remote Sens 58:945–949

Upreti DK (1998) Diversity of lichens in India. In: Agarwal SK, Kaushik JP, Kaul KK, Jain AK (eds) In perspectives in environment. APH Publishing Corporation, New Delhi, pp 71–79

Zhang J, Rivard B, Sánchez-Azofeifa A (2005) Spectral unmixing of normalized reflectance data for the deconvolution of lichen and rock mixtures. Remote Sens Environ 95:57–66

Lichens as Agents of Biodeterioration 9

M.R.D. Seaward

Contents

Abstract

One of the major roles lichens play in shaping the natural world, both physically and biologically, is as agents in soil development: formerly considered only in a geological context, recent research has shown that they are capable of biodeteriorating stone substrata within a relatively short timescale. Chemical alteration of the substratum is achieved by the disruptive action of many species, particularly those capable of producing an oxalate at the thallus–substratum interface. The oxalate contributes significantly to the bulk and composition of the thallus itself and persists as an obvious encrustation after the lichen's death. In the past, these disfiguring oxalate residues on ancient monuments have been variously interpreted as resulting from former mechanical/ chemical renovation treatments, atmospheric pollution, and climatic weathering. The origin and nature of oxalate accretions, the factors involved in pedogenesis, and the development of lichen mosaics are reviewed. Aesthetic disfigurement *versus* physical damage to stonework is considered, and various aspects of stonework conservation are discussed.

Keywords

Biological weathering · Pedogenesis · Calcium oxalate · Biofilms · Bioprotection · Stonework conservation

M.R.D. Seaward (✉)
Department of Archaeological, Geographical and Environmental Sciences (AGES), University of Bradford, Bradford BD7 1DP, UK
e-mail: m.r.d.seaward@bradford.ac.uk

D.K. Upreti et al. (eds.), *Recent Advances in Lichenology*,
DOI 10.1007/978-81-322-2181-4_9

9.1 Introduction

Although earlier literature provided an insight into the uniqueness of lichens, it gave little hint of the major role these organisms play in the shaping of the physical and biological environment of our planet and their importance in maintaining its equilibrium. Their role as biological weathering agents in the development of soils, for example, was formerly considered in a geological context only, but it has been shown that lichens are capable of biodeteriorating stone substrata within a relatively short timescale. Lichens also have the capacity to accumulate elements, such as nitrogen, phosphorus, and sulphur, thereby increasing their potential bioavailability to successive life forms. Organic material derived from lichen decomposition together with detached particles of the substratum and atmospherically derived dusts trapped by thalli all contribute to the development of primitive soils. As well as mineral cycling through biodeterioration processes (Wierzchos and Ascaso 1996; De los Rios et al. 2005), even in a stony littoral zone (Kulikova et al. 2009), lichens often contribute substantial biomass and support a high biodiversity of organisms, creating complex food webs and adding significantly to energy flow. The disappearance of lichens, due to many aspects of human interference in the natural world, leads inexorably to environmental impoverishment (Seaward 2008).

The role of lichens in biodeterioration processes has been investigated worldwide in habitats ranging from polar to desert, at low to high altitudes, in human influenced environments, and involving every conceivable rock type and substrate. However, little attention has been given to the impact of lichens in zones subjected to periodic inundation by marine or fresh water (Gómez-Pujol et al. 2007). In a natural setting, biodecomposition of rock is accepted as normal and even desirable, but in a landscape modified by human activity, biodeterioration of monuments, buildings, artwork, statues, and gravestones is regarded as a serious problem. Even in natural settings, culturally significant prehistoric and historic rock art is subject to the same processes of biodeterioration. However, there are often significant differences in the particular lichen assemblages, and therefore the biodeterioration processes, between those monuments which are composed of stones translocated to the study site (the classic example being Stonehenge in England) and those monuments created on in situ rocks or from locally occurring rocks.

In the case of translocation, studies which compare the differences between lichen colonization of the natural rock, as for example in quarries, and similar material when used in the construction of the monument, such as that demonstrated by Ascaso et al. (2004), are of relevance. In the case of in situ lichen colonization, the detailed study of El Morro Inscription Rock in New Mexico, USA, a large sandstone formation which has become a natural repository for more than 700 years of comments and notations recorded by passing travellers (Knight et al. 2004) is of particular interest. Rather interestingly, the lichen assemblages on monuments are often significantly different in composition to those on in situ rocks close to where they have been constructed, no doubt due to special, sometimes devotional, practices the latter have received over the centuries (Fig. 9.1).

Over the past half-century, considerable interest has been shown in the preservation of our cultural heritage from a biological standpoint, with an extensive and growing literature on the subject, particularly in respect of lichens (Nimis and Monte 1988; St. Clair and Seaward 2004a) which, unfortunately, as demonstrated even a decade ago is to be found in a wide variety of published sources, some aimed at a very specific readership (Koestler and Vedral 1991; Piervittori et al. 2004). It should be recognized that a wide range of organisms are implicated in biodeterioration processes (Paine et al. 1933; May et al. 1993; Scheerer et al. 2009). Of particular significance has been the proliferation of published material on the role of lichens in biodeterioration processes in terms of natural rocks, particularly pedogenesis (Syers and Iskandar 1973; Jones et al. 1981; Jones and Wilson 1985; Jones 1988), ancient monuments (Nimis et al. 1992; Piervittori et al. 2004),

Fig. 9.1 Prehistoric dolmen, Browne's Hill, Ireland—some of the lichen assemblages differ significantly from those on similar local in situ rocks, more particularly by the presence of a maritime species, *Anaptychia runcinata*, despite being more than 50 km from the sea (Seaward 1976)

and man-made materials such as cement, concrete, and brick (Brightman and Seaward 1977; Ariño and Saiz-Jimenez 2004).

Modern electron microscopy and chemical techniques have made it possible to recognize that many lichens contribute to the deterioration of a wide range of materials, particularly rocks and stonework, as a result of physical and/or chemical processes (Seaward 1997a). In the past, attention was drawn to the possible effect of dissolved carbon dioxide, derived from lichen respiration, attacking the substratum to produce pits and channels for easier penetration of hyphae, with attendant loosening of mineral particles and their incorporation into lichen tissue. Until recently, such effects, although important on a geological timescale, have been considered to be minimal in terms of the life of stone buildings and monuments.

Many lichen species create microclimatic effects at thethallus–substratum interface, particularly in terms of water retention, which undoubtedly lead to mechanical damage to stonework on a short timescale of ten, or even fewer, years; it has also been shown that raised temperatures are generated at the interface (Carter and Viles 2004). A wide variety of endolithic, crustose, and squamulose lichens are implicated, their aggressive behaviour no doubt promoted by particular man-made environmental conditions, but under special circumstances, foliose (Ascaso et al. 1976) and fruticose (Adamo and Violante 1991; Ariño et al. 1996) species are also effectiveweathering agents. The forces generated by climatic wetting and drying of lichen thalli cause them to expand and contract in conjunction with the chemical breakdown of substrata by lichen acids, which, despite their relatively low solubility (Iskandar and Syers 1971), are effective chelators (Schatz et al. 1954; Schatz 1963), forming metal complexes with silicates, etc., derived from the substratum. It would appear that lichens having the capacity to fix nitrogen, such as *Xanthoparmelia mexicana*, can enhance organic acid release, thereby leading to further solubilization of the mineral substrate (Seneviratne and Indrasena 2006). X-ray powder diffraction and transmission electron microscopy have clearly demonstrated the presence of characteristic alteration products at the interface between rocks and various lichens, and experiments involving pure lichen acids or lichen fragments incubated with different types of rock have confirmed these observations.

9.2 Biogeochemical Processes

The major chemical agent in biodeterioration processes is oxalic acid secreted by the mycobiont (Jones et al. 1980), which is extremely soluble in water and acts as a chelator of metal ions, the oxalates formed at the thallus–substratum interface being closely related to the chemical composition of the rock (Purvis 1996). Thus, species growing on serpentinite, mainly composed of magnesium silicate, form magnesium oxalate dihydrate at the interface (Wilson et al. 1981). Other alteration products have been shown to be incorporated into the thallus and/or precipitated at the lichen–substratum interface, such as manganese oxalate (Wilson and Jones 1984) and copper oxalate (Purvis 1984) on manganese-rich and copper-rich rocks, respectively.

However, the commonest oxalate found in lichens is that of calcium (Seitz and Schade 1976; Wadsten and Moberg 1985), the oxalic acid reacting with an underlying substratum containing calcium carbonate to form the insoluble compound calcium oxalate. The calcium oxalate can accumulate on the surface of thalli, often forming distinctive pruina, within lichen thalli, and/or at the thallus–substratum interface (Ascaso and Wierzchos 1995; Seaward and Edwards 1997); however, calcium carbonate derived from neighbouring substrata or atmospherically can also be sourced by the lichen. The thallial content of calcium oxalate ranges from 1 to 50 % according to the species and its underlying substratum (Syers et al. 1967; Edwards et al. 1994). It should be noted, however, that the compound exists in two hydrated forms, monoclinic monohydrate(whewellite) and tetragonal dihydrate(weddellite). The monohydrate form is the major biodeterioration product and the equilibrium (ratio between monohydrate and dihydrate forms) is dependent on various environmental factors (Edwards et al. 1992).

After the death of the lichen, residues of calcium oxalate often remain on such substrata, leaving significant and often unsightly white deposits. This phenomenon can be particularly serious in the case of delicate and intricate stone monuments, as calcium oxalate deposits often obscure the detail and historical significance of such structures (Seaward and Edwards 1995). Oxalic acid has also been shown to solubilize magnesium silicates (Jones 1988).

It has been shown that lichens are capable of producing calcium oxalate on a wide variety of substrata (Edwards et al. 1997); for example, *Dirina massiliensis* forma *sorediata* has the capacity to produce encrustations of various thicknesses, with different concentrations of calcium oxalate and a variable chemistry derived from the following range of substrata: gypsum/calcite underlying a fresco, stucco of a church wall, a Roman brick wall, a lead–glass interface of a church window, mortar, and acidic stone. However, it was also noticeable that the effectiveness of this lichen in degrading its particular substratum was determined by the various environmental conditions obtaining in the different habitats and geographical settings studied. The ability of lichens to produce significant levels of calcium oxalate, even when growing on non-calcareous substrata, is particularly interesting. FT-Raman spectroscopic studies of lichens on granitic monuments in Spain, for example, have shown that not all species are capable of producing calcium oxalate, and those that do have significant levels derive some of the calcium for its production from the atmospheric or from leachates of neighbouring substrata (Prieto et al. 2000).

Undoubtedly, many lichens known to contain calcium oxalate cause extensive damage to a range of rock substrata, but it is the production of calcium oxalate dihydrate by thalli, which is a measure of a lichen's capacity to biodeteriorate its substratum; its production is related to microclimatic conditions such as temperature and humidity of the air, as well as the chemical and physical nature of the substratum. It would appear, for example, that warmer and drier sites are more conducive to the production of calcium oxalate dihydrate by *Diploicia canescens* (Edwards et al. 1995) and by *Ochrolechia parella* (Prieto et al. 1999); this accords with the findings of Wadsten and Moberg (1985) who observed that lichens growing on humid sites produced calcium oxalate monohydrate, whereas those from drier sites produced a mixture of the monohydrate and dehydrate forms.

9.3 Biogeophysical Processes

The natural process of soil formation, pedogenesis, whilst inherently slow, is nevertheless relentless, usually operating on a geologic time scale and involving a complex and interactive combination of physical, chemical, and biological activities. Some aspects of the process are strictly abiotic, such as wetting/drying, heating/cooling, and freezing/thawing, whilst others are clearly biological in nature, such as the encroachment of roots, rhizines and hyphae into cracks and fissures in rock surfaces, the biomediated chemical erosion of cementing agents, and the biotransformation of the molecular structure of rock substrata.

Both biological and abiotic factors operate together, the effects of one accentuating the effects of the other, relentlessly breaking down rocks, large and small, to form the mineral component of soil. The nature and rate of pedogenesis are mainly dictated by the chemical and physical attributes of the rock (Favero-Longo et al. 2005; Rajakaruna et al. 2012), including elemental composition, molecular structure, cementing agents, density, porosity, pH and nature of origin, the composition and degree of development of the epilithic and endolithic biological community, and local and regional climatic patterns.

The dynamics of rock decomposition apply to both natural rocks and man-made 'stone' substrata, but the onset of human-induced changes to the environment, especially changes in air and water quality, have altered the dynamics and rates of rock degradation. More specifically, increasing levels of air pollutants accompanied by acidification of precipitation have dramatically altered, or in some cases eliminated, some biological communities, particularly in urban/industrial settings; therefore, although environmental degradation has reduced the impact of the biological components on rock decomposition, it has dramatically enhanced the impact of some abiotic factors.

It is highly probable that lichens were early colonizers of terrestrial habitats on our planet, coping with harsh environments and contributing to the evolution of atmospheres more suited to a much wider variety of life-forms. However, their major long-term role has been as biologicalweathering agents, with a pedogenic action which is both physical and chemical in nature (Syers and Iskander 1973; Jones 1988). Historically, their contribution to the breakdown of rock substrata was perhaps somewhat overrated; however, today it is often seriously underestimated (St. Clair and Seaward 2004b). Frequently, lichens are the first living things to occupy newly exposed rock surfaces, their invasion being dictated by several factors, including proximity of native lichen communities on similar substrata, reproductive strategies (sexual and/or asexual) of local lichens, and dispersal capacity of local lichens (Bailey 1976). The ability of reproductive propagules to succeed on what appears to be an inhospitable substratum such as a rock is remarkable. However, although the discharge, dispersal, and establishment of reproductive propagules in fungi have been extensively studied, these processes in lichenized fungi are still under-researched.

Efficient scattering of reproductive propagules in no way guarantees their successful establishment and development: suitable substrata and environmental conditions are necessary when propagules come to rest following one or more deposition processes involving sedimentation, gravitation, and wash-out by rain. Propagule impact on various surfaces leading to successful development of thalli is accounted for by efficient trapping, and indeed electrostatic charges generated by lichen ascospores have been suggested as the cause for this phenomenon in some species (Garrett 1972). The importance of aerobiology in understanding how different propagules are dispersed and transported in the air and how they impact on surfaces cannot be underestimated (Mandrioli et al. 2003; Piervittori 2004).

Lichens are capable of colonizing a wide range of substrata (Brodo 1973), including not only natural rock surfaces, but also stone imported for the construction of monuments and buildings, supplemented by other man-made materials, such as cement, concrete, asbestos-cement, brick, roofing tiles, metals and wood (Brightman and Seaward 1977), and even glass (Mellor 1923; Krumbein et al. 1991; Kossowska and Węgrzyn 2009), all of which are subject to biodeterioration. Field experiments have shown the importance of surface texture and micro-topography for the successful attachment and survival of propagules (Armstrong 1988). Although surface topography may provide a favourable environment for germination, it is likely that inter- and intra-specific competition as a result of a build-up of propagules in such niches will be intense, their interactions determining the ultimate composition of the local lichen flora.

Successful germination and establishment of the various types of lichen propagule are dependent not only on characteristics of the available substratum, such as texture, pH, nutrient status, and microtopography, but also on numerous other environmental factors such as humidity and temperature (Barkman 1958; Brodo 1973; Topham 1977). In addition, competition with established lichens and other organisms for space, light, and nutrients by the newly developing propagule can be considerable; such competition may arise as a consequence of mechanical action, inhibition of gaseous exchange, reduction in light intensity, or allelopathic or toxic chemicals. The possibility of intra-specific allelopathic effects determining the growth and spatial distribution of lichens has been postulated by Seaward (1982). All the above examples refer to mature thalli, but it is reasonable to suppose that actively germinating ascospores and other propagules are all capable of producing chemicals with similar inhibitory properties.

The factors which control morphogenesis are poorly understood; for example, little is known of the growth rate of thallus initials, a phase crucial to both the establishment and competitive ability of the new thalli, or how long a period must pass before lichens establish themselves on a particular substratum, how long before this can actually be detected, and how successful are the thallus initials. However, once established, their growth rates can be easily determined and a wealth of literature on this branch of lichenometry is available, the dating curves for particular species, for example, providing a guide to the date of their establishment and to their potential rate of colonization.

Many saxicolous lichen communities undergo regular patterns of successional change; for example, one assemblage of species may occupy a given rock surface for several years, steadily altering the substratum in ways that eventually better accommodate a new combination of species. Many species inhabit large areas for a considerably longer time than could be expected from the lifespan of a single thallus; this fact may be responsible for some exaggerated ideas about the age of individual thalli. With the passage of time, different lichen species and assemblages are subsumed by other lichen cover as a consequence of a particular succession, in many cases eventually giving way to bryophytes and vascular plants unless interrupted by a controlling factor, more than likely man-induced, that prevents the sequence. Although the ecology of the later phases of such successions has been widely studied and extensively cited in classic textbooks, changes in the development of the different lichen phases, spatially and temporally, have received less attention, particularly in respect of biodeterioration.

Lichen associations play an important role in energy-budgeting, modelling, etc., and should therefore not be underestimated. Energetics of lichen faunas are poorly understood due to the wide variety of data from disparate sources: information on ingestion, assimilation, egestion, respiration, growth and death rates, numbers and biomass needs careful co-ordination and interpretation (Seaward 2008). Of special interest to our understanding of biodeterioration processes is the pedogenetic study of Clarens sandstone, South Africa, (Wessels and Schoeman 1988) involving lichenophagous bagworm larvae (Wessels and Wessels 1991). The sandstone is weathered by endolithic lichens which dissolve the cementing material, thereby loosening quartz crystals which are then used by the larvae to construct the bags in which they live. It is estimated that the larvae contribute 4.4 kg of weathered sandstone per hectare per year to the area; furthermore, the larvae utilize the lichens as a food source, the resulting faeces providing 200 g of organic material per hectare per year and contributing to mineral cycling. In general, the impact of endolithic lichens on stonework has been underestimated (Matthews and Owen 2008) and there are relatively few detailed investigations on their role in biodeterioration and biomineralization (Bungartz et al. 2004; De los Rios et al. 2005; Favero-Longo et al. 2009a, b; McIlroy de la Rosa et al. 2012a, b). In their study of *Lecidea auriculata* thalli, Matthews and Owen (2008) found that the rapid initial weakening of rock surfaces is indicative of rates of biological weathering by endolithic lichens that may be 200–300 times faster than physico-chemical weathering alone.

Despite early controversy concerning the pedogenic significance of lichens, their effectiveness in the biodeterioration of rocks has been clearly demonstrated by recent research, which has revealed that substantial quantities of the substratum can be degraded even over relatively short periods of time. Furthermore, lichens have the capacity to accumulate elements such as N, P, and S, thereby increasing the letters' potential bioavailability to successive life forms which may replace lichens during soil development. Organic material derived from lichen decomposition, together with detached particles of the substratum and atmospherically derived dusts trapped by thalli, all contribute to the development of soils.

The weathering action of saxicolous lichens can be physical and/or chemical. Lichens contribute to mechanical weathering of rocks in four ways: (1) penetration of mycobiont hyphae (up to 15–20 mm) and rhizines into naturally occurring crevices and cracks in rock surface;, (2) expansion and contraction of lichen thalli with daily and seasonal changes in ambient temperature and humidity; (3) swelling action of organic salts produced by lichens; and (4) fracturing and incorporation of mineral fragments by lichen thalli (Chen et al. 2000).

Lichens chemically weather rocks by means of carbon dioxide (internally and/or externally derived), oxalic acid, and the complexing action of lichen substances. The latter have a low but significant solubility in water, forming soluble metal complexes under laboratory conditions when they react with minerals and rocks, particularly limestone (Syers and Iskandar 1973). Water is essential for many of the chemical reactions associated with the breakdown of rock substrata; since lichens are able to absorb water in either the liquid or vapour phase, chemical (and indeed) physical weathering processes are expedited. The mixing of respiratory carbon dioxide with water in lichen tissues results in the formation of carbonic acid, which also enhances the solubility of rock surfaces by lowering the pH of the substratum microenvironment adjacent to lichen thalli (Chen et al. 2000). Oxalic acid, formerly considered of minor importance in the biodeterioration process, has been proved otherwise; the nature of the thallus–substratum interface generated and the chemical disruption of the substratum are significant components of the weathering process. An ecological approach to our understanding of the genesis of calcium oxalate patinas on stone monuments is provided by Caneva (1993).

Detailed spectroscopic studies have demonstrated the highly destructive properties of calcium oxalate produced by lichen thalli, particular attention being directed towards the dramatic effects caused by the action of certain aggressive lichen species on historic monuments and other works of art, where biodeterioration processes have been shown to be devastatingly destructive within a surprisingly short timescale (Seaward and Edwards 1997). Such action on natural substrata is clearly of significance in a pedogenetic context, since lichens are usually regarded as weathering agents on a geological timescale.

The impact of lichen weathering of rocks on a global scale has been, and continues to be, important in terms of climatic consequences and the habitability of our planet: their disappearance from particular ecosystems would be critical over major areas. Lichens form such an important component of the complex web of life that their disappearance affects the balance of nature to a surprising degree, particularly in tundra zones, high altitudes, cold deserts, dune systems, semi-arid lands and deserts, and even urban areas, where they are, for example, important in community development and succession on rocks and soils.

There are areas of the globe where the results of lichen denudation are now being detected by means of remote sensing. Such losses may well have climatic repercussions and exert a measurable influence on global warming, as, for example, in the case of the disappearance of epilithic lichens over a very large area of the Canadian shield as a direct consequence of atmospheric pollution formerly emanating from the smelting operations at Sudbury (Seaward 1996); fortunately, this has subsided due to pollution abatement measures, allowing the barren rock surfaces to be recolonized by lichens, thereby restoring their light-absorbing ability (Rollin et al. 1994). It has been noted that if today's weathering were to take place under completely abiotic conditions, dramatic increases in global temperature would result (Schwartzman and Volk 1989).

9.4 Lichen Mosaics

Lichens are eminently successful and enjoy a worldwide distribution. Lichen species and assemblages are faithful to particular habitats and as such make ideal environmental indicators in biodeterioration studies. Their spatial andtemporal patterns are ecologically delimited by one or more factors. Lichen communities are often species rich and densely crowded with complex biotic interactions. In terms of determining the ecological processes in a community, spatial patterns need to be determined for habitats and ecosystems (e.g. an individual rock, a mountainside scree, and variously dated churchyard memorials). Lichens occur in almost every conceivable habitat and often play a key role in the operation of some natural systems, growing on a variety of substrata, including most natural ones as well as a host of human-manipulated or manufactured ones. Common natural substrata supporting lichens include all categories of rocks and common man-made substrata include fashioned stonework, asphalt, glass, concrete, cement, plaster, ceramic and terracotta tiles, bricks, processed wood products, and various types of metals (Brightman and Seaward 1977).

The presence of lichens on building materials, more particularly ancient monuments, is variously interpreted by the lay public and by specialists in different disciplines, whose attitudes are inevitably coloured by differing aesthetic and practical considerations. Undoubtedly, lichens are exceedingly sensitive to environmental change, and the diversity and composition of their assemblages can be a reliable indication of the level of atmospheric pollution, which in itself is one of the most serious factors in the deterioration of such stone surfaces.

The lichenologist regards the appearance of a lichen mosaic as a natural feature of ancient monuments, finding the diversity of species present aesthetically pleasing, besides being both taxonomically and ecologically interesting. Clearly, there is a direct correlation between the composition of the flora and the passage of time, the special technique of lichenometry being employed for dating surfaces (Innes 1985); furthermore, the different lichen communities established on stone surfaces not only reflect the various materials employed in their construction, but also can often be correlated with the chronology of successive building phases, thereby assisting in archaeological interpretation (Piervittori 2004). It should also be noted that many nationally or internationally rare lichen species that are to be found mainly or only on particular ancient monuments need to be considered in their own right as in need of conservation. In fact, some would argue that lichens can no longer be disregarded as aesthetical impairment or biodeterioration agents by architectural heritage preservation or cultural studies and that the 'conservation of a micro-flora is worthy of protection on historical monuments and its co-existence with the historically based elements' (Drewello and Drewello 2009).

It is ironic that in a bland, homogenous urban environment, where a lichen mosaic would be a welcome relief to the eye, the higher levels of air pollution prevent its establishment, only allowing the existence of a monotonous flora composed of a few algae and lichen crusts. Such a feature has proved problematic not only from an archaeological standpoint, but also architecturally in the development of modern buildings, in both cases being treated as a biodeterioration process. This lack of diversity and complexity in urban lichen communities is due to (a) local accumulation of high concentrations of interacting toxic airborne contaminants generated from a wide variety of urban and industrial activities and processes, (b) removal and/or replacement of natural substrata, and (c) alteration of natural hydrologic cycles. Prior to the onset of the industrial age, stone and wooden buildings, gates, fences and tombstones, as well as glass, bricks and tiles, in urban areas supported reasonably diverse lichen communities, as still witnessed in many rural settlements. As industrial activity developed in and around human population centres, lichen communities on natural substrata declined, but man-made substrata supported their own, albeit limited, lichen assemblages.

There is strong evidence that recent environmental changes have been conducive to increasingly detrimental invasion by certain aggressive lichens, as in the case of the establishment of nitrophilous species due to environmental hypertrophication (Deruelle 1988). Such evidence could help to explain why it is that monuments, undamaged for many centuries, appear in recent years to be vulnerable to lichen attack, in addition to the known problems resulting from air pollution (Pye and Schiavon 1989; Butlin 1991). Whilst acknowledging the recent effects of air pollution (especially acid rain) on monuments, it must also be recognized that *D. massiliensis* forma *sorediata*, for example, is a relative newcomer: its dramatic spread in Europe, and more particularly Britain, in recent years has been facilitated by new environmental regimes, including qualitative changes in air pollution, which have allowed it to dominate substrata in the wake of the rapid disappearance of other more pollution-sensitive species. *D. massiliensis* forma *sorediata* is by no means the only organism implicated in short-term deterioration processes: other lichens, and indeed other micro-organisms, capable of adapting to man-made environmental disturbances can be equally destructive when ecosystem equilibria are disrupted.

The urban ecosystem is of increasing interest to lichenologists, as well as ecologists and environmental scientists. For almost two centuries, urban areas, mostly in the developed world, have been of lichenological interest purely in terms of their lack or paucity of lichens, air quality being identified as the limiting factor, but the contributions of the many factors operating in a changing urban climate and complex landscape, although appreciated, were not fully evaluated. With the implementation of clean air policies in recent years, the urban environment is lichenologically unrecognizable: without the major limiting factor, a diverse landscape, in terms of habitats, substrata and chemistry (mainly nutrient-enrichment), is there to be exploited (Seaward 1997b). Some cities could soon be on a par with lichen-dominated ecosystems, since an urban landscape of man-made structures constructed from a very wide variety of natural and manufactured building materials can provide suitable substrata for the establishment of a diverse lichen mosaic, which in turn requires consideration from a biodeterioration perspective. Although it is pleasing to note that the undoubted improvement in air quality of many cities in recent years is reflected in the continued recovery of their lichen floras, there is no room for complacency, since other recent environmental changes brought about by human agencies have been conducive to increasingly detrimental invasion by certain aggressive micro-organisms, as in the case of the establishment of nitrophilous species due to hypertrophication (Seaward and Coppins 2004).

However, historians, archaeologists, and fine art specialists concerned with the conservation of ancient monuments in both rural and urban settings view the encroachment of lichens from a very different standpoint from that of the lichenologist: inscriptions and fine details may be obscured, and depending on the nature of the substratum, and in some cases the ambient conditions, serious physical damage is often caused through lichen-induced biodeterioration. The lichen floras vary considerably according to the spatial differences in the chemical properties of stone surfaces, the micro-environmental conditions, and the overall influence of air pollution. Clearly, with all these conflicting issues, it is necessary to establish whether the particular lichens observed are merely causing aesthetically unpleasing disfigurement or are actually causing physical and/or chemical damage to their substrata.

9.5 Aesthetic Disfigurement Versus Physical Damage

It has been shown above that the role of lichens as biological weathering agents in the development of soils should no longer be considered in a geological context only, since such organisms are capable of biodeteriorating stone substrata within a relatively short timescale. Expert diagnosis of those species mainly responsible for this, and the conditions under which they thrive, is essential for the conservation of stonework. It is essential that the real causes of deterioration in any given situation are scientifically established before well-meaning, but uninformed, action is taken; for example, cleaning techniques may well accelerate deterioration in situations where lichens have not been reliably established as the prime cause of damage. Indeed, some species may well afford a protective cover, shielding the cement work from external weathering agents. It is therefore necessary to determine which species are disfiguring but intrinsically harmless and which cause actual physical damage (Lallemant and Deruelle 1978; St. Clair and Seaward 2004a). To this end, several studies aimed at determining the nature of the disfigurement and/or physical damage by particular lichens to various substrata have been undertaken in various parts of the world, a few of which are detailed below. These studies demonstrate some of the dilemmas facing conservationists: namely, do lichens damage the particular substratum or merely create a 'patina of time' (cf. 'scialbatura' in Del Monte and Sabbioni 1987) which is aesthetically unpleasing, and furthermore, do lichens afford some protection to the substratum and if removed without consolidation of the exposed surfaces, will further, or indeed more intensive, weathering ensue (St. Clair and Seaward 2004a; McIlroy de la Rosa et al. 2012a).

The exposure of Roman mosaics through excavation after centuries of being buried beneath soil, etc. poses just such a dilemma. Pioneer colonization by lichens, sometimes with a predilection for particular coloured tesserae (Fig. 9.2), detracts from the detail or overall

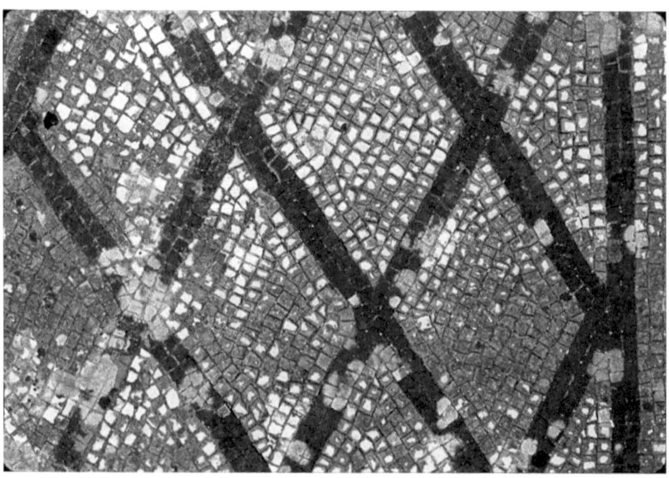

Fig. 9.2 An exposed Roman mosaic at Ostia Antica, Italy—the *light-coloured* lichens, such as *Aspicilia* spp. with a predeliction for the *dark-coloured* tesserae, clearly detracting from the overall effect of the original pattern

pattern of the mosaic, and it has been clearly demonstrated in a detailed study of exposed Roman mosaics in Italica, Spain that successional lichens, bryophytes, etc. have ultimately led to their complete destruction (Saiz-Jimenez et al. 1991). Common sense should prevail: clearly such archaeological exposures should either be re-covered after detailed recording, or suitably protected in situ againstweathering, or, if absolutely necessary, be carefully translocated to an appropriate museum setting.

In a study of archaeologically important monuments and artefacts constructed from a variety of materials in a grassed area surrounded by cloistered museum buildings of the Museo Nationale Romano in Rome, the role of the ambient urban climate and associated atmospheric pollutants had to be taken into consideration when interpreting the deterioration of these materials, particularly since the latter dramatically affects the lichen flora.

It was noted, for example, that several of the more toxi-tolerant lichen species (poleophiles) with an aggressive behaviour were actively colonizing stone monuments and other archaeological artefacts. A most dramatic case of lichen attack was observed in the case of several large terracotta pots scattered about the grassed area

Fig. 9.3 Section through *Lecanora muralis* to show the blister created on the shoulder of a terracotta pot at the Museo Nationale Romano—not only has pottery been detached from the surface to a depth of 2–4 mm, but a chamber to a depth of 3–5 mm beneath this has created a microenvironment which influences other biodeterioration processes (Seaward 1988)

(Seaward 1988). The rims and shoulders of these pots were lightly colonized by relatively innocuous species such as *Lecanora dispersa* and *Candelariella vitellina*, but here and there, thalli of *Lecanora muralis*, mainly 4–7 cm in diameter, probably representing 6–15 years growth, were causing demonstrable damage. A section through one of the thalli clearly reveals the results of such damage: a central blister, created by the crowding of apothecia, pulled away a fragment of the substratum, two or more mm in thickness, over an area almost 12 cm^2 (Fig. 9.3). *L. muralis* appears to be a highly successful lichen in urban environments into which it has spread dramatically in recent years, due in part to lack of competition from other species; a change in the nature of air pollution in Rome in that period may be a contributory factor to this aggressive behaviour.

In another study, the deterioration of frescoes by Zuccari painted in the 1560s on the inner walls on ground and first floor levels of a circular cloister at the Palazzo Farnese, a beautiful mansion at Caprarola in central Italy built by Vignola in 1547–1549, was examined in detail. It was clear that the relatively recent alarming signs of biodeterioration of the water-based paintwork were the result of a single lichen species, *D. massiliensis* forma *sorediata*, its attack being very pronounced in many places and in its early stages of colonization clearly demonstrating a predilection for the brown and yellow pigments, rather than the red pigment which contained one or more metals antagonistic to its growth. When viewed in 1986, the distribution pattern of this lichen was clearly dictated not only by the colour of the paintwork, since it was obvious that some recent cleaning activity had distributed lichen propagules from existing thalli to create distinctive areas of lichen invasion across the frescoes (Edwards and Seaward 1993). More recently, lichen colonization was less clearly demarcated, and cleaning removed not only the lichen but also the underlying paintwork (Fig. 9.4); however, it is pleasing to the note that since restoration, involving the use of paints impregnated with biocides after the cleaning procedure, lichen reinvasion has ceased—but for how long?

Fig. 9.4 Renaissance fresco in the Palazzo Farnese, Italy, attacked by *Dirina massiliensis* forma *sorediata*—its removal by light brushing from the central area has also removed the underlying paintwork

Fig. 9.5 Church wall, Fiskerton, England—the calcium oxalate coating generated by *Dirina massiliensis* forma *sorediata* suggesting a former surface rendering (Seaward and Edwards 1997)

Using FT-Raman spectroscopy, it was shown that *D. massiliensis* forma *sorediata* on these frescoes produced calcium oxalate encrustations at the thallus–substratum interface of almost 2 mm in thickness in less than 12 years. It was deduced that 1 m^2 of the fresco with a current 60 % obliteration by this lichen and the underlying substratum generated more than 1 kg of calcium oxalate over a similar time period; furthermore, with the incorporation of calcite and gypsum into the thallus encrustation, it is likely that more than four times this amount of the underlying substratum has been chemically and physically disturbed (Seaward and Edwards 1995, 1997).

The short-term biodeteriorative capacity of *D. massiliensis* forma *sorediata* is not specific to the above-mentioned frescoes and mural paintings (Nugari et al. 2009); for example, detailed studies of this lichen on exterior stonework of English churches, as demonstrated by a fourteenth-century church at Fiskerton in Lincolnshire, have shown its similarly destructive nature (Seaward and Edwards 1997). In the past, encrustations generated by this, and no doubt other lichen species, have been misinterpreted as the remaining traces of a whitish coating or rendering applied as a decorative or protective surface in a nineteenth- or twentieth-century restoration programme (Fig. 9.5). It is now clear that this is not so, since these 'renderings' consist essentially of calcium oxalate and more often than not evidence remains of the thalli producing them (Fig. 9.6). These encrustations are usually more than 0.5 mm in thickness and cover considerable areas of many church walls throughout England.

The impact of the paints and 'renderings' on the presence or absence of lichens in the above two examples raises an interesting issue: namely, would treatments of monuments in the past have intentionally or fortuitously discouraged lichen growth? Accepting the role of air pollutants in recent years, why, despite the ravages of various forms of weathering processes over centuries, has biodeterioration been shown to be of paramount importance in the deterioration of some of our cultural heritage, particularly in the Mediterranean region. A clue to this may be found on the Parthenon sculptures, the surviving traces of colour on them (Jenkins and Middleton 1988) representing an example of the widespread use of ancient polychromy, the applied paints, known to contain toxic metals, having biocidal properties. What other treatments have our monuments been subjected to by those who have revered either the structure or its (religious) function?

Undoubtedly, many of the major researchers in biodeterioration investigations have been, and continue to be, based in Italy. Their pioneer work and considerable output of published material have contributed significantly to our understanding of the characteristics of lichen assemblages on ancient monuments and their role in biodeterioration processes (for a review see Piervittori 2004). Valuable work has been undertaken in other Mediterranean countries, particularly Spain (Garcia-Rowe and Saiz-Jimenez 1991; Ascaso and Wierzchos 1995; Ascaso et al. 2004; Ariño and Saiz-Jimenez 2004; Silva and Prieto 2004).

Fig. 9.6 Calcium oxalate encrustation generated by in situ thalli of *Dirina massiliensis* forma *sorediata*

Contributors to our knowledge of biodeterioration processes based elsewhere, but not necessarily working solely on the cultural heritage of their own countries, include Antarctica, Australia, Britain, Canada, China, France, Germany, Holland, India, Israel, Japan, Korea, Norway, Poland, Portugal, South Africa, Sweden, Turkey, and USA. It is hoped that the review of lichens growing on historical monuments and buildings in India (Saxena et al. 2004; Bajpai and Upreti 2014) will stimulate others to investigate this under-researched subject in hot climates. Other than the occasional paper for Central and South America, south-east Asia and north-west Africa, relatively little is known of the impact of biodeteriorative processes on monuments in developing countries, much of the very considerable disfiguration observed in tropical and subtropical areas being due not only to lichens but also to cyanobacteria which create dark biofilms (Fig. 9.7).

9.6 Stonework Conservation: Practical Considerations

The above observations constitute a necessary first phase of any programme aimed at quantifying the actual role played by lichens in the deterioration of archaeological materials. Stone- and art-work in exposed and partially enclosed situations, giving rise to a variety of environmental conditions, need to be examined in detail. It is necessary to elucidate the relationships between particular lichen species and the physical and chemical nature of their substrata in order to determine the relative importance of those species in biodeterioration processes obtaining in specific circumstances. It is essential to determine, for example, which species are disfiguring but intrinsically harmless, and which cause actual physical damage (Fig. 9.8).

Fig. 9.7 Imam Bara Mosque, Lucknow, India—aesthetic disfiguration of building in a subtropical area, the dark biofilms created by both lichens and cyanobacteria

It is self-evident that baseline work is a prerequisite for laboratory research designed to establish the nature of the interface between problematic lichens and their substrata and field trials intended to test the relative effectiveness of differing techniques and treatments for the removal and discouragement of lichens from stonework. Any treatment should be selected with care, since although immediately effective, the long-term effects are highly likely to be deleterious. Other possibilities, mostly based on common sense, should also be investigated, as exemplified by the work 'towards a solution' on prehistoric rock art (Tratebas 2004). Recent work to quantify hyphal spread within the substratum (Gazzano et al. 2009a), for example, has shown that it may be possible to determine which particular lichens should be targeted in conservation measures. Currently, only tentative evaluations of lichen impact on stonework are available (Gazzano et al. 2009b).

Efforts to conserve our cultural heritage from deterioration are extensively practised worldwide. As would be expected, damage to stonework, more particularly limestones, is of international concern. Whilst accepting that both

Fig. 9.8 Limestone sculpture by F. Narvaez at the University of Caracas, Venezuela, showing lichen and cyanobacterial disfiguration about 40 years after its installation. (*Photo* M. Tabasso)

ancient monuments and more recent architectural structures of merit and/or importance are subjected to wind and water erosion, such damage is exacerbated by the chemical and mechanical effects of not only lichens but various other pioneer micro-organisms and mycota (Paine et al. 1933; Wee and Lee 1980; Danin and Caneva 1990; Ortega-Calvo et al. 1991; May et al. 1993; Burford et al. 2003), which in due course are succeeded by mosses and higher plants (Bech-Andersen 1984); each of these successional stages support other life forms as well as protista and invertebrates (Gerson and Seaward 1977), the complex ecosystem developed thereby creating a milieu for biodeterioration. Each of these successional stages generates its own problems for the conservationist to solve.

Various methods for controlling or eliminating lichen growth, and indeed other defacing micro-organisms, on rocks and stonework have been investigated, control measures ranging from mechanical removal to application of various types of biocides. Since many lichen species are able to regenerate from thallus fragments, mechanical removal of lichens may only temporarily reduce lichen coverage. Mechanical methods involving scraping and brushing, usually followed by washing, are tedious, damaging and often ineffective. Absorbed water may adversely affect the monument, particularly under fluctuating temperature regimes; although penetration can be minimized by the use of water repellents, entrapped water and rising damp can nevertheless prove highly destructive. Laser treatment of stonework to remove lichens has been tried, and although this could feasibly be adopted for small artefacts (such as those removed for display in a museum), it is too laborious for most of the cleaning tasks required. Lichen growth on buildings and monuments is often prevented, intentionally or indeed fortuitously, on stonework affected from downwash or adjacent to roofing, windows, guttering, and pipework (Fig. 9.9) employing metals, particularly lead, zinc, and copper.

Fig. 9.9 Church wall, Fiskerton, England—the absence of lichen assemblages to either side of a metal downpipe clearly demonstrated

Application of biocides has yielded mixed results, including: generally poor treatment response, changes in community dynamics, with surviving species aggressively exploiting open space, persistent dead thalli, which decompose slowly, especially in arid habitats, and good results on lichen removal but damage to substrate surfaces, with changes ranging from discoloration to structural issues requiring use of chemical consolidants (St. Clair and Seaward 2004a). A wide range of biocides have been tried (see literature cited in Martin and Johnson 1992), many of which have since been rejected due to side effects such as crystallization of soluble salts which have penetrated the stonework, staining, and discoloration of monuments where the chemicals used have interacted with particular metals present in the substratum, and the promotion of secondary biological growths, which may be even more unsightly than the primary growths.

Furthermore, regular treatments are likely to be necessary which are expensive both in terms of the chemicals used and the labour employed for the mechanical removal of only partially detached and brittle lichen growths which remain. As pointed out by Nascimbene et al. (2008), if restorations are not to be maintained, it would be better to retain a more historic, diverse, and complex lichen community than a simplified community of 'weedy' lichens that quickly cover almost the same area as the pre-restoration community did. The biocides employed may also be harmful to the operators and, not surprisingly, dangerous to wildlife. Some success has been achieved using Anatasa (Fonseca et al. 2010), Koretrel (Tretiach et al. 2007), and also organometallic compounds, quaternary ammonium compounds and borates, but the latter have proved problematic when used in air-polluted environments where, of course, many of the monuments it is desired to conserve are located. In a study by Prieto et al. (1995), it was found that biocides tested in the laboratory were highly effective in surface cleaning but remnants of crustose lichens and algae remained between the quartz and feldspar crystals; results obtained in the field, using the same substratum and biocides, proved to be unsatisfactory. Other approaches to solve the problems arising from standard biocides, such as the use of enzymes (Silver and Wolbers 2004), are under review.

After application of biocides, consolidants may be necessary, as in the case of eighth to eleventh-century structures at Xunanturick in Belize constructed of low strength, porous limestone composed of almost pure calcium carbonate. Clearly, the erosive forces of wind and water of the humid tropical climate have extensively weathered these monuments, but the impact of lichens and other organisms has also been significant (Kumar and Kumar 1999). To evaluate the effectiveness of possible stabilization treatments, tests were conducted on stone-penetrating consolidant solutions applied to the limestone under different environmental conditions over a period of 4 years. Several of these consolidants were capable of penetrating and stabilizing the moist limestone and, in combination with biocides, 'minimized' the growth of lichens and other local microflora (Ginell and Kumar 2004).

Consolidant testing was also undertaken to ascertain if lichen-damaged sandstone supporting rock art in Wyoming, USA, could be strengthened by chemical consolidation. To achieve this, sandstone samples with attached lichens were characterized by means of detailed laboratory analyses involving examination by reflected and transmitted light, water-absorption tests (i.e. porosity), ionic chromatography, X-ray diffraction, and environmental scanning electron microscopy (Silver and Wolbers 2004). One of the underlying objectives of this work was to determine if substrata exposed when lichens are removed by mechanical and/or chemical methods are far more vulnerable to weathering and therefore in need of consolidant treatment to restore their resistance to that of the adjacent hard-packed desert varnish surface (Silver and Wolbers 2004; Chiari and Cossio 2004; Tratebas 2004). It is clear that rock art, which depends upon an unaltered natural environment, presents different issues from the culturally constructed environment of historic buildings and monuments (Tratebas 2004).

The problem is further complicated by the fact that in some situations, lichen thalli may actually play an important role in consolidating and protecting the substrate surface. A protective effect by fungal hyphae for oolitic limestone buildings has also been noted (Concha-Lozano et al. 2011). However, when lichens are removed by mechanical and/or chemical methods, the structural integrity of the substratum can be reduced and is thereby far more vulnerable to weathering (Silver and Wolbers 2004; Tratebas 2004; McIlroy de la Rosa et al. 2012a). Furthermore, it has been shown that in certain situations lichens are bioprotective, shielding the underlying stonework from external weathering agents, such as rain and insolation (Viles and Pentecost 1994; Mottershead and Lucas 2000), and indeed from pollutants and salt deposition (Carter and Viles 2003, 2005). Increasing evidence has been assembled in recent years regarding the bioprotective role of lichens on stonework, particularly calcareous surfaces (McIlroy de la Rosa et al. 2012a). Therefore, there may be good reasons to retain the subtle colouration of a varied mosaic for its aesthetic appeal, provided it does not produce disfigurement or unduly obscure detail. Furthermore, in modifying the physical and chemical properties of, for example, asbestos-cement, lichens are acting as effective spontaneous bioattenuation agents (Favero-Longo et al. 2009a).

The ethical and conservation issues arising from remedial work aimed at combating biodeterioration of cultural monuments cannot be underestimated. Conflicting issues often result from the complexity of the problems involved in any restoration programme (Tratebas and Chapman 1996; Tratebas 2004). Recent work on the biodeterioration of ancient monuments has also shown that complex problems can arise which involve some combination of laws and ordinances enacted with the intent of protecting and maintaining the resource, altered local resource use patterns, which sometimes result in environmental conditions that enhance lichen growth and development on or near critical cultural resources, consideration of threatened and endangered and endemic lichen species, and religious and cultural concerns and expectations of aboriginal populations, which in some cases may run counter to government-mandated management programmes (St. Clair and Seaward 2004b).

There are instances where lichens are actually encouraged on new stone and cement surfaces, and various chemicals are employed to establish/promote their growth. Such practices are employed, for example, in the case of farm roofs (especially those constructed of asbestos-cement) where natural colonization by lichens can satisfy the design requirements regarding colour and reflectance specified by planning authorities and governmental agencies. Such 'blending' into the landscape of roofs, and sometimes other architectural features of new buildings and extensions to existing premises, such as garages, is not only aesthetically more pleasing but may be legally necessary in environmentally sensitive situations.

Restoration work on ancient monuments resulting in contrasting stonework may also require such chemical applications to tone down the differences (Fig. 9.10); furthermore, there are instances where new surfaces, such as mortar (=repointing), provide substrata conducive to lichen assemblages which contrast sharply with those on the original stonework and are the loci from which these alien lichens with a great potential for biodeterioration may spread onto the monument (Fig. 9.11). Furthermore, thalli established initially on the mortar subsequently invade the artefact in need of preservation such

Fig. 9.10 Lincoln Castle, England—the restoration of a wall demonstrating the need for remedial action through the promotion of a suitable lichen assemblage

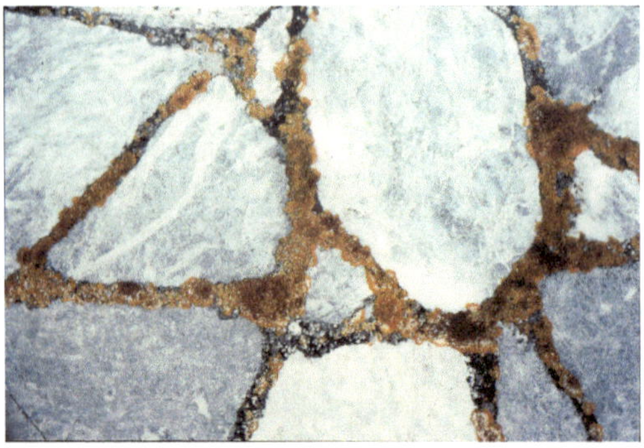

Fig. 9.11 Villa Adriano, Italy—the mortar of a reconstructed column providing loci for the colonization of distinctively coloured thalli of *Xanthoria* and *Caloplaca* spp. characteristic of calcareous substrata

Fig. 9.12 Ostia Antica, Italy—the mortared areas of a characteristically reticulated Roman wall clearly generating a distinctive lichen assemblage

Fig. 9.13 Bull's head monument, National Museum, Rome—the use of non-authentic material (cement) for the neck providing a locus for non-characteristic and unappealing lichens to spread onto the marble sculpture

as limestone tesserae (Saiz-Jimenez and Arino 2001) and reticulated Roman walls (Fig. 9.12). Why it is necessary to use such incongruent materials in the reconstruction of monuments (Fig. 9.13) when a supply of similar stonework is available?

On the other hand, there are instances where lichen recolonization should not be promoted by such practices; for example, where, for educational purposes, historical interpretation of building periods and architectural developments is highlighted or demarcated by the presence or nature of the lichen assemblages over time (Fig. 9.14).

It should also be noted that there are many instances where the actual lichens themselves are in need of conservation, since habitats generated by ancient monuments often represent ecological continuity. It is debatable whether habitats or species should be the primary concern: on the one hand, lichen conservation should aim at conserving habitats (their presence, for example, being a means of biomonitoring pollution) and, on the other, compiled regional, national, and international red lists of threatened species often include lichens known only from or mainly in such sensitive habitats (Fig. 9.15). Clearly, the maintenance of threatened species, lichen biodiversity, and our cultural heritage in general is dependent on good management practices.

Lichens are natural sensors of our changing environment and the presence or absence of particular species and the composition of assemblages on stonework are indicative of one or more identifiable factors. Information gained from our knowledge of how lichens respond to long-term perturbations and short-term upheavals

Fig. 9.14 Castle walls of Otranto, Italy—the area devoid of lichens demarcating the size and nature of a former earth embankment

Fig. 9.15 Gravestone, Baumber, England—*Ramalina capitata*, a very rare British lichen, currently found only in churchyards in two isolated areas of the country

in nature can be applied to the interpretation and monitoring of environmental changes and disasters brought about through a wide range of human activities. Lichens can often be used as an early warning system for other biota which without remedial action would subsequently suffer stress or indeed extinction through human mismanagement. Baseline information on lichen assemblages and ecosystems which are ecologically distinctive, such as those on prehistoric monuments (Aptroot and James 2002) and in churchyards, have proved invaluable in assessing human pressures on the environment. Clearly, long-term field investigations involving detailed inventories based on stringent ecological criteria are a necessary component of any programme aimed at interpreting environmental disturbances, both natural and man-made. Such work would be complementary to biodeterioration monitoring, many of the methods and measurement techniques for which are detailed in Mandrioli et al. (2003).

9.7 Conclusions

In the light of all of the above reasons, any decision to remove lichens from stonework must not be undertaken over-hastily or without very careful consideration of the wider implications of long-term effects. Unfortunately, it has to be acknowledged that the problem is under-researched and much of the work published to date is of a largely empirical nature, which has yet to be adequately substantiated by long-term experimentation. Biodeterioration of historical and culturally significant stone substrata is a complex problem: easy, risk-free solutions are simply not available. The dynamics of the problem are too complicated to address effectively with only one kind of expertise. Successful resolution of this issue will inevitably require a multidisciplinary effort, where lichenologists join with ecologists, geologists, geochemists, crystallographers, cultural conservators, archaeologists, and historians to recommend the most effective management scheme. The advantage of this approach is obvious: multidisciplinary management teams with good leadership can ask more appropriate questions whilst developing much more thoughtful and informed management decisions (St. Clair and Seaward 2004a).

References

Adamo P, Violante P (1991) Weathering of volcanic rocks from Mt. Vesuvius associated with the lichen *Stereocaulon vesuvianum*. Pedobiologia 35:209–217

Aptroot A, James PW (2002) Monitoring lichens on monuments. In: Nimis PL, Scheidegger C, Wolseley PA (eds) Monitoring with lichens—monitoring lichens. Kluwer Academic, Dordrecht, pp 239–253

Ariño X, Saiz-Jimenez C (2004) Lichens of different mortars at archaeological sites in southern Spain. In: St. Clair LL, Seaward MRD (eds) Biodeterioration of stone surfaces. Kluwer Academic, Dordrecht, pp 165–179

Arino X, Gomez-Bolea A, Hladun N, Saiz-Jimenez C (1996) *Roccella phycopsis* Ach.: colonization and weathering process on granite. In: Vicente MA, Delgado-Rodrigues J, Acevedo J (eds) Degredation and conservation of granitic rocks in monuments. Research report, protection and conservation of the European cultural heritage, European Commission, Brussels, pp 399–404

Armstrong RH (1988) Substrate colonization, growth, and competition. In: Galun M (ed) CRC handbook of lichenology. CRC Press, Boca Raton

Ascaso C, Wierzchos J (1995) Study of the biodeterioration zone between the lichen thallus and the substrate. Cryptogamic Bot 5:270–281

Ascaso C, Galvan J, Ortega C (1976) The pedogenic action of *Parmelia conspersa*, *Rhizocarpon geographicum* and *Umbilicaria pustulata*. Lichenologist 8:151–171

Ascaso C, Garcia del Cura MA, de los Rios A (2004) Microbial biofilms on carbonate rocks from a quarry and monuments in Vovelda (Alicante, Spain). In: St. Clair LL, Seaward MRD (eds) Biodeterioration of stone surfaces. Kluwer Academic, Dordrecht, pp 79–89

Bailey RH (1976) Ecological aspects of dispersal and establishment in lichens. In: Brown DH, Hawksworth DL, Bailey RH (eds) Lichenology: progress and problems. Academic Press, London, pp 385–406

Bajpai R, Upreti DK (2014) Lichens on Indian monuments: biodeterioration and biomonitoring. Bishen Singh Mahinder Pal Singh Publishers, DehraDun

Barkman JJ (1958) Phytosociology and ecology of cryptogamic epiphytes. Van Gorcum, Assen

Bech-Andersen J (1984) Biodeterioration of natural and artificial stone caused by algae, lichens, mosses and higher plants. In: Barry S, Houghton DR, Llewellyn GC, O'Rear CE (eds) Biodeterioration 6. CAB International, Slough, pp 126–131

Brightman FH, Seaward MRD (1977) Lichens of man-made substrates. In: Seaward MRD (ed) Lichen ecology. Academic Press, London, pp 253–293

Brodo IM (1973) Substrate ecology. In: Hale ME, Ahmadjian V (eds) The lichens. Academic Press, New York, pp 401–441

Bungartz F, Garvie LAJ, Nash TH (2004) Anatomy of the endolithic Sonoran Desert lichen *Verrucaria rubrocincta* Breuss: implications for biodeterioration and biomineralization. Lichenologist 36:55–73

Burford EP, Fomina M, Gadd GM (2003) Fungal involvement in bioweathering and biotransformation of rocks and minerals. Minerological Mag 67:1127–1155

Butlin RN (1991) Effects of air pollutants on buildings and materials. Proc R Soc Edinb 97B:255–272

Caneva G (1993) Ecological approach to the genesis of calcium oxalate patinas on stone monuments. Aerobiologia 9:149–156

Carter NEA, Viles HA (2003) Experimental investigations into the interactions between moisture, rock surface temperatures and an epilithic lichen in the bioprotection of limestone. Building Environ 38:1225–1234

Carter NEA, Viles HA (2004) Lichen hotspots: raised rock temperatures beneath *Verrucaria nigrescens* on limestone. Geomorphology 62:1016

Carter NEA, Viles HA (2005) Bioprotection explored: the story of a little known earth surface process. Geomorphology 67:273–281

Chen J, Blume H, Bayer L (2000) Weathering of rocks induced by lichen colonization—a review. Catena 39:121–146

Chiari G, Cossio R (2004) Lichens on Wyoming sandstone. Do they cause damage? In: St. Clair LL, Seaward MRD (eds) Biodeterioration of stone surfaces. Kluwer Academic, Dordrecht, pp 99–113

Concha-Lozano N, Gaudon P, Pages J, Billerneck G, Lafon D, Eterradossi O (2011) Protective effect of endolithic fungal hyphae on oolite limestone buildings. J Cult Heritage 13:120–127

Danin A, Caneva G (1990) Deterioration of limestone walls in Jerusalem and marble walls in Rome caused by cyanobacteria and cyanophilous lichens. Int Biodet 26:397–417

De los Rios A, Wierzchos J, Sancho LG, Sancho TG, Ascaso C (2005) Ecology of endolithic lichens colonizing granite in continental Antarctica. Lichenologist 37:383–395

Del Monte M, Sabbioni C (1987) A study of the patina called 'scialbatura' on imperial Roman marbles. Stud Conserv 32:114–121

Deruelle S (1988) Effets de la pollution atmospherique sur la vegetation lichenique des monuments historiques. Studia Geobot 8:23–31

Drewello R, Drewello UG (2009) Flechten auf Denkmälern: Indikatoren und Vermittler zwischen Denkmal- und Naturschutz. In: Ōkologische Rolle der Flechten. Rundgespräche der Kommission für Ōkologie 36. Bayerische Akademie der Wissenschaften (Herausgeber). Freiderich Pfeil, Munchen

Edwards HGM, Seaward MRD (1993) Raman microscopy of lichen–substratum interfaces. J Hattori Bot Lab 74:303–316

Edwards HGM, Farwell DW, Jenkins R, Seaward MRD (1992) Vibrational Raman spectroscopic studies of calcium oxalate monohydrate and dihydrate in lichen encrustations on Renaissance frescoes. J Raman Spectrosc 23:185–189

Edwards HGM, Edwards KAE, Farwell DW, Lewis IR, Seaward MRD (1994) An approach to stone and fresco lichen biodeterioration through Fourier Transform Raman microscopic investigation of thallus–substratum encrustations. J Raman Spectrosc 25:99–103

Edwards HGM, Russell NC, Seaward MRD, Slarke D (1995) Lichen biodeterioration under different microclimates: an FT-Raman spectroscopic study. Spectrochimi Acta Part A 51:2091–2100

Edwards HGM, Farwell DW, Seaward MRD (1997) FT-Raman spectroscopy of *Dirina massiliensis* forma *sorediata* encrustations growing on diverse substrata. Lichenologist 29:83–90

Favero-Longo SE, Castelli D, Salvadori O, Belluso E, Piervittori R (2005) Pedogenetic action of the lichens *Lecidea atrobrunnea, Rhizocarpon geographicum* gr. and *Sporostatia testudinea* on serpentinized ultramafic rocks in an alpine environment. Int Biodeterior Biodegradation 56:17–27

Favero-Longo SE, Borghi A, Tretiach M, Piervittori R (2009a) In vitro receptivity of carbonate rocks to endolithic lichen-forming aposymbionts. Mycol Res 113:1216–1227

Favero-Longo SE, Castelli D, Fabini B, Piervittori R (2009b) Lichens on asbestos-cement roofs: bioweathering and biocovering effects. J Hazard Mater 162:1300–1308

Fonseca AJ, Pina F, Macedo MF, Lean N, Romanowska-Deskins A, Laiz L, Gomez-Bolea A, Saiz-Jimenez C (2010) Anatase as an alternative application for preventing biodeterioration of mortar: evaluation and comparison with other biocides. Int Biodeterior Biodegradation 64:388–396

Garcia-Rowe J, Saiz-Jimenez C (1991) Lichens and bryophytes as agents of deterioration of building materials in Spanish cathedrals. Int Biodeterior 28:151–163

Garrett RM (1972) Electrostatic charges on freshly discharged lichen ascospores. Lichenologist 5:311–313

Gazzano C, Favero-Longo SE, Matteucci E, Piervittori R (2009a) Image analysis for measuring lichen colonization on and within stonework. Lichenologist 41:299–313

Gazzano C, Favero-Longo SE, Matteucci E, Roccardi A, Piervittori R (2009b) Index of lichen potential biodeteriogenic activity (LPBA): a tentative tool to evaluate the lichen impact on stonework. Int Biodeterior Biodegradation 63:836–843

Gerson U, Seaward MRD (1977) Lichen-invertebrate associations. In: Seaward MRD (ed) Lichen ecology. Academic Press, London, pp 69–119

Ginell WS, Kumar R (2004) Limestone stabilization studies at a Maya site in Belize. In: St. Clair LL, Seaward MRD (eds) Biodeterioration of stone surfaces. Kluwer Academic, Dordrecht, pp 19–44

Gómez-Pujol L, Stephenson WJ, Fornós JJ (2007) Two-hourly surface change on supra-tidal rock (Marengo, Victoria, Australia). Earth Surf Processes Land 32:1–12

Innes JL (1985) Lichenometry. Prog Phys Geogr 9:187–254

Iskandar IK, Syers JK (1971) Solubility of lichen compounds in water: pedogenetic implications. Lichenologist 5:45–50

Jenkins ID, Middleton AP (1988) Paint on the Parthenon sculptures. Ann Br Sch Archaeol Athens 83:183–207

Jones D (1988) Lichens and pedogenesis. In: Galun M (ed) Handbook of lichenology, vol 3. CRC Press, Boca Raton, pp 109–124

Jones D, Wilson MJ (1985) Chemical activity of lichens on mineral surfaces—a review. Int Biodeterior 21:99–104

Jones D, Wilson MJ, Tait JM (1980) Weathering of a basalt by *Pertusaria corallina*. Lichenologist 12:277–289

Jones D, Wilson MJ, McHardy WJ (1981) Lichen weathering of rock-forming minerals: application of scanning electron microscopy and microprobe analysis. J Microsc 124:95–104

Knight KB, St. Clair LL, Gardner JS (2004) Lichen biodeterioration at Inscription Rock, El Morro National Monument, Ramah, New Mexico, USA In: St. Clair LL, Seaward MRD (eds) Biodeterioration of stone surfaces. Kluwer Academic, Dordrecht, pp 129–163

Koestler RJ, Vedral J (1991) Biodeterioration of cultural property: a bibliography. Int Biodeterior 28:229–340 (special issue: Biodeterioration of Cultural Property)

Kossowska M, Węgrzyn M (2009) Lichens recorded on iron and glass in NE Poland. Polish Bot J 54:123–124

Krumbein WE, Urzi CE, Gehrmann CK (1991) Biocorrosion and biodeterioration of antique and mediaeval glass. Geomicrobiol J 9:139–160

Kulikova NN, Suturin AN, Boyko SM, Paradina LF, Timoshkin OA, Potemkina TG, Saybatalova EV (2009) The role of water lichens in the biogeochemical process in the Lake Baikal stony littoral. Inland Water Biol 2:144–148

Kumar R, Kumar AV (1999) Biodeterioration of stone in tropical environments. Research in conservation. Getty Conservation Institute, Los Angeles, pp 1–85

Lallemant R, Deruelle S (1978) Présence des lichens sur les monuments en pierre: nuisance ou protection? In: Proceedings of the international symposium on deterioration and protection of stone monuments. UNESCO, Paris

Mandrioli P, Caneva G, Sabbioni C (eds) (2003) Cultural heritage and aerobiology. Kluwer Academic, Dordrecht

Martin AK, Johnson GC (1992) Chemical control of lichen growths established on building materials: a compilation of published literature. Biodeterior Abstr 6:101–107

Matthews JA, Owen G (2008) Endolithic lichens, rapid biological weathering and Schmidt Hammer R-values on recently exposed rock surfaces: Storbreen Glacier foreland, Jotunheimen, Norway. Geogr Ann Ser A Phys Geogr 90:287–297

May E, Lewis FJ, Pereira S, Taylor S, Seaward MRD, Allsopp D (1993) Microbial deterioration of building stone—a review. Biodeterior Abstr 7:109–123

McIlroy de la Rosa JP, Warke PA, Smith BJ (2012a) Microscale biopitting by the endolithic lichen *Verrucaria baldensis* and its proposed role in mesoscale solution basin development on limestone. Earth Surf Process Land 37:374–384

McIlroy de la Rosa JP, Warke PA, Smith BJ (2012b) Lichen-induced biomodification of calcareous surfaces: bioprotection versus biodeterioration. Prog Phys Geogr 37:325–351

Mellor E (1923) Lichens and their action on the glass and leadings of church windows. Nature 25:299–300

Mottershead D, Lucas G (2000) The role of lichens in inhibiting erosion of a soluble rock. Lichenologist 32:601–609

Nascimbene J, Salvadori O, Nimis PL (2008) Monitoring lichen recolonization on a restored calcareous statue. Sci Total Environ 407:2420–2426

Nimis PL, Monte M (1988) Lichens and monuments. Studia Geobot 8:1–133

Nimis PL, Pinna D, Salvadori O (1992) Licheni e Conservazione dei Monumenti. Cooperativa Libraria Universitaria Ediatrice Bologna, Bologna

Nugari MP, Pietrini AM, Caneva G, Imperi F, Visca P (2009) Biodeterioration of mural paintings in a rocky habitat: the crypt of the original sin (Matera, Italy). Int Biodeterior Biodegradation 63:705–711

Ortega-Calvo JJ, Hernandex-Marine M, Saiz-Jimenez C (1991) Biodeterioration of building materials by cyanobacteria and algae. Int Biodeterior 28:165–185

Paine SG, Linggood FV, Schimmer F, Thrupp TC (1933) The relationship of micro-organisms to the decay of stone. Phil Trans R Soc Ser B 222:97–127

Piervittori R (2004) Lichens and the biodeterioration of stonework: the Italian experience. In: St. Clair LL, Seaward MRD (eds) Biodeterioration of stone surfaces. Kluwer Academic, Dordrecht, pp 45–68

Piervittori R, Salvadori O, Seaward MRD (2004) Lichens and monuments. an analytical bibliography. In: St. Clair LL, Seaward MRD (eds) Biodeterioration of stone surfaces. Kluwer Academic, Dordrecht, pp 241–282

Prieto B, Rivas T, Silva B (1995) Effect of selected biocides on granites colonized by lichens. In: Bousher A, Chandra M, Edyvean R (eds) Biodeterioration and Biodegradation 9. Rugby, Institution of Chemical Engineers, pp 204–207

Prieto B, Seaward MRD, Edwards HGM, Rivas T, Silva B (1999) Biodeterioration of granite monuments by *Ochrolechia parella* (L.) Mass.: an FT-Raman spectroscopic study. Biospectroscopy 5:53–59

Prieto B, Edwards HGM, Seaward MRD (2000) A Fourier Transform-Raman spectroscopic study of lichen strategies on granite monuments. Geomicrobiol J 17:55–60

Purvis OW (1984) The occurrence of copper oxalate in lichens growing on copper sulphide-bearing rocks in Scandinavia. Lichenologist 16:197–204

Purvis OW (1996) Interactions of lichens with metals. Sci Progress 79:283–309

Pye K, Schiavon N (1989) Cause of sulphate attack on concrete, render and stone indicated by sulphur isotope ratios. Nature 342:663–664

Rajakaruna N, Knudsen K, Fryday AM, O'Dell RE, Pope N, Olday FC, Woolhouse S (2012) Investigation of the importance of rock chemistry for saxicolous lichen communities of the New Idria serpentinite mass, San Benito County, California, USA. Lichenologist 44:695–714

Rollin EM, Milton EJ, Roche P (1994) The influence of weathering and lichen cover on the reflectance spectra of granitic rocks. Remote Sens Environ 50:194–199

Saiz-Jimenez C, Arino X (2001) Microbial corrosion of cultural heritage stoneworks. In: Stocker JG (ed) A practical manual on microbiologically influenced corrosion, vol 2. NACE Press, Houston, pp 11.25–11.33

Saiz-Jimenez C, Garcia-Rowe J, Rodriquez-Hidalgo JM (1991) Biodeterioration of polychrome Roman mosaics. Int Biodeterior 28:65–79

Saxena S, Upreti DK, Singh A, Singh KP (2004) Observations on lichens growing on artifacts in the Indian subcontinent. In: St. Clair LL, Seaward MRD (eds) Biodeterioration of stone surfaces. Kluwer Academic, Dordrecht, pp 181–193

Schatz A (1963) Soil microorganisms and soil chelation. The pedogenic action of lichens and lichen acids. J Agric Food Chem 11:112–118

Schatz A, Cheronis ND, Schatz V, Trelawny GS (1954) Chelation (sequestration) as a biological weathering factor in pedogenesis. Proc Pennsylvania Acad Sci 28:44–51

Scheerer S, Ortega-Morales O, Gaylarde C (2009) Microbial deterioration of stone monuments—an updated overview. Adv Appl Microbiol 66:97–139

Schwartzman DW, Volk T (1989) Biotic enhancement of weathering and the habitability of Earth. Nature 340:457–460

Seaward MRD (1976) The lichens of county Carlow, Ireland. Rev Bryol Lichénol 42:665–676

Seaward MRD (1982) Lichen ecology of changing urban environments. In: Bornkamm R, Lee JA, Seaward MRD (eds) Urban ecology. Blackwell Scientific, Oxford, pp 181–189

Seaward MRD (1988) Lichen damage to ancient monuments: a case study. Lichenologist 20:291–294

Seaward MRD (1996) Lichens and the environment. In: Sutton BC (ed) A century of mycology. British Mycological Society/Cambridge University Press, Cambridge, pp 293–320

Seaward MRD (1997a) Major impacts made by lichens in biodeterioration processes. Int Biodeterior Biodegradation 40:269–273

Seaward MRD (1997b) Urban deserts bloom: a lichen renaissance. Bibl Lichenologica 67:297–309

Seaward MRD (2008) Environmental role of lichens. In: Nash TH (ed) Lichen biology. Cambridge University Press, Cambridge, pp 274–298

Seaward MRD, Coppins BJ (2004) Lichens and hypertrophication. Bibl Lichenologica 88:561–572

Seaward MRD, Edwards HGM (1995) Lichen-substratum interface studies, with particular reference to Raman microscopic analysis. 1. Deterioration of works of art by *Dirina massiliensis* forma *sorediata*. Cryptogamic Bot 5:282–287

Seaward MRD, Edwards HGM (1997) Biological origin of major chemical disturbances on ecclesiastical architecture studied by Fourier Transform Raman spectroscopy. J Raman Spectrosc 28:691–696

Seitz W, Schade A (1976) Calciumoxalat—ein bemerkenswertes "Abfallprodukt" in Flechten (Lichenes). Beitr Biol Pflanzen 52:355–382

Seneviratne G, Indrasena IK (2006) Nitrogen fixation in lichens is important for improved rock weathering. J Biosci 31:639–643

Silva B, Prieto B (2004) Lichens as subversive agents of biodeterioration. In: St. Clair LL, Seaward MRD (eds) Biodeterioration of stone surfaces. Kluwer Academic, Dordrecht, pp 69–77

Silver CS, Wolbers R (2004) Lichen encroachment onto rock art in eastern Wyoming: conservation problems and prospects for treatment. In: St. Clair LL, Seaward MRD (eds) Biodeterioration of stone surfaces. Kluwer Academic, Dordrecht, pp 115–128

St. Clair LL, Seaward MRD (eds) (2004a) Biodeterioration of stone surfaces. Kluwer Academic, Dordrecht

St. Clair LL, Seaward MRD (2004b) Biodeterioration of rock substrata by lichens: progress and problems. In: St. Clair LL, Seaward MRD (eds) Biodeterioration of stone surfaces. Kluwer Academic, Dordrecht, pp 1–8

Syers JK, Iskandar IK (1973) Pedogenetic significance of lichens. In: Ahmadjian V, Hale ME (eds) The lichens. Academic Press, New York, pp 225–248

Syers JK, Birnie AC, Mitchell BD (1967) The calcium oxalate content of some lichens growing on limestone. Lichenologist 3:409–414

Topham PB (1977) Colonization, growth, succession and competition. In: Seaward MRD (ed) Lichen ecology. Academic Press, London, pp 31–68

Tratebas AM (2004) Biodeterioration of prehistoric rock art and issues in site preservation. In: St. Clair LL, Seaward MRD (eds) Biodeterioration of stone surfaces. Kluwer Academic, Dordrecht, pp 195–228

Tratebas AM, Chapman F (1996) Ethical and conservation issues in removing lichens from petroglyphs. Rock Art Res 13:129–133

Tretiach M, Crisafulli P, Imai N, Kashiwadani H, Moon KH, Wada H, Salvadori O (2007) Efficacy of a biocide tested on selected lichens and its effects on their substrata. Int Biodeterior Biodegradation 59:44–54

Viles HA, Pentecost A (1994) Problems in assessing the weathering action of lichens, with an example of epiliths on sandstone. In: Robinson DA, Williams RGG (eds) Rock weathering and landform evolution. Wiley, Chichester, pp 99–116

Wadsten T, Moberg R (1985) Calcium oxalate hydrates on the surface of lichens. Lichenologist 17:239–245

Wee YC, Lee KB (1980) Proliferation of algae on surfaces of buildings in Singapore. Int Biodeterior Bull 16:113–117

Wessels DJC, Schoeman P (1988) Mechanism and rate of weathering of Clarens sandstone by an endolithic lichen. South African J Sci 84:274–277

Wessels DJC, Wessels LA (1991) Erosion of biogenically weathered Clarens sandstone by lichenophagous bagworm larvae. (Lepidoptera: Psychidae). Lichenologist 23:283–291

Wierzchos J, Ascaso C (1996) Morphological and chemical features of bioweathered granitic biotite induced by lichen activity. Clays Clay Miner 44:652–657

Wilson MJ, Jones D (1984) The occurrence and significance of manganese oxalate in *Pertusaria corallina*. Pedobiologica 26:373–379

Wilson MJ, Jones D, McHardy WJ (1981) The weathering of serpentinite by *Lecanora atra*. Lichenologist 13:167–176

Lichen Growth and Lichenometry

10

Richard A. Armstrong

Contents

Abstract

Lichenometry is one of the most widely used methods available for dating the surface age of various substrata including rock surfaces, boulders, walls and archaeological remains. It depends on the assumption that if the lag time before colonisation of a substratum by a lichen is known and lichen age can be estimated, then a minimum date can be obtained by measuring the diameter (or another property related to size) of the largest lichen at the site. Lichen age can be determined by variety of methods including calibrating lichen size against surfaces of known age ('indirect lichenometry'), by constructing a growth rate–size curve from direct measurement of lichen growth ('direct lichenometry'), using radiocarbon (RC) dating and from lichen 'growth rings'. This chapter describes the following: (1) lichen growth rates and longevity, (2) methods of estimating lichen age, (3) the methodology of lichenometry and (4) applications of lichenometry. Despite its limitations, lichenometry is likely to continue to play an important role in dating a variety of surfaces and also in providing data that contribute to the debate regarding global warming and climate change.

Keywords

Lichen growth · Lichen growth rate–size curve · Direct lichenometry · Indirect lichenometry · Growth rings

R.A. Armstrong (✉)
Department of Vision Sciences, Aston University, Birmingham B4 7ET, UK
e-mail: R.A.Armstrong@aston.ac.uk

D.K. Upreti et al. (eds.), *Recent Advances in Lichenology*,
DOI 10.1007/978-81-322-2181-4_10

10.1 Introduction

Geological and climatic history is similar to many other disciplines, including history and politics, in the need to establish a timescale of events. In pre-history, timescales are often critical to understanding the evolution of biological, tectonic and climatic systems that have interacted to produce the contemporary surface of the earth. Many methods of establishing a timescale of events have been developed to allow elapsed time to be accurately estimated for timescales ranging from a few decades to millions of years.

One of the most widely used techniques to date surfaces over the last 500 years, a period in which radiocarbon dating is relatively inefficient, is 'Lichenometry', viz. the use of symbiotic fungi in the form of lichens, growing on rock surfaces or other suitable substrata, to obtain an approximate date of the deposition of the surface (Locke et al. 1979; Innes 1985; Matthews 1994; Benedict 2009). Lichenometry depends on the assumption that if the lag time before colonisation of a substratum by a lichen is known and lichen age can be estimated, then a minimum date can be obtained by measuring the diameter (or another property related to size) of the largest lichen present. This chapter describes various aspects of lichen growth and lichenometry including (1) lichen growth rates and longevity, (2) methods of determining the age of a lichen thallus, (3) the methodology of lichenometry and (4) discusses how the technique has contributed to the debate regarding global warming and climate change.

10.2 Lichen Growth Rates and Longevity

Many lichens, especially those which have a foliose or fruticose growth form, grow fairly rapidly and live for relatively short periods of time and are therefore less useful as lichenometric tools (Table 10.1). Hence, species such as *Xanthoparmelia conspersa* (Ehrh. ex Ach.) Hale may live up to 30–40 years and *Cladonia rangiferina* (L.) Web for 25 years (Hale 1967). Of considerably more interest are the slow-growing crustose species, especially those in areas of the world in which very low radial growth rates (RaGR) have been recorded and exceptionally large individual thalli have been described. One of the first to investigate lichen growth rates in such regions was Roland Beschel, the 'father of lichenometry'. In West Greenland, Beschel (1958, 1961) studied the yellow-green crustose lichen *Rhizocarpon geographicum* (L.) DC. (Fig. 10.1), one of the most useful lichenometric species, and found that for the first seven decades after colonisation of a rock surface, RaGR was slow and maximum size of thalli only occurred on moraines greater than approximately 1,000 years of age. At Disko Island in the same region, some individuals appeared to have not increased in size since 1896, whereas the fastest growing thalli had probably increased by a maximum of 1–4 mm in diameter. Beschel concluded from these data that some individuals of *R. geographicum* may be at least 4,500 years old. Similarly, Denton and Karlen (1973) studied

Table 10.1 Estimated ages for a selection of lichens

Site	Species	Estimated age (years)	Original data
North Canada	*Cladonia rangiferina*	10–25	Hale (1967)
Connecticut, USA	*Xanthoparmelia conspersa*	39	Hale (1967)
West Wales, UK	*X. conspersa*	30	Armstrong and Bradwell (2010a, b)
Utah, USA	*Caloplaca brachyphylla*	20	Hale (1967)
Cascade Mountains WA, USA	*Rhizocarpon geographicum*	1,428	Armstrong (2005a)
West Greenland	*R. geographicum*	4,500	Beschel (1958)
South Alaska	*Rhizocarpon alpicola*	9,000	Denton and Karlen (1973)

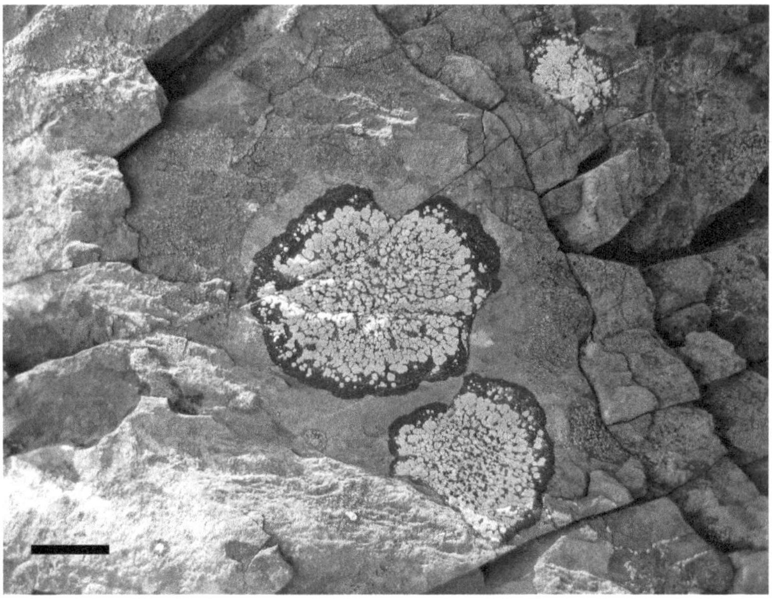

Fig. 10.1 A population of yellow-green thalli of the lichen *Rhizocarpon geographicum* on scree boulders in the Pacific Northwest, USA. The yellow-green islands of areolae can be clearly seen as well as the black fungal hypothallus at the margin (*bar* 10 mm)

Rhizocarpon alpicola Rabenh. at sites in the White River Valley, Skolai Pass, Southern Alaska. There was an initial, more rapid increase in diameter followed after a few centuries by a longer interval of nearly linear growth, a recorded RaGR of 3.4 mm 100 $year^{-1}$ being typical. Extrapolating from these data, an individual of diameter 15.5 cm would be approximately 3,700 years old, and the largest recorded *R. alpicola* thallus found in the region, viz. 480 mm in diameter, potentially 9,000 years old!

The range of RaGR recorded in *R. geographicum* in various regions is shown in Table 10.2. Hence, in a study in the Cascade Mountains of Washington State, USA, an average RaGR of 0.1 mm $year^{-1}$ was obtained for thalli of

Table 10.2 Annual growth rates (radial growth rate, RaGR, mm $year^{-1}$) for thalli of *Rhizocarpon geographicum* (L.) DC. from various habitats

Location	RaGR (mm $year^{-1}$)	Author
North Cascades, USA	0.07	Armstrong (2005a)
West Greenland	0.05–0.1	Beschel (1961)
South Orkney Islands	0.1	Hooker (1980a)
Alaska	0–0.18	Haworth et al. (1986)
West Greenland	0.1–0.2	Ten Brink (1973)
Maritime Antarctic	0.25	Sancho and Pintado (2004)
Canada	0.05–0.29	Rogerson et al. (1986)
Tierra del Fuego	0.31	Sancho et al. (2011)
Scotland	0.14– 0.32	Bradwell (2010)
South Iceland	0.09–0.37	Bradwell and Armstrong (2007)
British Columbia, Canada	0.26–0.41	McCarthy (2003)
South Norway	0.21–0.43	Trenbirth and Matthews (2010)
Switzerland	0.5	Proctor (1983)
North Wales	0.67–0.81	Winchester and Chaujar (2002)
North Wales	0.03–0.94	Armstrong (2005b)

Rhizocarpon section *Rhizocarpon* measured over six years (Armstrong 2005a), while Beschel (1958, 1961) in West Greenland reported RaGR of 0.05–0.10 mm year^{-1}. Hooker (1980a) studied 63 thalli of *R. geographicum* in South Orkney Islands and found no measurable growth over a period of 3 years. However, small amounts of growth were recorded during a subsequent three-year period of 0.1 mm year^{-1}. A higher RaGR was recorded for *R. geographicum* by Rogerson et al. (1986) in northern Labrador (0.17 mm year^{-1}) and by Haworth et al. (1986) in the central Brooks Range, Alaska (0–0.18 mm year^{-1}).

Significantly higher growth rates have also been recorded from maritime regions in Antarctica and especially the Antarctic Peninsula. Hence, on Livingston Island, an RaGR of 0.25 mm year^{-1} was reported (Sancho and Pintado 2004). Hence, exceptionally, large thalli of *R. geographicum* in this region may be much younger than previously supposed. In addition, in Tierra del Fuego, average growth rates of *R. geographicum* were estimated to be 0.32 mm year^{-1} (Sancho et al. 2011).

In more temperate regions, significantly higher RaGR has been recorded. Hence, McCarthy (2003) reported an RaGR for *R. geographicum* in British Columbia of 0.26–0.42 mm year^{-1}. Higher rates of growth have also been reported from a glacier in Switzerland (maximum 0.5 mm year^{-1}) (Proctor 1983). In a detailed long-term study of a single thallus of *R. geographicum* (max. diam. 46.5 cm) measured over 26 years, the mean RaGR was 0.48 mm year^{-1} (range 0.02–-1.54 mm year^{-1} in different years) (Mathews and Trenbirth 2011). In addition, in a study of 2,795 individuals of the subgenus *Rhizocarpon* measured at 47 sites from 18 glaciers in south Norway (Trenbirth and Mathews 2010), RaGR of 0.21–-0.43 mm year^{-1} was recorded at different sites reflecting local habitat differences.

Some of the highest recorded growth rates have been reported in Europe and especially in the UK. Hence, Bradwell (2010) measured *R. geographicum* at a maritime subpolar site in Scotland. In thalli greater than 10 mm in diameter, growth rates were 0.67 mm year^{-1}, while thalli less than 10 mm grew at 0.29 mm year^{-1}. Winchester and Chaujar (2002) studied the growth of *R. geographicum* spp. *prospectans* on gravestones in North Wales and estimated RaGR to be 0.74 mm year^{-1}. Comparable results have been reported by Armstrong (1983, 2005b) in North Wales in which RaGR was in the range 0.03–0.94 mm year^{-1}, depending on thallus size.

10.3 Determining Lichen Age

Essential to lichenometry is the estimation of the age of the largest and potentially oldest lichen thallus present on a substratum. This section describes four methods of obtaining such ages: (1) by calibrating lichen size against surfaces of known age, (2) by establishing a lichen growth rate–size curve from direct measurement of lichen growth, (3) radiocarbon (RC) dating and (4) in a selective group of species, by measuring lichen 'growth rings'.

10.3.1 Calibrating Lichen Size Against Age of Substratum

A variety of sources can provide information for dating a substratum and establishing a calibration curve of size versus age including gravestones (usually providing the most accurate dates), mine spoil heaps, abandoned farms or houses, stone walls and rock cairns (Innes 1983, 1985). In addition, natural deposits can be used that have been dated accurately from historical events, e.g. by RC dating or the use of tree growth rings (dendrochronology). Typical lichen growth curves relating lichen size and age are non-linear (Bradwell 2001a, b) and the relationship often fitted best by a third-order polynomial regression which describes the declining growth rates often apparent in larger thalli. This decline could be due to thallus senescence (Armstrong 1983, Armstrong and Smith 1997) or it could reflect the tendency for slower growth in the last 100 years.

10.3.2 Constructing a Growth Rate–Size Curve

A second method of obtaining an estimate of lichen age is by constructing a growth curve which describes how growth of a thallus may change with age. Because of very slow RaGR and consequent longevity of many crustose lichens, however, it has not been possible to measure the growth of any single individual throughout its life, even in relatively favourable environments. Hence, to obtain a growth curve, a 'cross-sectional' method has been employed, i.e. by studying individuals of different size and assuming a direct relationship between thallus size and age, the result being a growth rate–size curve. This type of study has three main limitations. First, there is no guarantee that the growth curve derived from a population of thalli of different size actually reflects the growth of any single individual throughout its life. Second, variation in RaGR between thalli of similar size, which are attributable in part to micro-environmental fluctuations and which can be considerable in foliose lichens (Fig. 10.2) (Armstrong and Smith 1992), can obscure changes due to size so that only major growth phases can be detected. Third, studies are often carried out over relatively short intervals of time, e.g. 1–3 years, and therefore ignore potential long-term effects of climate on growth rate by assuming 'average' climatic conditions over the lifespan of the thallus (Armstrong and Bradwell 2010a).

A number of studies have attempted to determine the growth rate–size curve of crustose lichens using a cross-sectional study (Bradwell and Armstrong 2007). In many of these studies, however, variability in RaGR combined with slow growth rates and small sample sizes has made it difficult to establish precisely the different growth phases (Proctor 1983; Haworth et al. 1986; Matthews 1994; McCarthy 2003; Armstrong 2005a). Furthermore, a major objective of this research has been to model the growth curve of crustose lichens as has been accomplished successfully for foliose species (Aplin and Hill 1979; Childress and Keller 1980). In foliose lichens, RaGR increases in smaller diameter

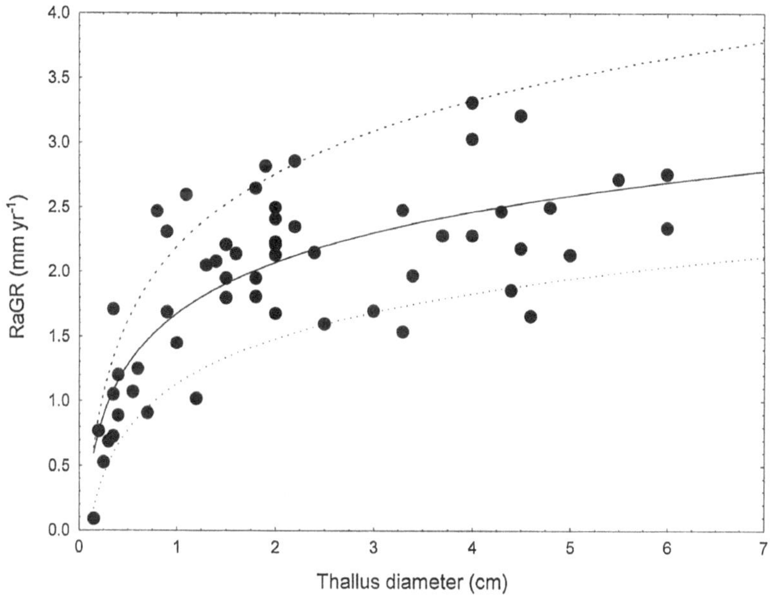

Fig. 10.2 Growth rate–size curve of the foliose lichen *Melanelia fuliginosa* ssp. *fuliginosa* (Fr. ex Duby) Essl. determined by a cross-sectional study. The *fitted curve* is a logarithmic function (data from Armstrong 1976). The *dashed lines* are logarithmic regressions fitted to the minimum and maximum RaGR values and indicate the range of variability in RaGR commonly observed in lichens

thalli becoming linear or asymptotic in larger individuals, the shape of the curve being approximately logarithmic (Fig. 10.2). There is no evidence for a declining growth phase in foliose lichens even during the later senescent phases when the thallus centre often degenerates and fragments (Armstrong 1973; Aplin and Hill 1979; Armstrong and Smith 1996, 1997; Armstrong and Bradwell 2011). This type of growth curve has been modelled based on rates of carbohydrate production within the thallus and the degree of translocation of materials between individual lobes and margin to the thallus centre (Aplin and Hill 1979; Childress and Keller 1980).

The earliest studies of the growth rate–size curves of crustose lichens often assumed a similarity to foliose lichens. Hence, Proctor (1977) studied the growth rate–size curve of the placodioid lichen *Buellia* (*Diploicia*) *canescens* (Dicks.) DNot. RaGR was assumed to be proportional to an area of thallus in an annulus of constant width at the growing margin, and therefore, the shape of the growth curve was essentially asymptotic. Similarly, in a study of *R. geographicum* on a glacier in Switzerland (Proctor 1983), the growth curve was similar to *B. canescens*, i.e. initial growth was logarithmic, rising asymptotically to a constant value. Proctor's data, however, did not include any thalli larger than 36 mm in diameter, and growth rates in thalli over 10 mm were assumed to be linear, the growth curve being extrapolated up to 60 mm without further measurement. Hence, although there was evidence that the growth rate–size curve of placodioid and foliose species were similar (Hill 1981, Benedict 2008), this may not be the case for many crustose lichens including the genus *Rhizocarpon* (Innes 1985).

In other growth studies (Bradwell and Armstrong 2007), variation in RaGR between individual thalli was often too great to establish the shape of the growth rate–size curve, and therefore, the data were not adequate to test the hypothesis that crustose and foliose lichens exhibited similar growth curves. Hence, Hooker (1980a) was unable to demonstrate any significant effect of thallus size on RaGR in South Orkney, while the studies of Rogerson et al. (1986) of seven thalli in northern Labrador shed little light on the relationship between growth and size. Similar conclusions were reported by Matthews (1994), Winchester and Chaujar (2002), Sancho and Pintado (2004) and Armstrong (2005a).

Brodo (1965) was one of the first to question whether crustose lichens grew differently to foliose species. He examined corticolous communities in Long Island, New York and observed lower growth rates of crustose thalli greater than about 40 mm in diameter compared with thalli 10–30 mm in diameter. Such a decline in RaGR in large thalli was also suggested by studies of the growth rate–size curve of *R. geographicum* in North Wales, UK (Armstrong 1983) (Fig. 10.3). Three phases of growth were identified as follows: (1) an early growth phase in which RaGR increased to a maximum, (2) a short phase in thalli 30–40 mm in diameter at which RaGR was maximal and (3) a phase in which RaGR declined in thalli greater than 50 mm in diameter. An inverse relationship between growth and size was also reported by Haworth et al. (1986) in Alaska. More substantial evidence for declining growth was obtained from a study of 134 thalli of four crustose lichen species growing in North Wales (Armstrong 2005b) and a study of *R. geographicum* in Iceland (Bradwell and Armstrong 2007). No evidence for an asymptotic growth rate–size curve was found in *R. geographicum*, *Rhizocarpon reductans*, Th. Fr. *Lecidea tumida* Massal. or *Buellia aethalea* (Ach.) Th. Fr. from North Wales or from *R. geographicum* in Iceland (Bradwell and Armstrong 2007). Instead, RaGR increased to the maximum and then declined in larger diameter thalli. In the absence of a growth model for crustose lichens, second-order polynomial curves have been fitted to growth data (Armstrong 2005b) (Fig. 10.3). More recently, Trenbirth and Matthews (2010) have proposed several models for crustose lichen growth in south Norway. At many of their sites, annual growth rates remained relatively constant or increased with size up to a diameter of 12 cm, the growth rate–size 'curve' being linear rather than parabolic or asymptotic,

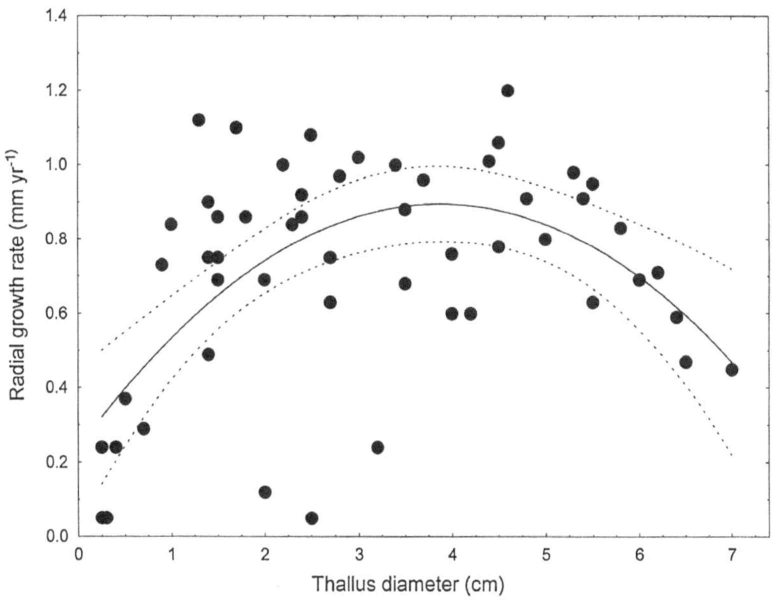

Fig. 10.3 The growth curve of the lichen *Rhizocarpon geographicum* (L.) DC. by direct measurement: a plot of radial growth rate (RGR) versus thallus diameter on rock surfaces in North Wales (data from Armstrong 2005b). Three phases of growth were identified as follows: (1) an early growth phase in which RaGR increased to a maximum, (2) a short phase in thalli 30–40 mm in diameter at which RaGR was maximal and (3) a phase in which RaGR declined in thalli greater than approximately 50 mm in diameter

raising the prospect that the growth curve of *R. geographicum* could vary at different sites or even within the same site in different years (Porter 1981).

That larger and presumably older crustose thalli may grow more slowly than younger individuals is also suggested by many traditional lichenometric studies. Porter (1981), for example, reviewed the lichen growth curves reported in the earliest lichenometric studies. Prior to 1980, growth curves were variously reported as linear (Andersen and Sollid 1971; Burrows and Orwin 1971), exponential (Mottershead and White 1972; Luckman 1977), exponential–linear (Benedict 1967; Denton and Karlen 1973) or 'irregular' (Curry 1969).

Lichenometric studies since 1980, however, have provided consistent evidence to support a declining phase of growth. Hence, Birkeland (1981) compared lichenometric and soil data from the Mount Cook region of New Zealand and found that the growth rate of *R. geographicum* decreased with time in a humid mountainous area. In addition, Bull and Brandon (1998) studied the growth of *Rhizocarpon* subgenus *Rhizocarpon* in the Southern Alps in New Zealand and observed that growth was non-linear. Hence, colonisation was followed by an exponential phase of growth for approximately the first 20 years succeeded by a phase of slower but more linear growth. The most similar type of growth rate–size curve of *R. geographicum*, provided by traditional lichenometry, to that obtained by direct measurement has been reported by O'Neal and Schoenenberger (2003). They revisited surfaces in the Cascade Range, USA originally measured by Porter (1981) and found that the growth curve was curvilinear with phases of increasing, constant and then decreasing RaGR, very similar in shape to the curve in Fig. 10.3.

10.3.3 RadioCarbon (RC) Dating

A number of authors have speculated whether RC dating could be used to age lichen thalli directly (Maguas and Brugnoli 1996; Clark et al. 2000; Bench et al. 2001). As a thallus grows radially, new tissue is produced at the margin which should be younger than the tissue in the centre. Hence, an RC date taken precisely at the centre could reveal the age of the thallus. Maguas and Brugnoli (1996) measured carbon isotope discrimination (d) in different individuals and with distance across thalli. Variation in 'd' was observed across the thalli, the highest values being observed at the margin and lowest in older, central tissues. It was concluded that variations in 'd' were caused by age-related changes in physiology in different regions of the thallus. In addition, the crustose lichen *Caloplaca trachophylla* (Tuck.) Zahlbr. was studied using C-14 accelerator mass spectrometry (AMS) (Clark et al. 2000). Sections of lichen were dated and exhibited a linear correlation with radial location, estimated ages also suggesting that RaGR was constant over long periods of time. The authors concluded that was potentially feasible to use AMS technology to date lichen tissues and to determine growth rates. However, Bench et al. (2001) measured radial C-14/C profiles across thalli *of R. geographicum* using AMS to determine whether C-14/C isotope ratio measurements could be used to estimate age. The C-14/C profile, however, was relatively flat, values of d-c-14 derived from modern RC standards for radial samples displaying a small spread. Hence, carbon cycling appears to occur within the thallus involving replacement and fusion of tissue as well as cellular and molecular replacement with age. It was concluded that lifespan estimates may not be routinely obtained from C-14/C profiles and that such methods were unlikely to provide an insight into lichen longevity or to be of use in lichenometry.

10.3.4 Lichen 'Growth Rings'

Certain species of the lichen genera *Pertusaria*, *Lecanora* and *Ochrolechia* have concentrically zoned margins (Fig. 10.4), and in some species, there are alternating light and dark bands at the margin, the lighter bands which may represent rapid summer growth and the narrow dark zones growth in winter (Hale 1973). Each zone or 'ring' could therefore represent growth in 1 year and, if this assumption was correct, in some thalli, growth could be traced back 3–7 years. In a preliminary study, the widths of successive marginal rings in 25 thalli of *Ochrolechia parella* (L.) Massal. growing at a maritime site in North Wales were measured (Armstrong and Bradwell 2010b). Mean ring widths of all thalli varied from a minimum of 1.02 mm (the outermost ring) to a maximum of 2.06 mm (the third ring from the margin) suggesting yearly variations in growth. Marginal ring width and thallus size were also positively correlated, growth rates increasing in larger thalli. In a further study on recently exposed bedrock adjacent to Breidalon, SE Iceland, the potential for using marginal growth rings to estimate thallus age of a lichen tentatively identified as of the genus *Rhizocarpon* [possibly *R. concentricum* (Davies) Beltram.] was investigated to confirm the timing of surface exposure (c. 50 years). Collectively, these results suggested the following: (1) the measurement of

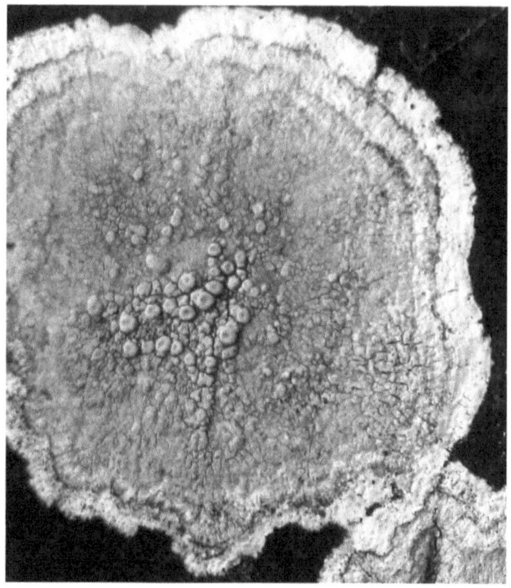

Fig. 10.4 Marginal rings in thalli of the crustose lichen *Ochrolechia parella* (L.) Massal. growing at a maritime site in North Wales, UK

marginal rings is a possible alternative method of studying the growth of crustose lichens, (2) *O. parella* may grow differently to other crustose species, exhibiting a rapidly increasing radial growth rate in thalli greater than 40 mm and (3) where lichens with marginal rings grow on recently exposed surfaces (<60 years), minimum age can be estimated using growth rings as an in situ indication of lichen growth rates.

A detailed study of marginal zonation, however, was carried out by Hooker (1980b) in Antarctica. He found that in *Buellia russa* (Hue.) Darb. zonation was present only in the non-lichenised hypothallus, and each concentric zone represented one year's growth. By contrast, *Buellia coniops* (Wahlenb.) Th. Fr. and *Caloplaca cirrochrooides* (Vainio) Zahbr. possessed 'pseudoannual rings' with each new ring representing two growth seasons. In addition, the rings of *Caloplaca* were not as distinct as those of the other species, and no new growth rings were formed during two subsequent growing seasons. Further research is therefore required before growth rings can be used as a reliable measure of growth in different habitats. Nevertheless, lichen growth rings potentially provide a previously unexploited, in situ lichenometric dating tool which could be used in areas where it is not possible to calibrate size and age (Armstrong and Bradwell 2010b).

10.4 Lichenometry

Lichenometry is a particularly useful method of dating surfaces in regions above and beyond the treeline and especially in Arctic–Alpine environments where lichens grow very slowly and have considerable longevity. In such environments, it may be possible to date surfaces and deposits up to thousands of years old, but in the majority of cases, the method has been most useful for dating over the past 500 years.

Although several species of lichen have been used in lichenometry, the majority have utilised a specific group of crustose lichens, viz. the yellow-green species of the genus *Rhizocarpon* (Fig. 10.1). This type of lichen is abundant in many environments from sea level to high elevations, grows very slowly (0.02–2 mm $year^{-1}$), especially in Arctic and Alpine environments (Table 10.2) and can live to a considerable age (Beschel 1958, 1961). Morphologically, the lichen comprises discreet areolae that contain the cells of the alga *Trebouxia*, located on a fungal medulla which is attached to the substratum and which extends into a black algal-free marginal zone around the thallus called the hypothallus (Armstrong and Smith 1987; Armstrong 2013). Within each areola, there is a cortical layer 15–80 μm thick, an algal layer and medullary tissue. The fungal hypothallus extends radially, and to grow, it relies on carbohydrate supplied from the areolae (Armstrong and Smith 1987). Primary areolae near the edge of the hypothallus may develop from free-living algal cells on the substratum that are trapped by the hypothallus, while secondary areolae may develop from zoospores produced within the thallus (Armstrong 2013). Different species of *Rhizocarpon* may also colonise a surface at different rates, e.g. species of the section *Rhizocarpon* may colonise a surface earlier than *R. alpicola*. However, *R. alpicola* may exhibit a faster RaGR than most members of the section *Rhizocarpon* and eventually may become the largest lichen on a substratum (Innes 1985).

10.4.1 Indirect Lichenometry

Traditional lichenometry, also known as 'indirect lichenometry', involves the dating of rock and stone surfaces using a calibration of size of lichen thalli present against known age of substratum (Locke et al. 1979; Innes 1985; Matthews 1994; Benedict 2009). Indirect lichenometry has been the most useful technique for dating and has the advantage that fluctuations in lichen growth that occur over long periods are taken into account. Several measurements of thallus size have been advocated as the most useful in indirect lichenometry. Hence, the use of the 'largest inscribed circle', i.e. the largest circle that can be drawn within an individual thallus, was suggested by Locke et al. (1979) and has been adopted by several workers and is equivalent to using the 'shortest diameter' of the thallus. By contrast, many workers have suggested that the largest diameter of the thallus is the most appropriate measure (Burbank 1981; Ellis et al. 1981; Proctor 1983). Other workers have averaged the longest and shortest axes or used both measurements to derive an estimate of the surface area of the thallus (Griffey 1977). In addition, various methods have been employed to sample a lichen population at a site. Either the single largest thallus at the site is measured and regarded as representative of age (Webber and Andrews 1973) or several of the largest thalli present are averaged (e.g. Mathews 1994), both methods being regarded as valid (Innes 1985). Some workers have used the frequency distribution of thallus size as an indicator of the age of the substratum (McCarroll 1993; Caseldine and Baker 1998; Dabski 2007), but is more controversial as size distributions can have different interpretations (Innes 1985).

A limitation of indirect lichenometry is that smaller lichen thalli could coalesce and therefore be recorded as a single individual (Armstrong 1984). In addition, thalli of the same and of different lichen species may fuse to form lichen mosaics, a feature of many lichen communities dominated by crustose species (Pentecost 1980; Armstrong and Welch 2007). The tendency for thalli to fuse without leaving evidence of the original thallus boundaries depends on species. For example, in North Wales, *R. geographicum* forms mosaics of sharply delimited individual thalli, while *R. lecanorinum* (Flörke ex Körb) Anders, thalli merge with one another at intraspecific contacts, leaving no visible trace of their initial boundaries (Clayden 1997). In *R. geographicum*, it is possible that there is somatic incompatibility between genotypes, whereas in *R. lecanorinum*, somatic compatibility may exist between genetically identical or closely related thalli allowing them to fuse more readily (Clayden 1997). Measurement of only circular or near-circular thalli has been advocated as one method of reducing this problem as fusion is likely to increase the degree of asymmetry of the thallus (Innes 1985).

10.4.2 Direct Lichenometry

In direct lichenometry, a lichen growth rate–size curve is constructed and then a lichenometric dating curve, from direct measurements of lichen growth (Armstrong 1976; Bradwell and Armstrong 2007; Trenbirth and Matthews 2010) and is potentially useful method where it is not possible to establish a calibration of size and age (Armstrong and Bradwell 2010a, b). Relatively few studies, however, have adopted the direct approach to lichenometry (Bradwell and Armstrong 2007, Trenbirth and Matthews 2010). Trenbirth and Matthews (2010) identified four main reasons for the neglect of the direct method: (1) high growth ratevariability, (2) lack of ecological data concerning the effects of environmental factors on growth, (3) small sample sizes and (4) the restricted number of years over which measurements were made.

Construction of a lichen growth curve usually involves dividing up the range of thallus size into suitable increments (Armstrong 1976). Hence, size classes of 1 cm were used to construct growth curves of *X. conspersa* (Armstrong and Smith 1996), *Melanelia fuliginosa* ssp. *fuliginosa* (Armstrong 1976) (Fig. 10.2) and *R. geographicum* (Armstrong 1983) (Fig. 10.3). RaGR of a sample of thalli is then measured within each size class, and RaGR plotted against thallus size to obtain the growth curve. The estimated age of a thallus of a diameter '*d*' cm is obtained by fitting a growth curve to the data and calculating the time taken to grow each size increment. A calibration of diameter against estimated age can then be constructed based on measured growth rates (Armstrong 1976). However, the within-thalli variation in RaGR results in large confidence intervals for estimated ages which accumulate with thallus size (Fig. 10.2). Nevertheless, it may be possible to model variation in RaGR on a single face using the normal distribution and, therefore, to estimate the number of thalli that should be measured in each size class to estimate RaGR within defined limits of error.

An example of an age/size curve for the foliose lichen *M. fuliginosa* ssp. *fuliginosa*

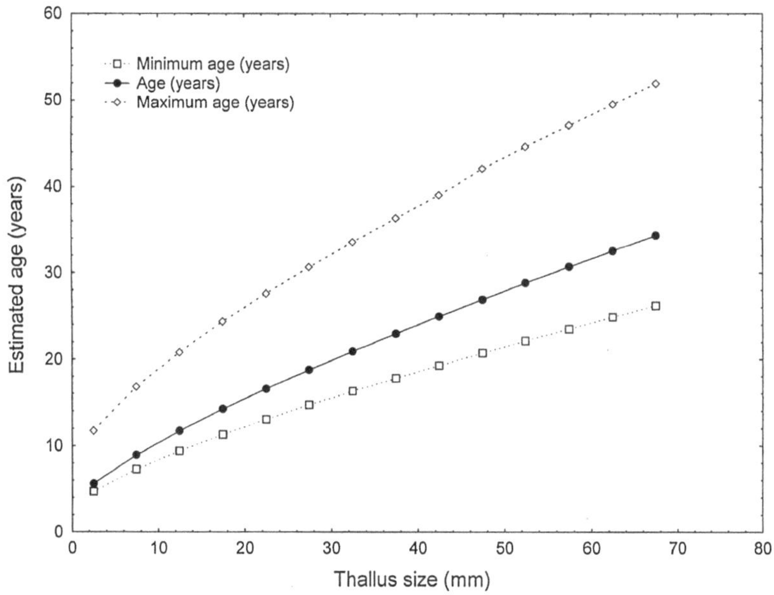

Fig. 10.5 Calibration of thallus age against size for the lichen *Melanelia fuliginosa* ssp. *fuliginosa* (Fr. ex Duby) Essl. in North Wales, UK by direct measurement for use in direct lichenometry. The data show the non-linear change in estimated age with thallus size and are best fitted by a second-order (quadratic) polynomial. In addition, the error limits for an estimated age are based on the logarithmic regressions fitted to the minimum and maximum values of RaGR in Fig. 10.2 and are asymmetrical, the upper error limit being greater than the lower. From this curve, a thallus of *M. fuliginosa* ssp. *fuliginosa*, 5 cm in diameter, would be estimated to be 28 years old (range 22–43 years)

derived from Fig. 10.2 is shown in Fig. 10.5. The data show a non-linear change in estimated age with thallus size which is best fitted by a second-order (quadratic) polynomial. In addition, the error limits for an estimated age, based on the maximum and minimum values of RaGR, are asymmetrical, the upper limit being greater than the lower (Fig. 10.2). From this curve, it was estimated that a thallus of *M. fuliginosa* ssp. *fuliginosa*, 5 cm in diameter, could be 28 years old (range 22–43 years), the large error reflecting the degree of variation in RaGR present in the original data. Hence, methods of controlling and reducing this variation are essential for the successful application of the direct method.

10.5 Applications of Lichenometry

Lichenometry has been used in many different environments and regions to date surfaces. However, it is the dating of the sequences of rocks that form glacial moraines that have been the most widely used application (Burrows and Orwin 1971). Porter (1981), for example, applied indirect lichenometry to the moraines of Mount Rainier in Washington State, USA and used historical evidence for the past marginal positions of the glaciers, including buildings, together with other structures in the national park, to provide surfaces of known age from 1857 to the present. The single largest circular *R. geographicum* thallus was used as a measure of lichen growth. The study indicated that the moraine probably stabilised 4 years earlier than had been previously suggested. Other applications of lichenometry include the dating of the stone images on Easter Island (Follman 1961), stone walls (Laundon 1980; Benedict 1967), river flooding (Gregory 1976), sea-level changes (Birkenmajer 1981) and the occurrence of landslides and avalanches (McCarroll 1993).

During the last 30 years, there has been increasing concern as to the environmental impact of global warming and climate change.

Lichenometric studies have played an important role in this debate (Bradwell 2001b). Hence, McCarroll (1993) used the size–frequency distribution of lichen populations to determine the ages of boulders resulting from avalanches in South Norway. Snow avalanches reflect periods of high winter snow and rapid spring melting rather than low temperatures. It was concluded that increasing avalanche activity in South Norway was attributable to global warming. Harrison and Winchester (2000), using a combination of lichenometry and dendrochronology, studied the nineteenth- and twentieth- century fluctuations of glaciers in South Chile. There was evidence for retreat of the glaciers following the 'little ice age' maximum between 1850 and 1880. Glacier retreat increased during the 1940s, and the degree of synchrony exhibited by different glaciers suggested a common climatic influence. This pattern has been repeated around the world. In Central Asia, there were small advances of the glaciers during 1908–1911, 1911–1934 and 1960–1977 (Narama 2002). However, significant recession also occurred in the 1900s, especially in the periods 1911–1934 and 1977–1978. Recession then accelerated after 1990 reflecting recent climatic warming in regions of inland Asia. Records of glacier fluctuation are now compiled by the World Glacier Monitoring Service and have derived estimates of global warming during the last 100 years (Oerlemans 1994). Examination of data from all over the world has confirmed that the retreat of glaciers is a common phenomenon over the globe and is likely to be explained by a linear warming trend of 0.66 K per century (Oerlemans 1994).

10.6 Conclusions

Several aspects of the biology and growth of lichens have implications for lichenometry (McCarthy 1999; Worsley 1981). First, there is uncertainty regarding the early stages of development of many lichens and especially the duration of a time lag between colonisation and the appearance of the first measurable thalli. In addition, small thalli may merge early in life to establish larger thalli than would be expected by their initial growth rates. There may also be substantial mortality of young thalli (Loso and Doak 2006) resulting in relatively few survivors of the first colonists on a mature rock surface. Therefore, studies of very early growth processes and mortality may enable a more accurate estimate of the time lag to be made and improve the accuracy of dating curves.

There is also uncertainty regarding the shape of the growth rate–size curve, especially in the *Rhizocarpon* group (Trenbirth and Matthews 2010). A major problem affecting most experimental growth data is the degree of between-thallus variability which is often larger than the growth–size fluctuations themselves. Different models have been proposed for the shape of the growth rate–size curve in *R. geographiucm* (Bradwell and Armstrong 2007; Trenbirth and Matthews 2010) including models suggesting parabolic, linear or increasing growth rates over time. It is possible that members of the *Rhizocarpon* group exhibit different growth curves at different sites or even at the same site in successive years. If this hypothesis is correct, it would limit the application of 'direct' lichen growth curves across different geographical regions.

R. geographicum is likely to exhibit seasonal patterns of growth (Armstrong 2006) and significant annual variation in RaGR. These variations may correlate with specific climatic factors (Armstrong 2006; Trenbirth and Matthews 2010), and further knowledge of such relationships could enable growth curves established in one region to be adjusted for use in others.

There has also been debate regarding the possible effects of competition in lichenometric studies. On older surfaces, there is likely to be increased competition leading to possible overgrowth and mortality, cessation of growth and the formation of lichen 'mosaics' (Armstrong and Welch 2007). Mature thalli of some species of *Rhizocarpon* may even merge imperceptibly, leading to the formation of much larger thalli than expected (Clayden 1997). Hence, Bradwell (2010) measured the growth of *R. geographicum* at two contrasting sites in Scotland and found that more intensive competition at one of the sites ruled out a meaningful comparison between the two. In addition, the rate of lichen growth may decline on older surfaces as a consequence of increased competition, a result which may have implications for dating older surfaces with considerable lichen cover (Gellatly 1982).

Lichenometry is one of a number of techniques now available for estimating the surface age of exposure of a substratum. Its advantages include an ability to date surfaces during the last 500 years, a time interval in which RC dating is least efficient and provides a quick, cheap and relatively accurate date for a substratum. Despite the continued lack of information regarding many aspects of lichen growth, lichenometry is likely to continue to play an important role in dating surfaces and especially in providing data that contribute to the debate regarding global warming and climate change.

References

Andersen JL, Sollid JL (1971) Glacial chronology and glacial geomorphology in the margin zones of the glaciers, Midtdalsbreen and Nigardsbreen, south Norway. Nor Geogr Tidsskr 25:1–28

Aplin PS, Hill DJ (1979) Growth analysis of circular lichen thalli. J Theor Biol 78:347–363. doi:10.1016/0022-5193(79)90335-7

Armstrong RA (1973) Seasonal growth and growth rate-colony size relationships in six species of saxicolous lichens. New Phytol 72:1023–1030. doi:10.1111/j.1469-8137.1973.tb02078.x

Armstrong RA (1976) Studies on the growth rate of lichens. In: Brown DH, Bailey RH, Hawksworth DL (eds) Progress and problems in lichenology. Academic Press, London, pp 309–322

Armstrong RA (1983) Growth curve of the lichen *Rhizocarpon geographicum*. New Phytol 94:619–622. doi:10.1111/j.1469-8137.1983tb04870.x

Armstrong RA (1984) Growth of experimentally reconstructed thalli of the lichen *Parmelia conspersa*. New Phytol 98:497–502. doi:10.1111/j.1469-8137.1984tb04143.x

Armstrong RA (2005a) Radial growth of *Rhizocarpon* section *Rhizocarpon* lichen thalli over six years at Snoqualmie Pass in the Cascade Range, Washington State. Arct Antarct Alp Res 37:411–415. doi:10.1657/1523-0430(2005)037[0411:RGORSR]:RGORSRJ2.0.CO;2

Armstrong RA (2005b) Growth curves of four crustose lichens. Symbiosis 38:45–57

Armstrong RA (2006) Seasonal growth of the crustose lichen *Rhizocarpon geographicum* (L.) DC. in south Gwynedd, Wales. Symbiosis 41:97–102

Armstrong RA (2013) Development of areolae and growth of the peripheral prothallus in the crustose lichen *Rhizocarpon geographicum*: an image analysis study. Symbiosis 60:7–15. doi:10.1007/s13199-013-0234-2

Armstrong RA, Bradwell T (2010a) Growth of crustose lichens: a review. Geogr Ann Ser A Phys Geogr 92A:3–17

Armstrong RA, Bradwell T (2010b) The use of lichen growth rings in lichenometry: some preliminary findings. Geogr Ann 92A:141–147

Armstrong RA, Bradwell T (2011) Growth of foliose lichens: a review. Symbiosis 53:1–16. doi:10.1007/s13199-011-0108-4

Armstrong RA, Smith SN (1987) Development and growth of the lichen *Rhizocarpon geographicum*. Symbiosis 3:287–300

Armstrong RA, Smith SN (1992) Lobe growth variation and the maintenance of symmetry in foliose lichen thalli. Symbiosis 12:145–158

Armstrong RA, Smith SN (1996) Factors determining the growth curve of the foliose lichen *Parmelia conspersa*. New Phytol 134:517–522

Armstrong RA, Smith SN (1997) Factors associated with degeneration of the thallus centre in foliose lichens. Symbiosis 22:293–302

Armstrong RA, Welch AR (2007) Competition in lichen communities. Symbiosis 43:1–12

Bench G, Clark BM, Mangelson NF, St. Clair LL, Rees LB, Grant PG, Southen JR (2001) Accurate lifespan estimates cannot be obtained from C-14 profiles in the crustose lichen *Rhizocarpon geographicum* (L.) DC. Lichenologist 33:539–542. doi:10.1006/lich.2001.0353

Benedict JB (1967) Recent glacial history of an alpine area in the Colorado Front Range, USA. 1. Establishing a lichen-growth curve. J Glaciol 6:817–832

Benedict JB (2008) Experiments on lichen growth. III. The shape of the age-size curve. Arc Antarct Alp Res 40:15–26. doi:10.1657/1523-0430(06-030)[BENEDICT]2.0.CO;2

Benedict JB (2009) A review of lichenometric dating and its applications to archeology. Am Antiq 74:143–172

Beschel RE (1958) Lichenometrical studies in West Greenland. Arctic 11:254

Beschel RE (1961) Dating rock surfaces by lichen growth and its application to the glaciology and physiography (lichenometry). In: Raasch GO (ed) Geology of the Arctic. University of Toronto Press, Toronto, pp 1044–1062

Birkeland P (1981) Soil data and the shape of the lichen growth curve for Mt Cook area. New Zeal J Geol Geophys 24:443–445

Birkenmajer K (1981) Lichenometric dating of raised beaches at Admirality Bay, King George Island (South Shetland islands, West Antarctica). Bull de l'Académie Pol des Sci, Série des Sci de la Terre 29:119–127

Bradwell T (2001a) A new lichenometric dating curve for southeast Iceland. Geograf Ann 83A:91–101. doi:10.1111/j.0435-3676.2001.00146.x

Bradwell T (2001b) Glacier fluctuations, lichenometry and climatic change in Iceland. PhD Thesis, University of Edinburgh, 365 pp

Bradwell T (2010) Studies on the growth of *Rhizocarpon geographicum* in northwest Scotland and some implications for lichenometry. Geogr Ann Ser A Phys Geogr 92:41–52

Bradwell T, Armstrong RA (2007) Growth rates of *Rhizocarpon geographicum* lichens: a review with new data from Iceland. J Quater Sci 22:311–320

Brodo IM (1965) Studies on growth rates of corticolous lichens on Long Island, New York. Bryologist 68:451–456

Bull WR, Brandon MT (1998) Lichen dating of earthquake-generated regional rockfall events, southern Alps, New Zealand. GSA Bull 110:60–84

Burbank DW (1981) A chronology of late Holocene glacier fluctuations on Mount Rainier, Washington. Arc Alp Res 13:369–386. doi:10.2307/1551049

Burrows CJ, Orwin J (1971) Studies of some glacial moraines in New Zealand. 1. The establishment of lichen-growth curves in the Mount Cook area. New Zea J Sci 14:327–335

Caseldine C, Baker A (1998) Frequency distributions of *Rhizocarpon geographicum*, modelling, and climate variation in Trollaskagi, Northern Iceland. Arct Alp Res 30:175–183. doi:10.2307/1552132

Childress S, Keller JB (1980) Lichen growth. J Theor Biol 82:157–165. doi:10.1016/0022-5193(80)90095-8

Clark BM, Mangelson NF, St. Clair LL, Rees LB, Bench GS, Southern JR (2000) Measurement of age and growth rate in the crustose saxicolous lichen *Caloplaca trachyphylla* using C-14 accelerator mass spectroscopy. Lichenologist 32:399–403. doi:10.1006/lich.2000.0279

Clayden SR (1997) Intraspecific interactions and parasitism in an association of *Rhizocarpon lecanorinum* and *R. geographicum*. Lichenologist 29:533–545

Curry RR (1969) Holocene climatic and glacial history of the central Sierra Nevada, California. Geol Soc Am 123:1–47 (Special paper)

Dabski M (2007) Testing the size-frequency-based lichenometric dating curve on Flaajokull moraines (SE Iceland) and quantifying lichen population dynamics with respect to stone surface aspect. Jokull 57:21–36

Denton GH, Karlen W (1973) Lichenometry: its application to Holocene moraine studies in southern Alaska and Swedish Lapland. Arct Alp Res 5:347–372

Ellis JM, Hamilton TD, Calkin PE (1981) Holocene glaciation of the Arrigetch Peaks, Brooks Range, Alaska. Arctic 34:158–168

Follman G (1961) Estudios liquenometriqos en los monumentos prehistoricos de la Isla de Pascua. Rev Univer; Anal de la Acad Chil de Cienc Nat 24:149–154

Gellatly AF (1982) Lichenometry as a relative-age dating method in Mt Cook National Park. NZ J Bot 20:343–354

Gregory KJ (1976) Lichens and the determination of river channel capacities. Erach Surf Proc 1:273–285. doi:10.1002/esp.3290010308

Griffey NJ (1977) A lichenometric study of the neoglacial end moraines of the Okstinden glacier, north Norway and comparisons with similar recent Scandinavian studies. Nor Geogr Tidsskr 31:163–172

Hale ME (1967) The biology of lichens. Edward Arnold, London

Hale ME (1973) Growth. In: Ahmadjian V, Hale ME (eds) The lichens. Academic Press, New York, pp 473–492

Harrison S, Winchester V (2000) Nineteenth and twentieth-century glacier fluctuations and climatic implications in the Arco and Colonia valleys Hielo Patagonico Norte, Chile. Arc Antarc Alp Res 32:55–63

Haworth LA, Calkin PE, Ellis JM (1986) Direct measurement of lichen growth in the central Brooks Range, Alaska USA, and its application to lichenometric dating. Arct Alp Res 18:289–296. doi:10.2307/2258323

Hill DJ (1981) The growth of lichens with special reference to the modelling of circular thalli. Lichenologist 13:265–287. doi:10.1017/S0024282981000352

Hooker TN (1980a) Factors affecting the growth of Antarctic crustose lichens. Brit Antarct Surv Bull 50:1–19

Hooker TN (1980b) Lobe growth and marginal zonation in crustose lichens. Lichenologist 12:313–323. doi:10.2307/2258324

Innes JL (1983) Development of lichenometric dating curves for highland Scotland. Trans Roy Soc Edin 74:23–32

Innes JL (1985) Lichenometry. Prog Phys Geog 9:187–254. doi:10.1177/030913338500900202

Laundon JR (1980) The use of lichens for dating walls in Bradgate Park, Leicestershire. Trans Leic Lit Pjil Soc 74:11–30

Locke WW III, Andrews JT, Webber PJ (1979) A manual for lichenometry. British Geomorphological Research Group. Tech Bull 26:1–47

Loso MG, Doak DF (2006) The biology behind lichenometric dating curves. Oecologia 147:223–229. doi:10.1007/s00442-005-0265-3

Luckman BH (1977) Lichenometric dating of Holocene moraines at Mount Edith Cavell, Jasper, Alberta. Can J Earth Sci 14:1809–1822

Maguas C, Brugnoli E (1996) Spatial variation in carbon isotope discrimination across the thalli of several lichen species. Plant Cell Environ 19:437–446. doi:10.1111/j.1365-3040.1996.tb00335.x

Mathews JA, Trenbirth HE (2011) Growth rate of a very large crustose lichen (Rhizocarpon subgenus) and its implications for lichenometry. Geogr Ann Ser A Phys Geogr 93A:27–39

Matthews JA (1994) Lichenometric dating: a review with particular reference to 'Little Ice Age' moraines in southern Norway. In: Beck C (ed) Dating in surface context. New Mexico Press, Albuquerque, pp 185–212

McCarroll D (1993) Modeling late-holocene snow-avalanche activity incorporating a new approach to lichenometry. Earth Surf Process Land 18:527–539. doi:10.1002/esp.329018606

McCarthy DP (1999) A biological basis for lichenometry? J Biog 26:379–386. doi:10.1046/j.1365-2699.1999.00272.x

McCarthy DP (2003) Estimating lichenometric ages by direct and indirect measurement of radial growth: A case study of *Rhizocarpon* agg. At the Illecillewaet Glacier, British Columbia. Arct Antarct Alp Res 35:203–213. doi:10.1657/1523-0430(2003)035[0203:ELABDA]2.0.CO;2

Mottershead DM, White ID (1972) The lichenometric dating of glacier recession, Tunsbergdalsbre, Southern Norway. Geogr Ann 54:47–52

Narama C (2002) Late holocene variation in the Raigorodskogo glacier and climate change in the Pamir-Alai, central Asia. Catena 48:31–37. doi:10.1016/S0341-8162(02)00007-3

Oerlemans J (1994) Quantifying global warming from the retreat of glaciers. Science 264:243–245. doi:10.1126/science.264.5156.243

O'Neal MA, Schoenenberger KR (2003) A *Rhizocarpon geographicum* growth curve for the Cascade Range of Washington and northern Oregon, USA. Quatern Res 60:233–241. doi:10.1016/S0033-5894(03)00104-2

Pentecost A (1980) Aspects of competition in saxicolous lichen communities. Lichenologist 12:135–144. doi:10.1017/S0024282980000060

Porter SC (1981) Lichenometric studies in the Cascade Range of Washington: establishment of *Rhizocarpon geographicum* growth curves at Mount Rainier. Arc Alp Res 13:11–23. doi:10.2307/1550622

Proctor MCF (1977) The growth curve of the crustose lichen *Buellia canescens* (Dicks) De Not. New Phytol 79:659–663. doi:10.111/j.1469-8137.1977.tb02250.x

Proctor MCF (1983) Sizes and growth-rates of thalli of the lichen *Rhizocarpon geographicum* on the moraines of the Glacier de Valsorey, Valais, Switzerland. Lichenologist 15:249–261. doi:10.1017/S0024282983000389

Rogerson RJ, Evans DJA, McCoy WD (1986) Five-year growth of rock lichens in a low-arctic mountain environment, Northern Labrador. Géographie physique et Quaternaire XL:85–91

Sancho LG, Pintado A (2004) Evidence of high annual growth rate for lichens in the maritime Antarctic. Polar Biol 27:312–319. doi:10.1007/s00300-004-0594-4

Sancho LG, Palacios D, Green TGA, Vivas M, Pintado A (2011) Extreme high lichen growth rates detected in recently deglaciated areas in Tierra del Fuego. Polar Biol 34:813–822. doi:10.1007/S00300-010-0935-4

Ten Brink NW (1973) Lichen growth rates in West Greenland. Arct Alp Res 5:323–331

Trenbirth HE, Matthews JA (2010) Lichen growth rates on glacier forelands in southern Norway: preliminary results from a 25-year monitoring programme. Geogr Ann Ser A Phys Geogr 92A:19–40

Webber PJ, Andrews JT (1973) Lichenometry: a commentary. Arc Alp Res 5:295–302

Winchester V, Chaujar RK (2002) Lichenometric dating of slope movements, Nant Ffrancon, North Wales. Geomorphology 47:61–74. doi:10.1016/S0169-555X(02)00141-1

Worsley P (1981) Lichenometry. In: Goudie A (ed) Geomorphological techniques. Allen and Unwin, London, pp 302–305

Analysis of Lichen Metabolites, a Variety of Approaches

11

Pierre Le Pogam, Gaëtan Herbette and Joël Boustie

Contents

P. Le Pogam · J. Boustie (✉)
Faculté des Sciences Pharmaceutiques et Biologiques, Institut des Sciences Chimiques de Rennes, PNSCM-UMR 6226, Université Européenne de Bretagne, Univ. Rennes 1, 2 Av. Du Pr Léon Bernard, 35043 Rennes Cedex, France
e-mail: joel.boustie@univ-rennes1.fr

P. Le Pogam
e-mail: pierre.lepogam-alluard@univ-rennes1.fr

G. Herbette
Spectropole, Aix-Marseille Université, Campus Scientifique de Saint Jérôme, Service, 511, 13397 Marseille Cedex 20, France
e-mail: gaetan.herbette@univ-amu.fr

D.K. Upreti et al. (eds.), *Recent Advances in Lichenology*,
DOI 10.1007/978-81-322-2181-4_11

Abstract

Lichens produce secondary metabolites which have been first considered as a chemical support fully involved in lichen taxonomy. As a consequence, analytical methods were developed and applied to these organisms from a long time, some of them being standardized. Then, lichen analysis benefitted from new developments and techniques applied for isolation and identification of secondary metabolites which are exposed and discussed herein. Some ancient techniques for lichen taxonomy are still used as spot tests, which involve application of specific reagents directly on the lichen thallus. TLC is also still extensively used with standardized protocols affording more accurate information on lichens' metabolic profiles. Identification of lichen compounds from the shape of crystals observed under microscope was facilitated as some major lichen metabolites are extracted in high yields. X-rays are now used in some cases where classical spectroscopic, UV, IR, MS, and NMR techniques do not allow unambiguous assignments. Using such techniques for isolated lichen compounds, some characteristic identification patterns of these substances are presented here. Hyphenated techniques, coupling separation and identification, are more and more used and broaden the analysis facilities of lichen compounds. They enable early dereplication and subsequent focusing on bioactive or original compounds. Other trends in lichenology involve in situ analyses thanks to specific and innovative NMR or MS techniques that yield valuable information directly from the natural complex matrix. Several advantages can be expected from those approaches: no extraction steps, qualitative and quantitative information in a few minutes or hours, and direct analysis of genuine compounds, avoiding chemical artefacts associated with extraction and purification processes (Table 11.2). Moreover, some of these methods pave the way for the development of imaging techniques that might help correlating metabolites to their specific ecological environment. Altogether, such enhancements might enable harnessing of lichens' unique chemo-diversity.

Keywords
Secondary metabolites · Characterization · Identification · Spectroscopic analysis · Phytochemistry · Natural structure

11.1 Introduction

The lichenous lifestyle is maintained by approximately 18,900 known species (Feuerer and Hawksworth 2007) resulting from the association between a fungus and an alga (chlorolichens). For 10 % of lichens, green alga is replaced or accompanied by a cyanobacterium, which can absorb atmospheric nitrogen due to specific structures named heterocysts. The fungus plays a role in water and mineral supply and in mechanical protection of the whole organism. Algal or cyanobacterial partners realize photosynthesis and provide their fungal associate organic compounds to be metabolized.

Irrespective of the shape and the size of the lichen thallus, the symbiotic thallus yields unique metabolites among all life forms (Fig. 11.1). Most lichen substances are phenolic orcinol and β-orcinol derivatives: dibenzofurans and usnic acids (pannaric acid and usnic acid), depsides (thamnolic acid and atranorin), depsidones (norstictic acid), depsones (picrolichenic acid), diphenylethers (leprolomin), benzyl esters (alectorialic acid), and tridepsides (gyrophoric acid). Other than specific and unique lichen metabolites, various other structural families can be recognized: aliphatic acids (roccellic acid) and related lactones, xanthones (lichexanthone), chromones (siphulin), pulvinic acid derivatives (pulvinic acid dilactone), quinones (haemoventosin, parietin), terpenes (zeorin), steroids (ergosterol peroxide), and carotenoids (astaxanthin). Cyanolichens mostly contain sugars and amino acid derivatives although original compounds such as mycosporines were detected recently (Torres et al. 2004; Roullier et al. 2011).

Unusual compound families can also be encountered as polypropionates, cyclic depsipeptides, brominated depsidones, and brominated acetylenic fatty acids, along with some glycosides (Boustie and Grube 2005). One should find a recent update of lichenic chemodiversity in the catalogue established by Elix (2014) (http://www.anbg.gov.au/abrs/lichenlist/Chem%20Cat%203.pdf). This diversity is also extended through the increasing number of metabolites isolated from the cultivated fungi and microflora associated with the lichen thallus (Grube et al. 2012).

Lichen metabolites considered in this chapter are mainly produced by the fungal partner and often released into the extracellular space to subsequently crystallize on or near the hyphal surface (Honegger 1986; Ozenda and Clauzade 1970; Hale 1983). Thus, large amounts of secondary metabolites can be detected in a lichen thallus. Extraction yields can rate from 1 to 25 % of dried lichen material, and composition is frequently characterized by one to three metabolites accumulated in high yields (over 80 % ratio) (Boustie et al. 2011).

Lichen secondary compounds are mainly arising from four pathways of biosynthesis: polyketide path, mevalonic acid path, shikimic acid path, and photosynthetic products of the phycobionts as presented in Fig. 11.2.

Phenols are the most frequently produced lichen metabolites and generally found accumulated in high yields as extrolites, crystallized

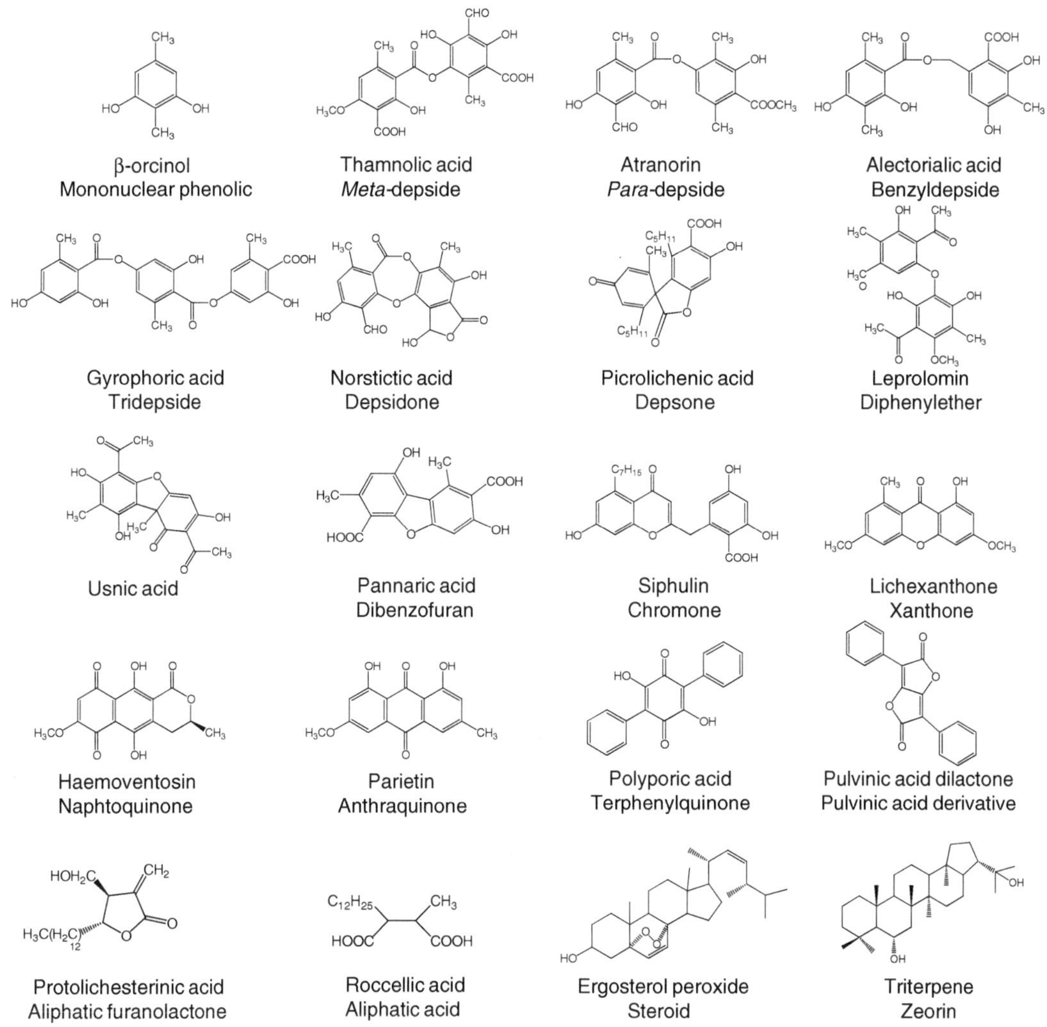

Fig. 11.1 Some typical secondary lichen metabolites illustrating a variety of structural classes

outside of the fungal hyphae. Oxidative coupling of orcinol and beta-orcinol derivatives result in the most typical lichen metabolites corresponding to depsides, depsidones, and dibenzofurane derivatives (Fig. 11.3). These lichen substances have indeed several possible biological roles that may impact biotic and abiotic interactions of lichens with their environment (Rikkinen 1995; Molnár and Farkas 2010).

Lichen secondary metabolites can act as allelochemicals, affecting the development and growth of lichens (Culberson et al. 1977; Whiton and Lawrey 1984), mosses (Heilman and Sharp 1963; Lawrey 1977), vascular plants (Pyatt 1967; Fisher 1979; Marante et al. 2003), and microorganisms that compete for space and light on a variety of substrates (Ranković et al. 2008; Schmeda-Hirschmann et al. 2008; Halama and Van Haluwin 2004). Strong experimental data support protection of lichens from herbivores grazing granted by lichen secondary metabolites (Asplund and Gauslaa 2008; Nimis and Skert 2006; Pöykkö et al. 2005; Dailey et al. 2008).

Ertl (1951) first observed that cortical accumulation of lichen compounds increased opacity of the upper cortex, restraining solar irradiance

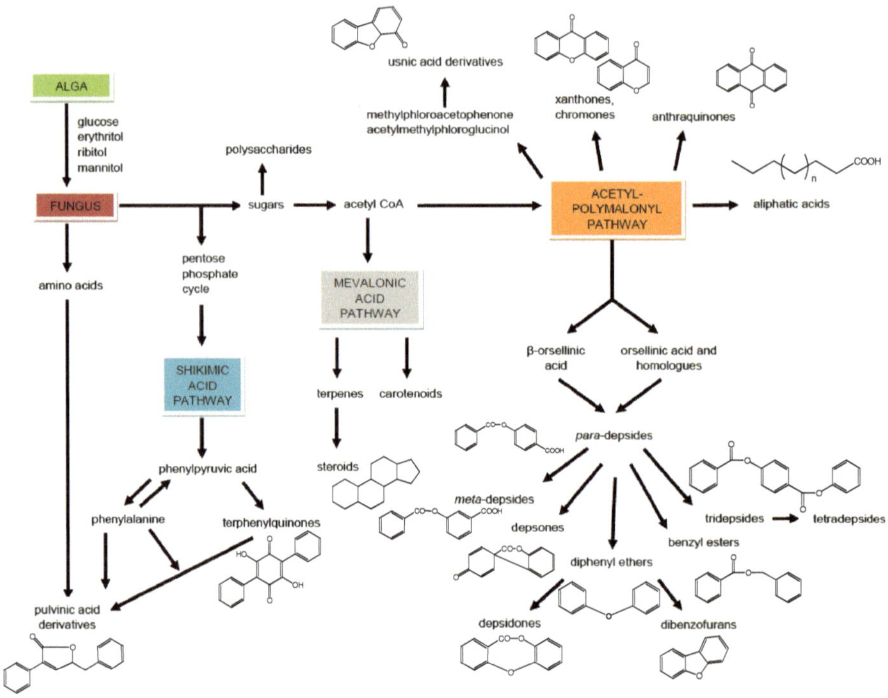

Fig. 11.2 Main metabolic pathways recognized in lichens [adapted from Elix (1996) and Stocker-Wörgötter (2008)]

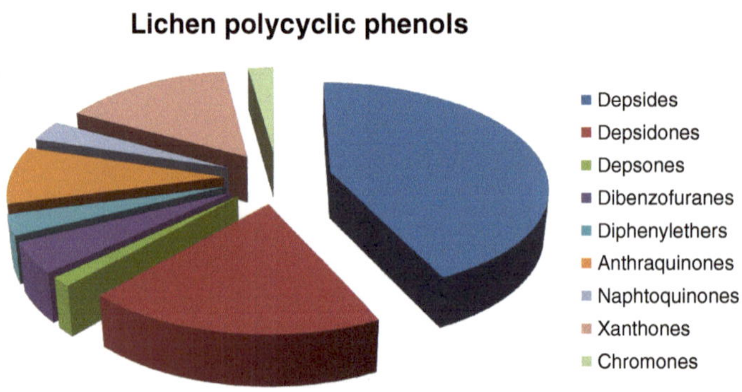

Fig. 11.3 Distribution of the main polycyclic lichen phenols arising from the acetyl-polymalonyl pathway (based on the 600 lichen polycyclic phenols compiled by Huneck and Yoshimura 1996)

reaching the light-sensitive algal layer. Various light-screening pigments display strong UV absorption abilities and might function as filters in order to prevent excessive UV-B irradiation (Nguyen et al. 2013). Some lichen phenolics have high ability to scavenge toxic free radicals generated by UV light (Russo et al. 2008; Odabasoglu et al. 2006; Karakus et al. 2009). Some other metabolites step in metal homeostasis and pollution tolerance (Hauck and Huneck 2007a, b; Hauck et al. 2009). Secondary metabolites quantification revealed tremendous changes in the concentrations of secondary metabolites from *Hypogymnia physodes* when thalli were transplanted to areas polluted with heavy metals and acidic inorganic sulfur compounds (Bialonska and Dayan 2005). As a result, the value of lichen metabolites in chemotaxonomy and systematics is questionable, but their isolation and identification is mandatory to recognize new active compounds and/or original physicochemical properties.

Most common biological activities deal with cytotoxicity and antibiotic activities (for a review see Shukla et al. 2010; Shrestha and St. Clair 2013). Additionally, miscellaneous other properties were described within several families of lichen products including antiviral or antiretroviral (Yamamoto et al. 1995; Pengsuparp et al. 1995), anti-inflammatory (Vijayakumar et al. 2000; Freysdottir et al. 2008), and analgesic activities (Okuyama et al. 1995). As natural plant-derived products have a less detrimental impact on the environment than synthetic chemicals, lichen metabolites also represent good candidates for new pesticides (Cetin et al. 2008) (Dayan and Romagni 2001). However, although studies carried on lichens to value their secondary metabolites are increasing, difficulties encountered in identification of the species, collection of bulk quantities, and the isolation of pure substances limit the number of tested compounds. Moreover, detection and isolation of minor compounds is generally complicated by the high abundance of redundant lichen compounds and techniques designed to bypass such a difficulty should allow findings of more specific and interesting compounds.

This chapter aims at presenting successive analytic methods developed during the history of lichens' secondary metabolites analysis. Arising from spot tests used in the first lichen taxonomy approaches, thin-layer chromatography (TLC) is still extensively used and standardized protocols have been established. Crystallography from crystal shapes to X-rays accurate measurements has been developed with lichen metabolites. The main classical spectroscopic techniques, UV, IR, MS, NMR used to identify isolated lichen compounds, are presented herein with a focus on some characteristic identification patterns of these substances. Early and refined dereplication approaches to eliminate already known compounds and to focus on species containing original or bioactive compounds have now to be used first. Accumulated spectroscopic data and hyphenated techniques, coupling separation and identification, are more and more used and broaden the analysis facilities of lichen compounds. Subsequent trends in lichenology involve in situ analyses due to more specific and innovative NMR or MS techniques that yield valuable information directly from the natural complex matrix.

Some of these methods are described in a second part as well as expected results they should yield. A special insight is given to some imaging techniques with high potential to dramatically improve our understanding of the metabolic activities revealed at the very root of the lichen symbiosis.

11.2 Oldest Analytical Methods

11.2.1 Spot Test (Reagents Directly Applied on the Lichen Thallus)

Color reactions should be considered as useful hints for the presence of functional groups or elements within lichen molecules and thus remain rather limited compared with classical spectroscopic analysis methods. Nevertheless, spot tests, involving application of tiny amounts of specific reagents directly to the lichen thallus or parts of this thallus (i.e., cortex, medulla, apothecia), have been applied since the nineteenth century (Nylander 1866) and can lead to distinguish varieties and even sometimes species according to any associated color change or not. This is particularly useful to distinguish some closely shaped species which could not be unambiguously discriminated on the basis of morphological characters. For instance, species of the *Ramalina siliquosa* complex are tricky to be separated and display a remarkable chemical plasticity, leading to six different chemotypes according to different thallus reactions related to the main compound as assessed with LC-DAD-MS analysis (Parrot et al. 2013). So, such colors obtained with given reagents are found in herbarium specimens since the method has been standardized.

Nowadays, spot tests remain either helpful or necessary to confidently discriminate between some lichens and are still met as a very important criterion in determination keys. However, color intensities which are to be interpreted are not easy to standardize, attribution to a given major compound is often hazardous, and this technique is poorly informative in analysis of pure compounds. Noteworthy, because different compounds are present in specific parts of the lichen, reference tests usually describe parts of the thallus which should be tested, which corresponds to a first attempt toward histo-localization (Fig. 11.4). Three main reagents used in color reactions are potassium hydroxide (commonly called K), sodium hypochlorite (abbreviated C), and paraphenylenediamine (mainly known as P or PD).

K (consisting of 10 % potassium hydroxide in water) is a useful reagent to differentiate between quinones and pulvinic acid derivatives as it turns quinones into a bright red to deep purple color. It also turns yellow and then red with most *ortho*-hydroxy aromatic aldehydes.

C (saturated aqueous $Ca(OCl)_2$ or commercial laundry bleach) is a reagent for *meta*-dihydroxy phenols, except for those substituted between the hydroxy groups with a –CHO or –COOH. This explains why arthothelin and thiophanic acid are C+, whereas lichexanthone does not react with C (Fig. 11.5). Different colors can be obtained ranging from light yellow to blood red. C turns green with dihydroxy dibenzofurans (Nash 2008).

In some cases, successive reagent application is required, and KC+ is pretreatment by KOH before having a coloration with C reagent. KC turns red with C– depsides and depsidones which undergo rapid hydrolysis to yield a *meta*-dihydroxy phenolic moiety. Blue color is obtained with dihydroxy dibenzofurans, whereas a yellow reaction appears with usnic acid (Nash 2008).

Although highly toxic, last routinely used macrochemical reagent is ethanolic solution of para-phenylenediamine (2–5 %) that gives yellow to orange colors with aldehyde-containing depsides and depsidones turned into a Schiff base according to the following reaction:

$$\begin{aligned} &R{-}CHO + H_2N{-}C_6H_4{-}NH_2 \\ &\quad \rightarrow R{-}CH{=}N{-}C_6H_4{-}NH_2 + H_2O \end{aligned}$$

Various other reagents can be found in literature, but their use is not common compared with those previously described. $FeCl_3$ can be used to

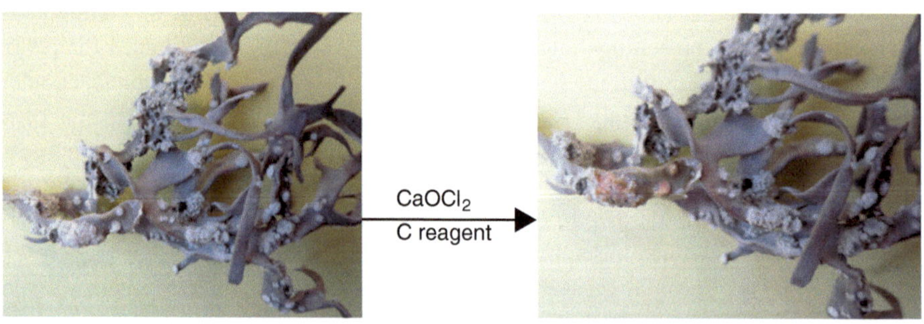

Fig. 11.4 Spot test with $Ca(OCl)_2$=C reagent drop on a thallus of the fruticose *Roccella fuciformis*. On the *right picture*, the C+ bright red reaction is limited to soralia (= white parts, corresponding to medulla as the gray thallus cortex is disrupted). This test is also confirming the distinction with the related species *Roccella phycopsis* where only the gray cortex is turning rose red C+

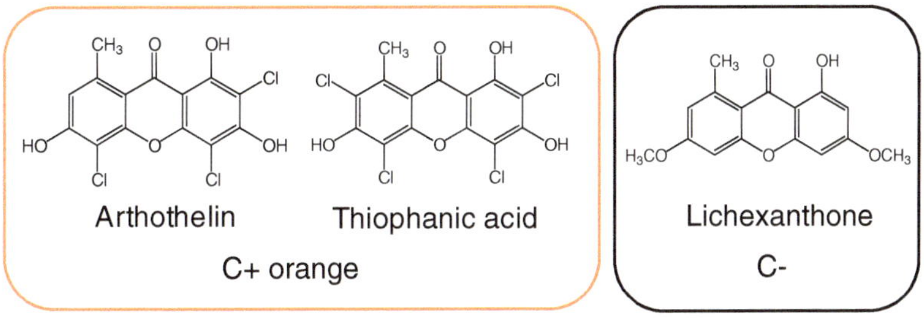

Fig. 11.5 Structures of chromones exemplifying the marked *orange* color obtained with $Ca(OCl)_2$ reagent =C spot test, indicating the presence of *m*-dihydroxyphenols in arthothelin and thiophanic acid compared to lichexanthone

detect phenolic groups (leading to red to violet coloration), quinones having free hydroxyl group in β-position are revealed with both 2 % magnesium acetate ethanolic solution (red or violet) and Dimroth's reagent, e.g., boric acid solved in acetic anhydride giving red to violine colors. Chlorinated metabolites can be detected thanks to Beilstein test which consists of bringing one milligram of the molecule on the tip of a glowed copper wire introduced into the flame of a Bunsen burner, halogenated compounds then transiently turn the flame into a deep green color (Huneck and Yoshimura 1996). Spot tests corresponding to a set of typical lichen products can be found below (Table 11.1). Chromones, xanthones, anthraquinones, and pulvinic acids are naturally colored compounds.

11.2.2 Crystal Tests

Microcrystallization was once a very if not the most important way to identify lichen metabolites in times without TLC and high-performance liquid chromatography (HPLC). Microcrystallization of lichen substances was mainly developed by Asahina and Shibata since 1950 (Shibata 2000). This approach was based on the formation of distinctive crystals from lichen extracts (Fig. 11.6).

The general procedure can be summarized as follows: the extract of a small piece of lichen is first filtered before being evaporated. Acetone is the most used solvent as it extracts compounds with a wide range of polarity. Dried residue is then transferred to a microscope slide, and a drop of proper microcrystallization reagent is added on it before capping with a cover glass. Different microcrystallization solutions were described, the most commonly used being GAW and GE, corresponding to H_2O/glycerol/ethanol 1:1:1 (v/v/v) and to acetic acid/glycerol 1:3, respectively. Slides using GE or GAW are further gently heated over microflame (other solvents do not need this heating step) and then let aside for progressive cooling, enabling the crystallization process to occur. Once formed, crystals are best observed under polarized light with a 200–1,000-fold magnification and present incredible diversity in both shapes and colors that was used to recognize the major metabolite in a given specimen. Although now outstood by modern analysis methods, this concept is still used nowadays after compound purification with great performances thanks to modern X-ray crystallography (Deschamps 2010).

This simple and rapid technique allowed to characterize the major metabolites in hundreds of lichen species but was obviously useless to detect minor substances and to analyze too complex mixtures of lichen substances.

11.2.3 Melting Point

Melting point was once an important criterion for identification of a pure lichen substance. It can be determined with a few crystals under a glass capillary or with a Köfler device. It should be noted that some compounds do not have a sharp melting point but decompose over a wide range

Table 11.1 Thalline reactions corresponding to a set of common lichen metabolites (Brodo et al. 2001)

Compound	K	C	KC	PD
Depsidones				
Psoromic acid	–	–	–	Light yellow
Salazinic acid	Dark red	–	–	Orange
Stictic acid	Yellow	–	–	Orange
Virensic acid	–	–	–	Orange/red
Benzyldepsides				
Alectorialic acid	Pale yellow	Red	Red	Dark yellow
Barbatolic acid	Yellow	–	–	Yellow
Dibenzofurans and related				
Didymic acid	–	Green	Green	–
Pannaric acid	–	Olive green	Olive green	–
Strepsilisin	–	Green	Green	–
Usnic acid	–	–	Yellow	–
Chromones				
Siphulin	Yellow brown	Violet	–	–
Xanthones				
Arthothelin	–	Orange	Orange	–
Lichexanthone	–	–	–	–
Thiophanic acid	–	Orange	Orange	–
Anthraquinones				
Nephromin	Red purple	–	–	–
Parietin	Red purple	–	–	–
Skyrin	Red purple	–	–	–
Triterpenoids				
Zeorin	–	–	–	–
Pulvinic acid derivatives				
Calycin	–	–	–	–
Rhizocarpic acid	–	–	–	–
Vulpinic acid	–	–	–	–

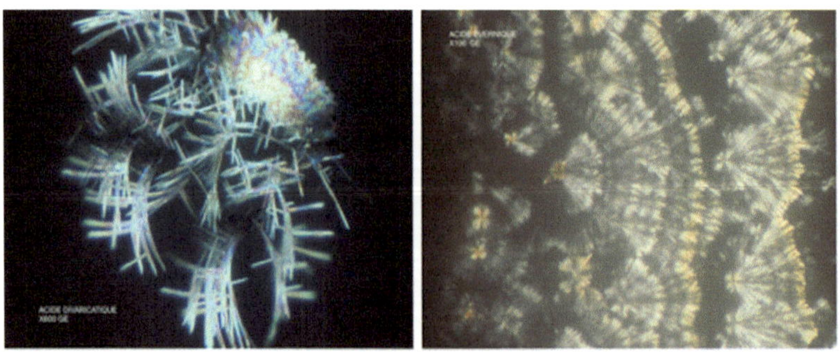

Fig. 11.6 Microcrystallizations of secondary lichen metabolites: divaricatic acid and evernic acid (both crystallized in GE from an acetone lichen extract then magnified 100-fold and observed through polarized light) (pictures from Jacques Lagabrielle)

of temperature. Decomposition of lichen substances may be accompanied with characteristic odors that can differ between aliphatic molecules, phenolic compounds, or triterpenes. Obviously, melting points can only serve as a hint to further confirm structural elucidation granted by other means. One could find a list of lichen substances' melting points in Hüneck and Yoshimura's reference book (Huneck and Yoshimura 1996).

11.3 Structure-elucidating Methods

General chemical procedure to study secondary metabolites is done from lichen extracts in organic solvent and further purification through crystallization or chromatography separation. Isolated compounds are then analyzed in most cases by NMR, MS, IR, and UV spectroscopy plus stereochemistry studies, if required.

11.3.1 NMR Spectroscopy (NMR, MAS-NMR, Innovative Methods)

As for any organic compound, NMR is an essential method for the structural determination of lichen secondary metabolites. Structural determinations for lower milligram samples could be made with data obtained in proton and carbon NMR. The assignments of signals can be achieved via crosspeak correlation observed in homonuclear 2D spectra of scalar couplings (COSY, TOCSY ...) or dipolar couplings (NOESY, ROESY) and heteronuclear ^{1}J type (HSQC/HMQC ...) and $^{2,3}J$ type (H2BC, HMBC ...), 2D or 3D sequences to detect heteronuclear scalar couplings as HSQC-TOCSY, and HSQC-NOESY generally for complex structures.

However, most secondary lichen metabolites isolated so far belong to a limited variety of structural families as illustrated in Fig. 11.1. Some dedicated NMR studies related to several structural families were published: depsides (Cheng et al. 2013), tridepsides (Narui et al. 1998), depsidones and depsones (Sundholm and Huneck 1980, 1981), xanthones (Sundholm 1978, 1979), and triterpenes (Wenkert et al. 1978; Wilkins et al. 1990). A compilation of the published NMR data of lichen compounds is reported in the Huneck and Yoshimura (1996).

Solubilization of pure lichen compounds is often a challenge, but most of the NMR spectra are recorded after dissolution in $CDCl_3$ or DMSO-d_6. Some characteristic NMR features commonly found in NMR spectra of lichen compounds are selected here for monoaromatic or rearranged phenols corresponding to depsides, depsidones, depsones, or dibenzofuranes along with some NMR landmarks for aliphatic and related paraconic acids.

11.3.1.1 Phenolic Compounds

Chemical shifts are greatly influenced by inductive and mesomeric effects originating from other substituents that depend on their respective position. Some publications reported on the effect of substituents nature and respective position to predict corresponding chemical shift with the reference being the benzene chemical shift of 128.5 ppm (Ewing 1979). For orcinol, phenol-bearing carbons are deshielded (140–160 ppm) as well as the carbon that has two *meta*-hydroxy moieties (140 ppm), whereas ortho-hydroxy-located carbons are shielded (100–110 ppm). Aromatic methyls are usually found around 20 ppm. Noteworthy, for the closely related β-orcinol, the second aromatic methyl can be easily distinguished since it is strongly shielded due to two ortho-hydroxy groups (8 ppm). Side chains are evidenced through a variable number of signals arising in the 10–40 ppm range (divarinol, olivetol). Carbonyl functions are easily observable in ^{13}C NMR where shifts around 170–200 ppm have to be distinguished between a methyl ester from an acid function, e.g., methyl orsellinate at 170.3 ppm compared to orsellinic acid at 173.3 ppm, while shifts of aldehyde function are found around 194 ppm as of haematommic acid.

Depending on the substitution on the aromatic ring, ^{1}H NMR spectrum shows sometimes only one aromatic signal (singlet) due to the symmetry of the molecules with a signal at 6.09 ppm in ^{1}H NMR and at 106.7 ppm in ^{13}C NMR (β-orcinol). According to moieties present on the aromatic ring, other signals are methyl groups at 2.14 ppm (orcinol), 2.08 and 1.89 ppm (β-orcinol), 2.47 and 2.00 ppm (orsellinic acid), variable hydroxyl and acid shifts, or alkyl chains between 0.85 and 2.47 ppm as the pentyl moiety of the olivetol. Signals can also be found at 3.82 ppm indicating a *O*-methyl group (4-*O*-methylolivetolcarboxylic) or at 3.92 ppm for acid or ester (methyl-β-orcinolcarboxylate). A signal around 10.3 ppm is indicative for an aldehyde function (haematommic acid).

11.3.1.2 Dibenzofuranes

Most true dibenzofuranes exhibit two aromatic protons (6.6 and 6.9 ppm) while related compounds such as the commonly found and related usnic acid display a shielded aromatic signal at 5.92 ppm. Additional ^{1}H NMR signals correspond to the substitution of the dibenzofuran skeleton by a variety of substituents such as methyl, alkyl chain, hydroxyl, acid, ester, or ketone. The ^{13}C NMR of the aromatic ring signals is observed between 98.3 and 179.4 ppm, and those corresponding to the ketone function are at around 200 ppm and to methyl groups between 7.7 and 32.0 ppm (usnic acid).

A fourth cycle from a gamma-lactone (e.g., strepsilin) or a delta-lactone (e.g., haemophaein) may also occur with a ^{13}C NMR chemical shift around 170 ppm. Such an additional ring, which may be substituted by methyl (e.g., alectosarmentin), hydroxyl groups, and/or alkyl chains (e.g. haemophaein), is also found for some depsides and depsidones.

11.3.1.3 Depsides, Depsidones, and Depsones

Depsides are phenolic compounds bounded by ester linkage of two units (didepside), three units (tridepside), or four units (tetradepside). Barbatic acid, a didepside, shows two signals in ^{13}C NMR at 171.4 and 174.8 ppm for ester and acid function, respectively. Aromatic rings may be substituted by different groups: hydroxyl, aldehyde, methoxy, methyl, saturated or unsaturated alkyl chain with a ketone function, and halide (chlorine or bromine). Aromatic methyl groups' chemical shifts range from 7 to 30 ppm according to the neighboring substituents as previously discussed. Methoxyl groups can be easily recognized with chemical shifts ranging from 50 to 60 ppm, methylenic groups of alkyl chains have a chemical shift between 26 and 40 ppm, and methyls are encountered around 15 ppm. Noteworthy, one may recognize patterns of signals corresponding to the constitutive monomers within some depsides (e.g., isolecanoric acid and orsellinic acid, divaricatic acid and divaricatinic acid). Tri- and tetra-depsides obviously contain more carbon signals and can thus be easily distinguished from classical didepsides.

Benzyldepsides have an additional characteristic chemical shift at 5.60 ppm in ^{1}H NMR and 66.0 ppm in ^{13}C NMR of the methylenic group (alectorialic acid) due to a strong deshielding of the ester function and the aromatic ring in *alpha* position.

Depsidones are characterized by an ether linkage bridging the two aromatics moieties through a third ring including the C_7 carbonyl of the ether linkage which is shielded around 161 ppm while found around 169 ppm in depsides. The number of aromatic protons and substituents is consecutively reduced with regard to depsides, but substitutive patterns are often similar.

Depsones are usually picrolichenic acid derivatives which are cyclized depsides by a C_1–C_5 bond thereby forming a gamma-lactone with a signal at 170 ppm in ^{13}C NMR. A quaternary C_1 at 59 ppm and a cyclic ketone at 188 ppm are in most cases accompanied with signals of two alkyl chains.

11.3.1.4 Aliphatic Acids and Paraconic Acids

Aliphatic acids contain a long carbon chain whose chemical shifts are comprised between 10 and 40 ppm in ^{13}C NMR, and a variable number of carboxylic acid moieties located around 180 ppm. Identification of carbons belonging to the carbon chain is tricky since most of them are very close to one another. Double bond-containing compounds display a distinct chemical shift around 130–140 ppm (lichesterylic acid, fumaric acid), whereas molecules containing hydroxyl or methoxyl groups possess chemical shifts in the 60–80 and 50 ppm range, respectively.

Such aliphatic acids can cyclize, thus forming a gamma-lactone ring. The carbon bearing the lactone function displays a typical ^{13}C NMR chemical shift located between 174 and 178 ppm, whereas the carbon linked to the oxygen of the cycle is slightly deshielded at around 50 ppm. NMR is useful to characterize the length of the alkyl chain in C_{11} (nephrosterinic acid), C_{13} (lichesterinic acid), or C_{15} (muronic acid) as well as the substitution by a ketone (220.3 ppm in ^{13}C NMR, muronic acid), hydroxyl function (68.4 ppm in ^{13}C NMR, neuropogolic acid), a terminal acid group (168 ppm in ^{13}C NMR, protopraesorediosic acid), or acetoxy group (174.7 ppm in ^{13}C NMR, 19-acetoxylichesterinic acid). α, β-unsaturated-γ-lactones (e.g., neuropogolic acid) give signals in the range of ethylenic carbons (120–160 ppm) and deshields neighboring carbon (commonly found at 80 ppm). γ-lactones displaying an exocyclic methylenic function (e.g., protolichesterinic acid) are more shielded (125 ppm) than those comprising the double bond inside the lactone ring (135–145 ppm) with two characteristic signals in ^{1}H NMR at 6.39 and 6.03 ppm as doublets (J 3 Hz).

The gamma-lactone cycle may be substituted by a methyl group and an acid function as respectively observed at 14.4 and 177.3 ppm for (−)-dihydropertusaric acid. Structure of unsaturated lactones like isomuronic acid with two ethylenic carbons at 139.0 and 146.6 ppm in ^{13}C NMR or of lactones with an exocyclic ethylenic group can be resolved with NMR ^{1}H NMR characteristic signals (6.03 and 6.39 ppm as two doublets (J 3 Hz) in ^{1}HNMR and signals at 125.9 and 132.6 ppm in ^{13}C NMR).

Solid-state MAS-NMR (magic angle spinning) techniques enable monitoring the chemical composition in the solid state of intact biological samples (Miglietta and Lamanna 2006).

Identification of molecules from 5 *Cladonia* lichens, hardly differentiated on the basis of botanical characters, could be achieved by comparison of the ^{13}C CP-MAS-NMR (CP: Cross Polarization) spectrum of solid crushed lichens with the reference powder spectra of the three major secondary metabolites (fumarprotocetraric acid, perlatolic acid, and usnic acid) that enabled attributing the chemical content to different unambiguous NMR signals (Chollet-Krugler et al. 2008).

Likewise, ^{1}H HR-MAS-NMR (HR standing for high resolution) provided specific fingerprints that could be reliably assigned to different carbohydrates detected from several intact lichens (Alcantara et al. 2007).

Recent NMR studies describe ^{1}H NMR spectra to achieve analysis under metabolomic conditions. Overall spectra from complete extract, with no further workup, display specific fingerprints enabling direct characterization of major compounds. Eisenreich et al. (2011) submitted methanol-D4 extract of chlorolichen *Xanthoria parietina* and cyanolichen *Peltigera horizontalis* to ^{1}H NMR. Those spectra comprise lichen-specific features as well as strong signals standing for mannitol, while those corresponding to sucrose are missing. Such methods seem interesting for both comparative studies (developing fingerprinting for given lichen species) or to assess metabolite modulations in response to environment changes (see Sect. 11.4).

11.3.2 Mass Spectrometry

The formula of an organic compound can be calculated from HR mass spectrometry data. Fractionations and skeletal rearrangements are also of particular significance in the application of mass spectrometry to structure elucidation. As an example, the fractionation of divaricatic acid is given below (Fig. 11.7).

Mass spectrometry also provides characteristic peaks regarding signature elements or specific

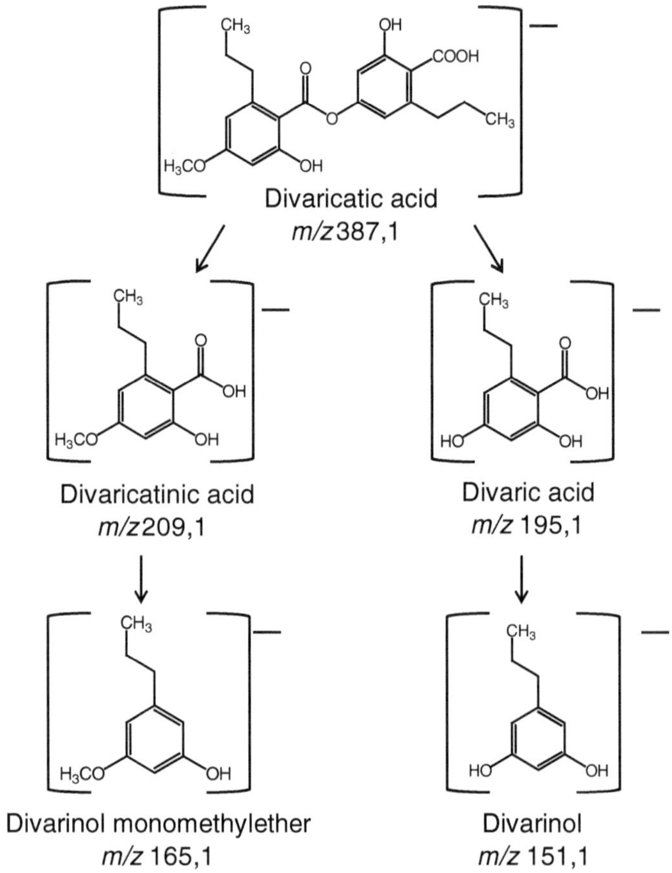

Fig. 11.7 Fractionation of divaricatic acid in negative-mode electrospray mass analysis

organic moiety, which hints regarding chemical structure of analyzed compounds. For instance, substances containing chloride give characteristic isotope pattern in their mass spectrum, indicating the number of chlorine atoms per molecule.

Huneck et al. (1968) first reported on mass spectra in both positive and negative mode corresponding to numerous depsides, depsidones, depsones, dibenzofuranes, and diphenylbutadienes. This paper described for the first time the main fragmentation route of depsides, consisting in the cleavage of its ester bond, enabling structural assignments to the carboxylic acid or phenolic parts of the molecule (Huneck et al. 1968). The specific cleavage of the ester bond as a typical process of didepsides can be followed by loss of water as the result of *ortho* interaction of the 2-hydroxybenzoic acid residue (Elix and Norfolk 1975; Addison 1985; Dmitrenok et al. 1987; Holzmann and Leuckert 1990). The classical fragmentation patterns of depsidones include neutral loss of CO_2 on their central ring to yield a dibenzofurane moiety alongside different elimination processes depending on the nature of the substituents (Parrot et al. 2013). Mass spectrum behavior of dibenzofuran and usnic acid derivatives was thoroughly discussed (Kutney et al. 1974; Schmidt et al. 1981). Hüneck and Schmidt (1980) described the fragmentation behavior of 16 usnic acid derivatives and established elimination mechanisms using deuterated analogues in both positive and negative modes. Characteristic rearrangements were also reported with some pulvinic acid derivatives (Letcher and Eggers 1967), polyporic acid derivatives (Grigsby et al. 1974), di- and tri-oxygenated stictane triterpenoids, and their trimethylsilyl derivatives (Holland and Wilkins 1979).

Non-hyphenated mass spectrometry was also applied to the screening of lichen compounds. Santesson developed a special technique, the so-called lichen mass spectrometry by introducing small lichen samples (sometimes less than 50 ng) into the direct inlet system of a mass spectrometer aiming to study xanthone pigments. The xanthones sublime as the temperature is increased (150 °C) under very low pressure, and the mass spectrum is recorded. Xanthones generally give prominent molecular ions, and the spectrum of the mixtures is often seen as additive of the individual components. However, this method remained limited to lichens containing only a few compound, otherwise giving spectra that are too complex to be interpreted (Santesson 1969).

Other mass spectrometry techniques relied on combination of soft-ionization negative-mode fast atom bombardment yielding intense $[M-H]^-$ ions that were subsequently analyzed by MS/MS. This enabled investigation of individual $[M-H]^-$ ions originating from crude extracts of lichen material and the selective study of the structure of components (Holzmann and Leuckert 1990). Tabacchi et al. (1991) also directly analyzed *Pseudevernia furfuracea* and *Lobaria pulmonaria* by tandem MS.

Recently, NMR and HRMS are the most commonly used structure-elucidating methods. Hyphenations of mass spectrometry with chromatographic techniques will be developed in Sect. 11.3 on dereplicative tools as well as recent innovations in the field of ambient and imaging mass spectrometry.

11.3.3 Infrared Spectroscopy

Like Raman spectroscopy, infrared spectroscopy is a technique based on the vibrations of the atoms of a molecule. The infrared spectrum is obtained by passing infrared radiation through a sample, in virtually any physical state, and further determining what fraction of the incident radiation is absorbed at a particular energy. Thus, infrared spectroscopy gives ability to associate patterns of absorption peaks with their corresponding functional groups (Stuart 2004). For example, carboxylic acid might be identified by a broad peak centered around 3,000 cm^{-1}, a strong peak around 1,720 cm^{-1}, and a broad medium intensity band around 920 cm^{-1} (Woodruff and Munk 1977). One might find pattern of absorption peaks corresponding to a wide variety of functional groups in Silverstein et al. (1998).

Infrared spectrometry is part of the classical set of structure-elucidating methods and IR data are thus most of the time available in Huneck and Yoshimura (1996).

Further enhancements in infrared spectroscopy now enable collecting space-resolved information, which has been discussed in detail in the Sect. 11.3.

11.3.4 UV Spectroscopy

As previously described, lichens are miniature ecosystems that can survive under very harsh conditions. Therefore, these stress-resistant organisms developed adaptive mechanisms to cope with such high light exposure including synthesis of UV-protectant metabolites (Nguyen et al. 2013). As most lichen compounds have chromophores, UV is useful for their detection and identification but remains most of the time associated with TLC or HPLC (see further).

Yoshimura et al. (1994) first divided lichen secondary metabolites into three groups according to their UV spectral pattern.

- depsides and monocyclic compounds
- depsidones
- dibenzofurans, pulvinates, xanthones, and anthraquinones exhibit typical UV spectra and are therefore easy to distinguish.

Then, mycosporines, more recently reported in lichens (Torres et al. 2004), are characterized through their UV spectrum absorbing in the 310–365 nm range with a strong molar extinction coefficient and a symmetrical shape. Detection through analytical techniques hyphenated with a PDA spectrophotometer offers a simple and easy method to screen them from lichen extracts

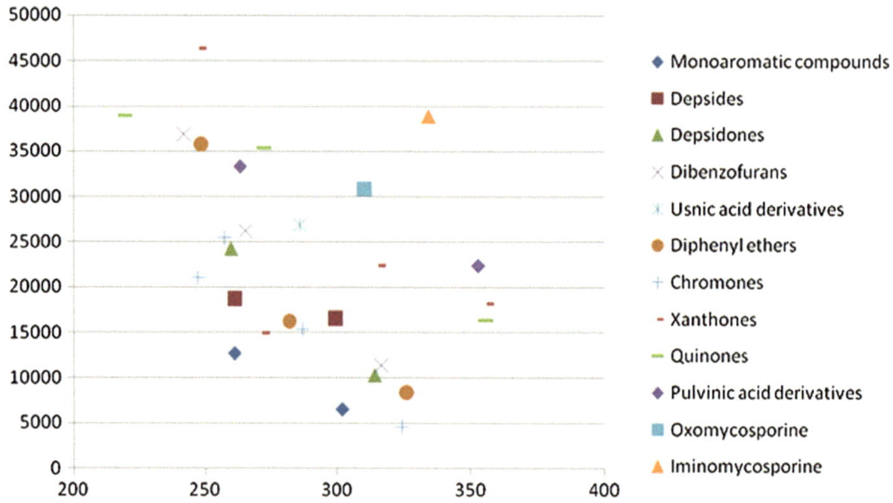

Fig. 11.8 UV properties of main structural families met within lichens (for quinones, additional absorptions are observed in visible range with medium λ_{max} of 458 and 529 nm) [adapted from Huneck and Yoshimura (1996)]

(Roullier et al. 2011). As an attempt, UV data of most lichen structural families are visualized in the Fig. 11.8:

Some relationships between chemical structures and UV spectra of depsides have been considered (Fig. 11.9) (Yoshimura et al. 1994).

First, it was noted that molecular weight affects UV spectra. For instance, comparison of orsellinic acid with lecanoric and gyrophoric acids, containing two and three orsellinic acid units, respectively, showed quite a similar spectrum, but both lecanoric and gyrophoric acids

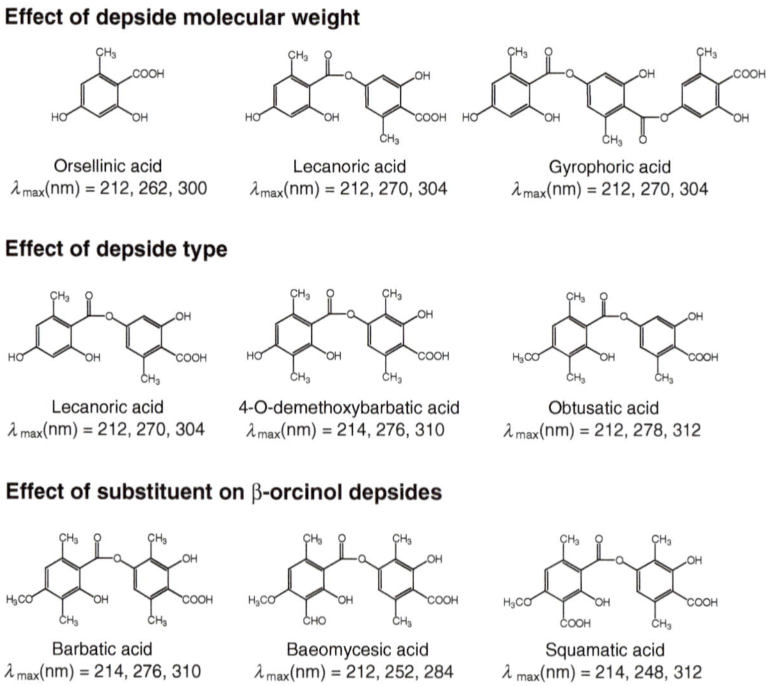

Fig. 11.9 Some relationships between chemical structure and UV features of depsides [adapted from Yoshimura et al. (1994)]

displayed bathochromic shifts of long wavelength bands and lower absorbance. Moreover, monocyclic compounds (except 2-*O*-methyl derivatives) have broader UV second bands ($\lambda_{max} \sim 260$ nm).

Spectral differences were also linked to the depside type. Indeed, depsides are classically divided into three groups: orcinol type (e.g., lecanoric acid), β-orcinol type (having C_1 unit at 3 position of benzene ring, e.g., 4-*O*-demethylbarbatic), and mixed β-orcinol/orcinol type (e.g., obtusatic acid). Attachment of the methyl moiety has an influence on UV spectra as the second band of 4-*O*-demethylbarbatic acid and obtusatic acid is bathochromically shifted (+10 and +20 nm in comparison with lecanoric acid, respectively).

Influence of substitutions of β-orcinol depsides has also been studied. β-orcinol depsides may have CH_3, CH_2OH, CHO, or a COOH substituent at the 3 position of a benzene ring. Using barbatic acid (CH_3), baeomycesis acid (CHO), and squamatic acid (COOH), it was demonstrated that changing from CH_3 to CHO or COOH allows conjugation with the other ring and thus complicates the corresponding UV spectrum.

Hydroxyl group at the 2 position of depsides and monocyclic compounds is seldom methylated. Such 2-*O*-methylated compounds have shorter wavelength and lower intensity for their second band in comparison with their 4-*O*-methylated derivatives.

Methylation and alkyl chain length modifications of depsides do not affect their UV spectrum.

11.3.5 Stereochemistry Determination

Interests in determining the absolute stereochemistry of a chiral organic molecule originate from the widely admitted fact that stereochemistry determines important features regarding both chemical, biological, and physical properties.

X-ray crystallography remains the primary method to determine the absolute configuration of a molecule and originates from crystallographic studies that grant unambiguous, accurate, and reliable three-dimensional structural parameters (Deschamps 2010).

In a number of aromatic lichen substances, fully unsaturated structures do not justify the measurement of molecular optical activity. However, numerous lichen substances are optically active. Most associated optical rotations could be found in reference books (Huneck and Yoshimura 1996) and generally expressed through $[\alpha]^{DT}$ values according to the Biot formula. Nevertheless, accurate measurements require full solubilization and solutions in g/100 mL which are sometimes difficult to achieve.

Alternatively, optical rotatory dispersion (ORD) corresponding to the variation of circular birefringence as a function of wavelength (Brittain 1998) has been determined for lichen products, including different classes of compounds: macrolactone glycosides (Rezanka and Guschina 2001a), aliphatic acid derivatives (Rezanka and Guschina 2001b), aliphatic lactones (Aberhart et al. 1970; David et al. 1990), triterpenoids (Huneck 1976), quinones (Ejiri et al. 1975), and carotenoids (Czygan 1976).

Circular dichroism is another option when experimental circular dichroism of the molecule can be compared to either (i) a set of experimental spectra from related compounds, (ii) data from literature or (iii) calculated theoretical curves (Berova et al. 2007). Circular dichroism can also grant information regarding conformational features and was applied to a wide array of lichen compounds including aliphatic acids (Bodo and Molho 1980), cycloaliphatic compounds (Huneck and Himmelreich 1995; Polborn et al. 1995; Aberhart et al. 1970; Huneck and Höfle 1980; Huneck et al. 1986; Huneck and Takeda 1992), quinones (Mathey et al. 1980), chromones (Huneck 1972), carotenoids (Czeczuga et al. 1988), and dibenzofuran-related compounds (Millot et al. 2013).

Assignment of absolute configuration by NMR represents another approach. Most of such NMR techniques rely on previous derivatization of the substrate prior to its analysis into two different species. This is performed with enantiomeric substances known as chiral derivatizing agents (CDA). Most used CDA is Mosher reagent (also known as methoxy trifluoromethyl phenyl acetic acid—MTPA) (Seco et al. 2004).

Some reports stand with application of Mosher derivatization to elucidate stereochemistry of lichen metabolites (Rezanka and Dembitsky 1999a; Rezanka and Guschina 2001a, b).

11.4 Dereplicative Tools

Over the last three decades, considerable attention has been paid to the development of hyphenated techniques, which combine different separation techniques (TLC, LC, GC) with structure-elucidating methods (mass spectrometry, NMR, UV/Vis spectroscopy).

11.4.1 Thin-layer Chromatography (TLC)

Study of lichen metabolites moved one step forward with the advent of commercially available TLC at the end of the 60s leading to a marked decrease of Asahina's crystal tests.

In 1952, Wachtmeister first introduced paper chromatography for the separation and characterization of lichen substances (Wachtmeister 1952, 1956). A few relationship between chemical structure and chromatographic behavior has been studied (Mitsuno 1953).

Ramaut (1963a, b) applied TLC in the field of lichenology to separate depsides and depsidones. Then, because different authors used different solvent systems and chromatographic conditions (Bendz et al. 1967; Santesson 1967a), procedured methods for routine identification of lichen products by TLC were first attempted by Santesson who reported R_f values of 80 lichen metabolites (Santesson 1967b). Culberson and Kristinsson (1970) first described a TLC-based method to cope with R_f variations ensuring higher reproducibility and associating other identification methods to curtain identification possibilities. Indeed, to overcome R_f varying problems, internal standards atranorin and norstictic acid were chromatographed alongside samples of interest, and 8 R_f classes were defined according to the relative positions of each spot compared with atranorin and norstictic acid. TLC was developed in three solvent systems ("A": benzene/dioxane/acetic acid 90/25/4; "B": hexane/ethyl ether/formic acid 5/4/1; "C": toluene/acetic acid 85/15). Further identification possibilities among compounds that share same R_f classes for both 3 solvents were further narrowed according to (i) reaction after spraying H_2SO_4 and heating (ii) color reactions (as previously described), (iii) appearance of the spot under visible light, short- and long-wave UV light, and (iv) any other pertinent data available as solubility in different solvents. One hundred and four compounds were listed into R_f classes according to this approach (Culberson and Kristinsson 1970) Table 11.2. Due to diethyl ether fast evaporation, it was later replaced in solvent system B by a higher boiling ether with nearly identical chromatographic properties: Methyl tertiary butyl ether (MTBE) and corresponding changes in R_f of lichen metabolites were also published (Culberson and Johnson 1982). Another solvent system (referred to as solvent G: toluene/ethyl acetate/formic acid 139/83/8) was later introduced to separate polar compounds such as β-orcinol depsidones (Culberson et al. 1981).

Further enhancements of this previous method were later published. In particular, it was reported that acidic hydrolysis (with concentrated sulfuric acid) of a number of compounds difficult to distinguish under their intact form might yield fragments that chromatograph very well in standard solvent systems and thus enable identification of the unhydrolyzed fragments (Culberson 1972). Another simple derivatization step

Table 11.2 Various reagents used to detect the lichen substance

Reagent	After treatment	Identified compounds	References
UV light (254, 365 nm)	None	Most lichen substances	
		Specific set of wavelengthes can enable detection of compounds with characteristic UV profiles, e.g., mycosporines 310 nm	
Iodine vapor	Place the chromatogram into a chamber in which some crystals of iodine have been placed. Iodine vapor is more quickly generated by gently warming the chamber	General detection reagent	Brante (1949)
		All lichen compounds, brown spots	Munier and Macheboeuf (1949)
Sulfuric acid	Spray the chromatogram with one of these reagents, Allow to dry for 15 min in the air and heat to 110 °C until maximal visualization of the spots	All lichen compounds, gray, brown or blue spots	Jones et al. (1963)
A: Mix equal parts of 95 % sulfuric acid and methanol with cooling			Metz (1961)
B: 5 % ethanolic solution of 95 % sulfuric acid			
C: 5 % solution of 95 % sulfuric acid in acetic anhydride			
D: Mix equal parts of 95 % sulfuric acid and glacial acetic acid			
Chlorosulfonic acid dissolve 5 ml chlorosulfonic acid in 10 ml glacial acetic acid with cooling	Heat 5–10 min at 130 °C	All lichen compounds, wide array of colors, terpenoids violet	Tschesche et al. (1961)
			Tschesche and Wulff (1961)
			Takeda et al. (1963)
Sulfuric anisaldehyde	Heat to 100–105 °C until maximum visualization of the spots	Depsides, depsidones, diphenylethers, monophenolic compounds, pink, red, blue-violet	Stahl and Kaltenbach (1961)
A: anisaldehyde + 1 mL sulfuric Acid + 8.5 mL methanol or B: 0.5 ml anisaldehyde in			Lisboa and Diczfalusy (1963)
50 ml glacial acetic acid and 1 ml 97 % sulfuric acid			
5 % p-phenylene diamine		Aromatic aldehydes, yellow to red	Huneck and Yoshimura (1996)
Ninhydrin	Heat at 110 °C until maximal visualization of the spots	Aminoacids, peptides, proteins, yellow (secondary aminoacids) to purple (primary aminoacids)	Patton and Chism (1951)
			Fahmy et al. (1961)
0.2 % ninhydrin in ethanol or 0.3 mg ninhydrin in 100 mL 1-butanol and add 3 mL glacial acetic acid			Friedman (2004)

(continued)

Table 11.2 (continued)

Reagent	After treatment	Identified compounds	References
0.5 % magnesium acetate in methanol	Heat 5 min at 90 °C	Anthraquinones glycosides and related aglycones (pink to violet)	Shibata et al. (1950)
0.2 % primulin solution	Dip and wait for 30 min	Lipids, aliphatic compounds. White fluorescent spots at 366 nm	
Sulfuric thymol Dissolve 0.5 g thymol in 95 ml ethanol and add 5 ml 97 % sulfuric acid with caution	Heat 15 min at 120 °C	Sugars. Pink spots	Adachi (1965)
Diazotized benzidine + NaOH (solution A: 2.5 g benzidine + 7 mL conc.HCI in 500 ml H2O; solution B: 50 g $NaNO_2$ in 500 mL H_2O; equal volumes of A and B are mixed immediately before the application of the reagent)	The reagent is stable for a few hours Colors can appear quickly or develop long after depending on the nature of the phenol	Depsides, depsidones, monophenolic compounds, red	Sherma and Hood (1965)

described is *O*-methylation using diazomethane that can be used as another identification hint. Eventually, acidic hydrolysis of *O*-methyl derivatives can provide further evidence for structure identification and to establish the order of linkage of phenolic acid units in simple depsides. As a consequence, R_f of hydrolyzed and *O*-methyl derivatives began being compiled, while an increasing number of lichen metabolites was also referenced (Culberson 1972). A useful tool for separation of complex mixtures is two-dimensional TLC as successfully applied for the separation of β-orcinol depsides from *Parmelia loxodes* and *P. verruculifera* (Culberson and Johnson 1976). TLC separation of (+) and (−) usnic acid was enabled by reaction with brucine as performed by Bendz et al. (1967).

Later, the use of high-performance thin-layer chromatography (HPTLC) in screening lichen substances was developed. HPTLC is more sensitive, allows the running of more samples in a shorter period of time, and requires smaller amounts of solvent (Arup et al. 1993).

As similar R_f value does not mean single compound, therefore, reliability to only R_f value remains a challenging and risky task. However, standardized methods listed above enable collecting valuable clues regarding chemical structures and might curtain identification possibilities.

Image-giving format of open, stationary phase of TLC can lead to numerous hyphenations. In particular, TLC can grant valuable information dealing with bioactivity and thus guide fractionation to focus on biologically active compound. For instance, bioautography is a microbial detection method developed on a planar chromatography plate and designed to recognize compounds with antibacterial and antifungal properties (Choma and Grzelak 2011). Thus, TLC bioautography was used to assess antibiotic properties of metabolites from *Ramalina farinacea* (Tay et al. 2004), anti-*Helicobacter* activity from various species of lichens (Luo et al. 2011) as well as fungitoxic compounds from *Lobaria kurokawae* (Wang et al. 2009). Biochemical detection can also be used, and for instance, screening of radical scavenging molecules can be evaluated using TLC by spraying stable free radical DPPH, which has already been performed on several lichen extracts (Bhattarai et al. 2008).

To sum up, TLC was one of the way to perform multiple detection in situ for (i) physical detection, (ii) microchemical detection, and

(iii) effect-directed analysis (either biochemically or microbiologically). Recent innovations in TLC allow successful hyphenations with different kinds of spectroscopic tools enabling proper in situ identification. Without any doubt, most developed hyphenated techniques are those combining TLC with mass spectrometry either online (Morlock and Schwack 2010) or off-line (Cheng et al. 2011). However, associating TLC with Fourier transform infra-red (FTIR) or Raman spectroscopy remain possible approaches. Furthermore, development of zone-collecting techniques (Luftmann et al. 2007) permits selective desorption and collection of molecules from a single TLC spot and further achievement of any structure elucidation technique as NMR (Morlock and Schwack 2010) (Fig. 11.10).

11.4.2 High-performance Liquid Chromatography

HPLC has become an important and efficient tool for the separation and identification of lichen substances, especially for compounds that cannot be studied by gas chromatography (GC) either because they are not volatile enough or are too unstable at high temperatures. First attempts were undertaken for chemotaxonomy with reversed-phase columns using methanol–water–acetic acid solvent system (Culberson and Culberson 1978).

Although such isocratic methods gave excellent results, gradient elution methods were later developed that better fit with complex lichen extracts containing a wide range of hydrophobic compounds (Culberson and Elix 1989). Gradient

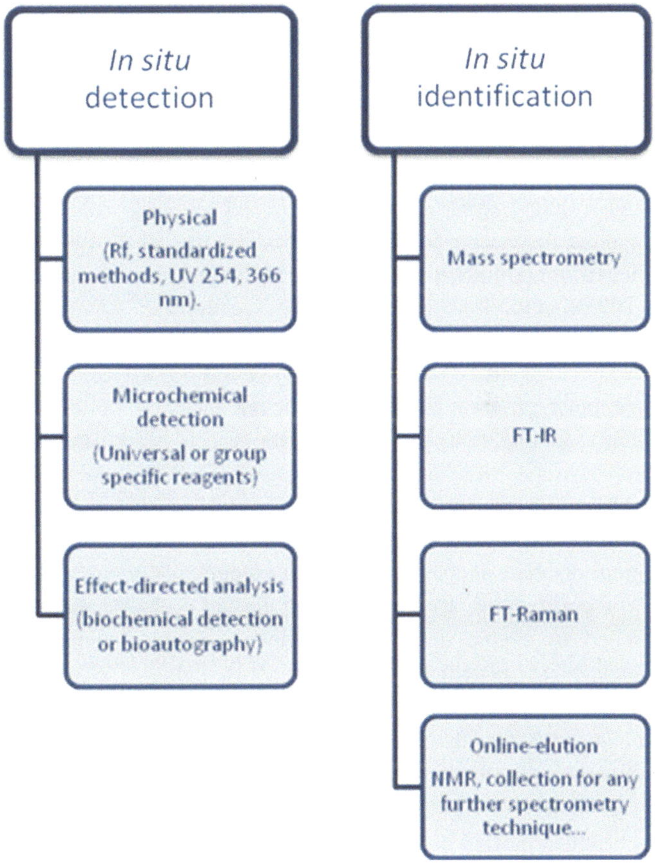

Fig. 11.10 Possible hyphenations with thin-layer chromatography [modified from Morlock and Schwack (2010)]

elution in lichenology was introduced by Strack et al. (1979) that separated 13 different lichen products (depsides, depsidones, dibenzofurans, and pulvinic acids) using a reversed-phase column with a linear gradient based on a water/methanol system. A first standardized method for separation of lichen aromatic products was described using different kinds of reversed-phase column on the basis of a methanol/orthophosphoric acid gradient system and including two internal standards: benzoic acid (low retention time) and bis-(2-ethyl-hexyl)phtalate (high retention time) (Huovinen 1987) according to which retention indices are defined ensuring better reliability. However, the latter standard was further replaced by solorinic acid (more hydrophobic) making the method suitable for the identification of lichen extracts containing chloroxanthones or long chain depsides as well. This second standardized reversed-phase HPLC method for identification of aromatic secondary metabolites from lichens was published by Feige et al. (1993) and is based on a retention index *I* calculated from the elution time of the appropriate peak compared with the standards benzoic and solorinic acids.

Such a relative index corresponding to 331 lichen compounds was determined by Feige et al. (1993), and this protocol was successfully applied to several genera of lichens (Lumbsch et al. 1993). Mietzsch et al. (1994) developed a computer program dedicated to lichen metabolites identification named Wintabolites. This software made use of HPLC-R_I values, TLC-R_f values, UV/visible colors of TLC spots, and lichen spot tests to identify molecules. The program database included 550 lichen compounds.

Derivatization techniques were further published to enable detection of aliphatic lichen acids (e.g., roccellic, rangiformic acids, lichesterinic acids) using the same method. Such metabolites are previously transformed into their phenacyl, para-nitrophenyl, or benzyl esters that display strong absorption in the UV (λ_{max} ca. 240 nm) (Huneck et al. 1994).

Later, hyphenation with photodiode array spectrophotometers (DAD), leading to complete UV–Vis spectra of corresponding peaks, greatly helped for the identification of molecules. Yoshimura et al. described detection of lichen metabolites according to their UV spectra features and relative retention time. HPLC was achieved using a reversed phase with an isocratic method ($MeOH/H_2O/H_3PO_4$). Retention times were expressed according to 2 standards: acetone and 1-phenyl-1-dodecanone.

However, relying on retention times only (even normalized with internal standard) and UV spectrum is not always sufficient to unambiguously identify a given compound. Nowadays, most liquid chromatography strategies rely on LC-DAD-ESI-MS that enable both (i) collection of UV–Vis spectrum and (ii) determination of the molecular weight associated with each peak of the chromatogram.

Hyphenated techniques are now commonly used as powerful tools of dereplication to focus on original or biologically relevant metabolites.

As an example, screening of mycosporines within different lichens was first based on their characteristic UV symmetrical shape while parallel ESI-MS helped focusing on undescribed molecules according to their molecular weight. Therefore, purification of a new molecule mycosporine hydroxyglutamicol from *Nephroma laevigatum* was achieved from 8 g of dried lichen material (Roullier et al. 2011). However, since the last standardized method described by Feige et al. and Yoshimura et al. in the nineties, no further publication updated extensively the behavior of lichen metabolites with hyphenated techniques such as LC-DAD-ESI/MS. Such a standardized method of reversed-phase LC-DAD-ESI/MS is described for the conspicuous reign of microfungi in which are described retention time, UV/Vis spectrum, and monoisotopic mass for 474 fungal metabolites (Nielsen and Smedsgaard 2003). Noteworthy, this database includes some fungal metabolites that can also be met within lichens (e.g., emodin, lichexanthone, skyrin …).

As previously discussed, NMR is arguably the most versatile analytical platform for complex mixture analysis (Corcoran and Spraul 2003). Indeed, successful interfacing of high-pressure liquid chromatography with parallel NMR and mass spectrometry were reported, yielding extensive structural data on each chromatographed compound (Exarchou et al. 2005). Wolfender et al. (2001) stated that coupling of HPLC with NMR is the one of the most powerful methods for the separation and structural elucidation of unknown compounds in mixture. However, to the best of our knowledge, no LC-NMR analysis was reported yet in the field of lichenology.

11.4.3 Gas Chromatography

GC found some limited applications in lichenology, since most lichen substances are polar and present low volatility. To cope with this problem, prior derivatization is usually required in order to yield volatile enough compounds to be analyzed by GC, most of the time as their trimethylsilyl, acetyl, methyl, or trifluoroacetyl derivatives. Derivatization procedures have been described by Nishikawa et al. (1973).

Using this method, Ikekawa et al. (1965) separated trimethylsilyltriterpene derivatives belonging to ursane, oleanane, and lupane groups (including compounds found within lichens as friedelin, taraxerol, taraxerone, and zeorin among others). Different C_{27}, C_{28}, and C_{29} sterols from the moss tree lichen *Pseudevernia furfuracea* were also characterized and quantified by GC-MS (Wojciechowski et al. 1973) after acylation from low polar organic solvent extractions. A few studies described GC analyses carried out on other structural families among which Furuya et al. (1966) demonstrated that retention time increases with the number of hydroxyl on the anthraquinone skeleton.

Low molecular weight carbohydrates from eight species of lichens were analyzed and quantified by GC as their acetyl, trifluoroacetyl, and trimethylsilyl derivatives in comparison with standard specimens. Arabitol, fructose, glucose, glycerol, erythritol, mannitol, sucrose, and trehalose were commonly found within tested samples alongside more specific sugars (as umbilicin, among others) (Nishikawa et al. 1973). Later publications reported on GC-MS characterization of *Ramalina celastri*'s mono and polysaccharides as their acetate derivatives (Stuelp et al. 1999). In the same way, acetylated saccharides from *Sticta* sp. were analyzed by GC-MS (Corradi da Silva et al. 1993) as well as carbohydrates from *H. physodes* and *Platismatia glauca* (Dahlman et al. 2003).

Fatty acid compositions of different *Parmelia* species were examined by capillary GC-MS. This enabled detection of 68 fatty acids as their methyl ester, trimethylsilyl, or oxazoline ethers (Dembitsky et al. 1992c). Fatty acid and phospholipid compositions were also reported within lichens growing in the Volga river basin using GC-MS (Dembitsky et al. 1992a), in the Tian Shan Mountains (Rezanka and Dembitsky 1999b), with different species of *Cladonia* (Dembitsky et al. 1991) or different lichens belonging to Lecanorales (Dembitsky et al. 1992b). More recently, fatty acid components—either free or combined—of the tropical lichen *Teloschistes flavicans* were analyzed by GC-MS of derived methyl esters and compared to those of separately cultivated symbionts. It was evidenced that the fatty acid compositions of the whole lichen strikingly differ from those of isolated symbionts. Moreover, changes were also observed according to the temperature. Likewise, capillary GC-MS demonstrated that seasonal changes also occurred in the fatty acid composition of two species of *Physcia* as well as in *Xanthoria parietina* (Bychuk 1993; Dembitsky et al. 1994).

In some cases, GC-MS analyses were performed for biomonitoring purposes as lichens are known to accumulate some given pollutants in their tissues (especially epiphytic lichens). Therefore, GC-MS air pollution assessment techniques were applied to detect polycyclic aromatic hydrocarbons within different Pyrenean lichens (Blasco et al. 2008) or within Italian alps-dwelling lichens (Nascimbene et al. 2014).

11.4.4 In Situ Detection

As discussed earlier, having access to the distribution of secondary metabolites could help understanding both fundamental facts about the symbiosis itself and enlightening relationships existing between the lichen and its environment. In many lichens, some metabolites are only present or occur in higher concentration in reproductive compared to somatic structures, giving support for optimal defense theory in lichens (Hyvärinen et al. 2000). Accordingly, it was shown that grazing pattern of snails inversely reflects the partitioning of the secondary compounds that have a documented deterring effect, avoiding reproductive parts of the lichen due to internal defense allocation (Asplund et al. 2010). Moreover, in situ characterization of compounds has always been considered as a very interesting approach since it eliminates complications and artifacts caused by time-consuming isolation procedures that are still required when structural analysis is achieved using conventional spectroscopic methods (Mathey et al. 1987).

11.4.4.1 Histochemical Reactions

First attempts of mapping metabolites' distribution within lichens used spot tests as described above. Indeed, specific pattern of distribution of some metabolites make thalline reactions organ specific (Kirschbaum and Wirth 1997). Honegger (1986) described ultrastructural studies of various *Parmeliaceae* using enhanced scanning electron microscopy sample preparation techniques to avoid dissolution of lichen substances. SEM images showed varying amounts of crystals covering the surface of both fungal and algal cells within different parts of the lichen thallus. Moreover, histochemical reactions (K, C, KC, PD) performed with thallus fragments or cryostats allowed localizing different secondary metabolites within the different layers of the thallus. The presence of mycobiont-derived crystalline lichen products on the surface of algae raises questions. One possible reason may be that such substances might slow down the metabolism of the photobiont to keep an harmonious growth pattern with a mycobiont whose growth abilities are limited (Kinraide and Ahmadjian 1970; Honegger 1986).

11.4.4.2 Laser Microprobe Mass Spectrometry

Along with thalline reactions, the first general method used to in situ analyze lichen substances on a microscale was laser microprobe mass spectrometry (LMMS also known as LAMMA) (Mathey 1981). Physical principle of this mass spectrometry method relies upon the irradiation of inorganic or organic samples with pulsed power UV laser under vacuum. Resulting ions are further analyzed using a time of flight or Fourier transform mass spectrometer (Van Roy et al. 1996). Microscale detection is enabled thanks to a concomitant light microscopic observation of thin cryosection that allows to focus laser beam on any sample detail (Wink et al. 1984). UV irradiation enables selective desorption and ionization of products with a chromophoric group. Indeed, a severe limitation of LMMS is its requirement for UV absorbing moieties.

As most lichen compounds have conjugated rings, this analysis method was the first showing ability to correlate analysis to given structures or substructures of the microscopical image. A large variety of lichen metabolites (especially lichen pigments) sublime when submitted to very low pressure, such as the vacuum existing in LAMMA-MS, especially lichen pigments (Mathey 1981). In such conditions, a mild elevation of temperature leads to vaporization of lichen metabolites that further yield mass spectra displaying very prominent parent ions, often corresponding to the base peak of the spectrum.

The tropical lichen *Laurera benguelensis* is characterized by yellow thallus warts that present a marked yellow fluorescence under UV radiation, originating from microcrystals located in the cortex (Mathey 1979). LAMMA spectrum obtained from single microcrystals displayed a characteristic lichexanthone profile while physcion and xanthorin, orange anthraquinone pigments, were distinctly found in the warts containing the fruiting bodies in additional *Laurera* species (Mathey et al. 1980, 1987, 1994).

LMMS was also successful in detection of a variety of lichen pigments belonging to different structural families (Mathey et al. 1987; Van Roy et al. 1996). However, LAMMA analysis only grants low mass resolution. Hence, unambiguous detection of compounds requires additional data from either NMR or HR mass spectroscopy.

11.4.4.3 Raman and FTIR Spectroscopy

First developed Raman analyses mostly dealt with biodeterioration of monuments and frescoes due to oxalic acid produced by encrusting microlichens, aiming at characterizing the physical and chemical nature of the lichen/substratum system (Edwards et al. 1991). Further works then moved to ecophysiological studies of lichen itself.

In situ Fourier transform Raman spectroscopy was used for identification of pigments and biodegradative calcium oxalate directly within thalli of viable epilithic lichens (*Acarospora* spp.) harvested in contrasted sites from Antarctic and Mediterranean (Holder et al. 2000). Raman-specific fingerprints could be used to recognize two photoprotective pigments: rhizocarpic acid and β-carotene. Near IR excitation (1,064 nm) eliminated possible fluorescence originating from pigments and insensitivity of Raman techniques to water enabled analysis of field fresh samples with a spatial resolution of about a 20 μm spot diameter. Hydration states of calcium oxalate could be differentiated with different signals between monohydrate (whewelite) and dihydrate (weddelite). Edwards et al. (1998) compared parietin contents of *Xanthoria* lichens harvested in Antarctic and temperate habitats using Raman spectroscopy. Other publications reported on scytonemin detection by Raman spectroscopy from cyanobacteria (*Nostoc* sp., *Chlorogloeopsis* sp., *Scytonema* sp., *Lyngbya* sp.) or from cyanolichen *Collema* species (Wynn-Williams et al. 1999).

Raman spectra of lichen samples colonizing active volcanic environment (Kilauea volcano, Kona, Hawaii) were acquired to decipher biomolecular protective strategies adopted in such extreme environments (Jorge-Villar and Edwards 2010). Raman profiles were acquired from 16 lichen specimens belonging to eight genera, growing on basaltic lava or on wood substrates. It was shown that chlorophyll and a carotenoid (most likely lutein or astaxanthin) were ubiquitous, but a wide range of protective pigments could be identified as atranorin, gyrophoric acid, parietin, pulvinic acid dilactone, and usnic acid (as well as calcium oxalate mono- and dihydrate in some cases) using specific patterns (Edwards et al. 2003a, b).

Lecanoric acid was characterized by FT-Raman spectroscopy in *Parmotrema tinctorium*, highlighting reliability of Raman spectroscopy for in situ detection of this metabolite from the whole lichen (de Oliveira et al. 2009).

FT-IR spectromicroscopy fits well to the examination of the conspicuous reign of filamentous fungi providing spatially resolved information on cellular biochemical content (Szeghalmi et al. 2007; Jilkine et al. 2008; Isenor et al. 2010).

More recently, Raman spectroscopy and FPA-FTIR were used to image the distribution of usnic acid within *Cladonia arbsucula*, *C. sulphurina,* and *C. uncialis* (Liao et al. 2010). This dibenzofuran-related metabolite was shown to be present in pycnidia and younger branches of *C. arbuscula* and *C. uncialis*, in *C. sulphurina*'s soredia as well as in the spore-forming region of *C. uncialis*' apothecia.

11.4.4.4 Ambient Mass Spectrometry

Recent technical innovations in mass spectrometry now allow detection of molecules in ambient conditions dealing with a large number of compounds in complex matrices and play an important role in various fields such as drug discovery, doping control, forensic identification, and food safety. Open-air sources especially revolutionized the way samples are introduced into the ion source for mass analysis and result in rapid acquisition, without special sample preparation enabling quick and sensitive dereplication. Among these open-air sources is especially worth quoting direct analysis in real-time mass spectrometry (DART-MS) that is based on the interactions of a sample in a water-containing atmosphere with a gas in a metastable state heated to a temperature up to 350 °C (Cody et al. 2005). DART's detection limit is comprised in the picomolar range and showed satisfying results for chemical fingerprinting of many plants (Kim and Jang 2009). Moreover, a number of lichenic compounds were shown to accumulate in upper cortex (Ozenda and Clauzade 1970). Such a distribution in the most superficial layers enables their detection using ambient mass spectrometry techniques as shown on the cyano-lichen *Lichina pygmaea* using DART-MS (Le pogam et al. 2014).

Ambient mass spectrometry methods have to be tried with lichens to measure defense reactions as responses to environmental challenges as successfully applied to the red algae *Callophycus serratus* using DESI *imaging* (Lane et al. 2009). Other DESI imaging experiments reported on hydrolysis of hydroxynitrile glucosides along leaves wounds of *Lotus japonicus* (Li et al. 2013).

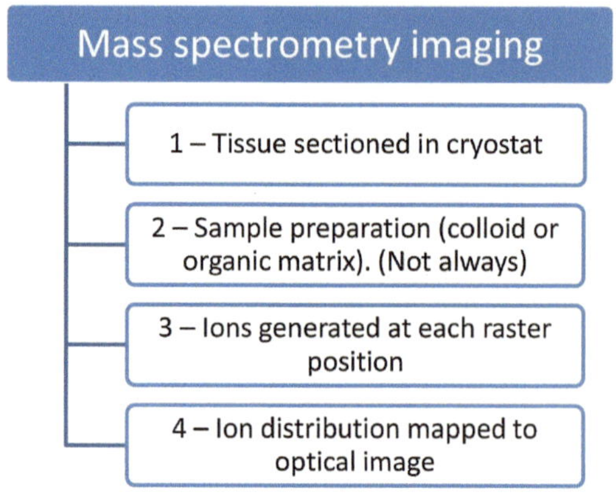

Fig. 11.11 Schematic outline of the workflow in a typical imaging mass spectrometry experiment [adapted from Greer et al. (2011)]

11.4.4.5 Imaging Mass Spectrometry

Imaging mass spectrometry allows to monitor the spatial distribution and abundance of metabolites. Experimental scheme for mass spectrometry imaging experiments can be defined as follows (Fig. 11.11). IMS involves rastering a laser or any other ionization source across a thin section of biological tissue by moving the sample stage in predefined *x–y* coordinates to generate thousands of position-dependent mass spectra. Sample preparation may be assisted to aid the ion ionization or can remain unassisted depending on both the mechanism of the ion source and the nature of the compounds to be analyzed. Then, data processing is accomplished by displaying ions of interest from a traditional *m/z* plot to a false color image eventually superimposed on a photograph of the analyzed sample (Greer et al. 2011).

Recently, three-dimensional imaging of metabolites in plant tissues was achieved thanks to LAESI-MS. LAESI-MS relies on tissue ablation with a single laser pulse of 2.94 μm wavelength. O–H vibrations of the native water molecules of the tissue samples absorb the laser pulse energy, leading to ejection of microscopic volumes of the most superficial layers of the sample (as neutral particulates and molecules) that enables establishing depth profiles. This plume is further intercepted by an electrospray that postionizes the ablated material (Nemes and Vertes 2012). 3D-LAESI-MS revealed specific distribution pattern of several molecules within the plant *Aphelandra* in correlation with tissue specificity (Nemes et al. 2009). Common ionization methods applied in imaging mass spectrometry are listed above (Table 11.3).

Although achieved on a wide range of organisms [for review see Esquenazi et al. (2009)], no ambient or imaging mass spectrometry was published until now on lichens to the best of our knowledge.

11.5 Deciphering Metabolic Networks and Partners' Cooperation at the Very Heart of Symbiotic Systems

Systematic analysis of the metabolite composition caused by cellular reactions is called metabolomics (Oliver et al. 1998). A metabolomic approach was developed on *Xanthoria elegans* in 2007 (Aubert et al. 2007). In this study, analysis

Table 11.3 Characteristics of most commonly used imaging mass spectrometry sources

Ionization source	Probe beam	Pressure regime	Spatial resolution (μm)
MALDI	Laser beam	Vacuum (most of the time) or ambient (AP-MALDI)	10–100
SIMS	Ion beam	Vacuum	0, 2–3
DESI	Solvent stream	Ambient	200
NIMS	Laser beam	Vacuum	15–20
LAESI	Infrared laser + electrospray	Vacuum	Lateral: 300–500 Depth: 30–40

Adapted from Lee et al. (2012), Greving et al. (2011)

of polar metabolites was achieved from crude perchloric acid extracts using ^{13}C and ^{31}P NMR spectroscopy to compare chemical profiles of both dry and wet thalli. Such analysis methods afforded the detection of 30 metabolites from the raw extract with no further sample workup (mostly sugars, sugar phosphates, polyols, phosphoenolpyruvate, and nucleotides). The data implied that metabolite composition was affected by stress conditions. Aubert also described increased concentrations of ribitol and mannitol under dry conditions. Sugars are known protectors of cellular substructures when lichens have to survive under dryness (Crowe et al. 1992).

As previously discussed, lichens are able to synthesize unique secondary metabolites. However, the relative contribution of its partners in providing such original compounds remains quite unstudied. This question is especially interesting when both partners possess the enzymatic machinery required to biosynthesize one metabolite [e.g., mycosporine and cyanolichens (Balskus and Walsh 2010)]. Moreover, knowledge about biosynthesis pathway is also a prerequisite for natural compounds production by biotechnological means.

Most current methods used to decipher metabolic pathways are based on stable isotope labeling (Eisenreich and Bacher 2007). Typically, organisms are grown with labeled structures that are most of the time very general precursors (e.g., glucose, acetyl CoA, CO_2, NH_4, aminoacids) that will spread in all their possible metabolic fates. Subsequent analysis of enriched biomolecules by either quantitative NMR spectroscopy or mass spectrometry has shown to be a powerful method for unraveling the dynamics of metabolic networks (Fan et al. 2012). Indeed, according to the detected profiles, metabolic history of the studied metabolite can be traced and then compiled to a comprehensive map about metabolic pathways and fluxes occurring in the network.

Although initially focused on single organisms such as plants or bacteria [for review see Eisenreich and Bacher (2007) and Heinemann and Sauer (2010)], techniques of stable isotope labeling were later applied to multi-organismic systems (Götz et al. 2010).

To our knowledge, only a few labeling studies for biosynthetic-unraveling purposes were reported on lichens. Lines et al. (1989) investigated polyol metabolism in *Xanthoria calcicola* using a ^{13}C NMR spectroscopy approach. More recently, a proof on concept was published in 2011 on the chlorolichen *X. parietina* (Eisenreich et al. 2011). Photobiont cells (*Trebouxia decolorans*) were first harvested alone in a medium containing ^{13}C-glucose, which enabled satisfying labeling of most if not all metabolites according to the authors. In a second step, algae were combined with unmarked mycobiont to reconstitute lichen symbiosis. Therefore, detection of ^{13}C-labeled fungal metabolites is indicative for prior transfer and usage of algal biomass for biosynthetic purposes by the mycobiont. ^{13}C NMR of crude methanol extracts of marked photobionts alone and reconstituted symbiosis by cocultivation with unlabeled fungus bear striking differences that demonstrate substrate transfer and further metabolism by the mycobiont. Closer investigation of those signal patterns shall provide information regarding (i) the nature of transferred compounds, (ii) the rate of this transfer and eventually, (iii) subsequent transformations occurring within the second partner (Eisenreich et al. 2011).

Such technologies took advantage of tremendous advances in mass spectrometry that now enable microscale mapping of metabolites directly within producing organisms (Esquenazi et al. 2009). Coupling of stable isotope profiling with imaging mass spectrometry enables both tracing and mapping of molecules within its biological system. This approach is especially promising for interfaces or symbiotic systems in order to know the compartment in which a molecule undergoes biotransformation (Jones et al. 2013; Pernice et al. 2012).

11.6 Conclusion

The evaluation of lichen metabolites for therapeutic purposes is generating an increasing interest as their activities are varied and often significant (Boustie et al. 2011). Very few lichen compounds are commercially available, and the thousand known metabolites are only partly or not at all investigated for their medicinal potential (Stocker-Wörgötter 2008). Moreover, lichens appear to harbor a diversity of culturable microorganisms from which original active compounds were recently isolated (Grube et al. 2012).

As lichens grow slowly, both in nature and axenic conditions, analysis must deal with low amounts of crude material. The past few years have witnessed major developments in purification, dereplication, and structure elucidation of lichen compounds, enabling much faster access to sufficient amounts of pure compounds.

Although not described here, innovative strategies were also developed for extracting and separating compounds which facilitate detection and identification of lichen compounds. New extraction methods developed on lichens include supercritic fluid extraction (SFE) involving the so-called supercritic fluids (SCF) (Lisickov et al. 2002; Zizovic et al. 2012) that appears to be a powerful strategy. Although a specific device is required, the main advantages of these methods are that (i) they display solvating powers similar to liquid organic solvents with higher diffusivity and lower viscosity, (ii) they can provide quantitative or complete extraction (by forcing continuously the fluid to flow through the sample), (iii) the solvating power of the fluid can be adjusted by changing the pressure and/or the temperature, (iv) adding modifiers to a SCF can change its polarity, (v) they avoid solvent residuals and last but not least, (vi) they are chemically inert which avoid chemical artifacts that can be dread when using classical separation procedures (Lang and Wai 2001).

A light fluorous catch and release approach was another option used to separate three closely related paraconic acids from *Cetraria islandica* allowing the isolation and identification of roccellaric acid from this lichen for the first time (Horhant et al. 2007).

The recent innovations in hyphenated techniques combining separation (HPTLC, HPLC, GC) and detection tools (currently MS, UV, IR) have already had a substantial impact in shortening the timeline for dereplication, isolation, and structure elucidation of the lichen products present in crude extract. Once hampered by methods limitations, lichen study might benefit from tremendous advances in analytical chemistry (MAS-NMR, DART-MS, LC-NMR...) described herein. Consequently, sensitive dereplication of lichen chemodiversity should facilitate focusing on minor, and possibly unknown, compounds. Standardized methods to identify recurrent products and to remove them should be developed in this way. As illustrated, various methods can be performed on solid samples then bypassing chemical artifacts associated with purification processes, giving in some cases direct field information. Furthermore, improvements in imaging techniques pave the way for future investigations into metabolic interactions taking place in lichen miniature ecosystem and might help deciphering ecological relevance of lichens metabolites.

Coupling these approaches with advances in purification, structure elucidation, and streamlining of the screening processes, the timeline for lichen products drug discovery is shortened similar to that expected for synthetic compounds, now being compatible with the present regime of blitz screening campaigns.

Acknowledgments Jacques Lagabrielle is greatly acknowledged for his support in providing crystal pictures.

References

Aberhart DJ, Overton KH, Huneck S (1970) Portentol: an unusual polypropionate from the lichen *Roccella portentosa*. J Chem Soc C 11:1612–1623

Adachi S (1965) Thin-layer chromatography of carbohydrates in the presence of bisulfite. J Chromatogr A 17:295–299

Addison JB (1985) Application of pyrolysis-chemical ionisation mass spectrometry to lichenology. Analyst 110:933–935

Alcantara GB, Honda NK, Ferreira MMC, Ferreira AG (2007) Chemometric analysis applied in 1H HR-MAS NMR and FT-IR data for chemotaxonomic distinction of intact lichen samples. Anal Chim Acta 595:3–8

Arup U, Ekman S, Lindblom L, Mattsson J-E (1993) High performance thin layer chromatography (HPTLC), an improved technique for screening lichen substances. Lichenologist 25:61–71

Asplund J, Gauslaa Y (2008) Mollusc grazing limits growth and early development of the old forest lichen *Lobaria pulmonaria* in broadleaved deciduous forests. Oecologia 155:93–99

Asplund J, Solhaug KA, Gauslaa Y (2010) Optimal defense: snails avoid reproductive parts of the lichen *Lobaria scrobiculata* due to internal defense allocation. Ecology 91:3100–3105

Aubert S, Juge C, Boisson A-M, Gout E, Bligny R (2007) Metabolic processes sustaining the reviviscence of lichen *Xanthoria elegans* (link) in high mountain environments. Planta 226:1287–1297

Balskus EP, Walsh CT (2010) The genetic and molecular basis for sunscreen biosynthesis in cyanobacteria. Science 329:1653–1656

Bendz G, Bohman G, Santesson J (1967) Studies en the chemistry of lichens. 5. Separation and identification of the antipodes of usnic acid by thin layer chromatography. Acta Chem Scand 21:1376–1377

Berova N, Di Bari L, Pescitelli G (2007) Application of electronic circular dichroism in configurational and conformational analysis of organic compounds. Chem Soc Rev 36:914–931

Bhattarai HD, Paudel B, Hong SG, Lee HK, Yim JH (2008) Thin layer chromatography analysis of antioxidant constituents of lichens from Antarctica. J Nat Med 62:481–484

Bialonska D, Dayan FE (2005) Chemistry of the lichen *Hypogymnia physodes* transplanted to an industrial region. J Chem Ecol 31:2975–2991

Blasco M, Domeño C, Nerín C (2008) Lichens biomonitoring as feasible methodology to assess air pollution in natural ecosystems: combined study of quantitative PAHs analyses and lichen biodiversity in the Pyrenees Mountains. Anal Bioanal Chem 391:759–771

Bodo B, Molho D (1980) Structure des acides isomuronique et neuropogolique, nouveaux acides aliphatiques du lichen *Neuropogon trachycarpus*. Phytochemistry 19:1117–1120

Boustie J, Grube M (2005) Lichens—a promising source of bioactive secondary metabolites. Plant Genet Resour Charact Util 3:273–287

Boustie J, Tomasi S, Grube M (2011) Bioactive lichen metabolites: alpine habitats as an untapped source. Phytochem Rev 10:287–307

Brante G (1949) Iodine as a means of development in paper chromatography. Nature 163, 651–651

Brittain HG (1998) Applications of chiroptical spectroscopy for the characterization of pharmaceutical compounds. J Pharm Biomed Anal 17:933–940

Brodo IM, Sharnoff SD, Sharnoff S (2001) Lichens of North America. Yale University Press, New Haven

Bychuk IA (1993) Change in the lipid composition of lichens in the spring period. Chem Nat Compd 29:582–585

Cetin H, Tufan-Cetin O, Turk AO, Tay T, Candan M, Yanikoglu A, Sumbul H (2008) Insecticidal activity of major lichen compounds,(−)- and (+)-usnic acid, against the larvae of house mosquito, *Culex pipiens* L. Parasitol Res 102:1277–1279

Cheng S-C, Huang M-Z, Shiea J (2011) Thin layer chromatography/mass spectrometry. J Chromatogr A 1218:2700–2711

Cheng B, Cao S, Vasquez V, Annamalai T, Tamayo-Castillo G, Clardy J, Tse-Dinh Y-C (2013) Identification of anziaic acid, a lichen depside from *Hypotrachyna* sp., as a new topoisomerase poison inhibitor. PLoS ONE 8:e60770

Chollet-Krugler M, Le-Floch M, Articus K, Millot M, Boustie J (2008) Carbon-13 CP-MAS NMR studies of some lichens of the Genus *Cladonia* section Cladina. In: Planta Medica. Georg Thieme Verlag Stuttgart, Germany, p 1118

Choma IM, Grzelak EM (2011) Bioautography detection in thin-layer chromatography. J Chromatogr A 1218:2684–2691

Cody RB, Laramée JA, Durst HD (2005) Versatile new ion source for the analysis of materials in open air under ambient conditions. Anal Chem 77:2297–2302

Corcoran O, Spraul M (2003) LC–NMR–MS in drug discovery. Drug Discov Today 8:624–631

Corradi da Silva ML, Iacomini M, Jablonski E, Gorin PA (1993) Carbohydrate, glycopeptide and protein components of the lichen *Sticta* sp. and effect of storage. Phytochemistry 33:547–552

Crowe JH, Hoekstra FA, Crowe LM (1992) Anhydrobiosis. Annu Rev Physiol 54:579–599

Culberson CF (1972) Improved conditions and new data for identification of lichen products by standardized thin-layer chromatographic method. J Chromatogr A 72:113–125

Culberson CF, Culberson WL (1978) β-Orcinol derivatives in lichens: biogenetic evidence from *Oropogon loxensis*. Exp Mycol 2:245–257

Culberson CF, Elix JA (1989) Lichen substances. Methods Plant Biochem 1:509–535

Culberson CF, Johnson A (1976) A standardized two-dimensional thin-layer chromatographic method for lichen products. J Chromatogr A 128:253–259

Culberson CF, Johnson A (1982) Substitution of methyltertbutyl ether for diethyl ether in the standardized thin-layer chromatographic method for lichen products. J Chromatogr A 238:483–487

Culberson CF, Kristinsson H-D (1970) A standardized method for the identification of lichen products. J Chromatogr A 46:85–93

Culberson CF, Culberson WL, Johnson A (1977) Nonrandom distribution of an epiphytic Lepraria on two species of Parmelia. Bryologist 80:201–203

Culberson CF, Culberson WL, Johnson A (1981) A standardized TLC analysis of β-orcinol depsidones. Bryologist 84:16–29

Czeczuga B, Cifuentes B, Reynaud PA (1988) Carotenoids in lichens from the Canary Islands. Biochem Syst Ecol 16:117–118

Czygan F-C (1976) Carotinoid-Garnitur and-Stoffwechsel der Flechte *Haematomma ventosum* (L.) Massal. s. str. and ihres Phycobionten. Z Fuer Pflanzenphysiol 79:438–445

Dahlman L, Persson J, Näsholm T, Palmqvist K (2003) Carbon and nitrogen distribution in the green algal lichens *Hypogymnia physodes* and *Platismatia glauca* in relation to nutrient supply. Planta 217:41–48

Dailey RN, Montgomery DL, Ingram JT, Siemion R, Raisbeck MF (2008) Experimental reproduction of tumbleweed shield lichen (*Xanthoparmelia chlorochroa*) poisoning in a domestic sheep model. J Vet Diagn Invest 20:760–765

David F, Elix JA, Binsamsudin MW (1990) Two new aliphatic acids from the lichen *Parmotrema praesorediosum*. Aust J Chem 43:1297–1300

Dayan F, Romagni J (2001) Lichens as a potential source of pesticides. Pestic Outlook 12:229–232

De Oliveira LFC, Pinto PCC, Marcelli MP, Dos Santos HF, Edwards HGM (2009) The analytical characterization of a depside in a living species: spectroscopic and theoretical analysis of lecanoric acid extracted from *Parmotrema tinctorum Del Ex Nyl.* lichen. J Mol Struct 920:128–133

Dembitsky VM, Rezanka T, Bychek IA, Shustov MV (1991) Identification of fatty acids from Cladonia lichens. Phytochemistry 30:4015–4018

Dembitsky VM, Rezanka T, Bychek IA (1992a) Lipid composition of some lichens. Phytochemistry 31:1617–1620

Dembitsky VM, Rezanka T, Bychek IA (1992b) Fatty acids and phospholipids from lichens of the order lecanorales. Phytochemistry 31:851–853

Dembitsky VM, Rezanka T, Bychek IA, Shustov MV (1992c) Fatty acid composition of Parmelia lichens. Phytochemistry 31:841–843

Dembitsky VM, Rezanka T, Bychek IA (1994) Seasonal variability of lipids and fatty acids in the tree-growing lichen *Xanthoria parientina* L. J Exp Bot 45:403–408

Deschamps JR (2010) X-ray crystallography of chemical compounds. Life Sci 86:585–589

Dmitrenok PS, El'kin YN, Stepanenko LS, Krivoshchekova OE (1987) Fast-atom ionization mass spectra of some lichen metabolites. Chem Nat Compd 23:249–250

Edwards HGM, Farwell DW, Seaward MRD (1991) Raman spectra of oxalates in lichen encrustations on Renaissance frescoes. Spectrochim Acta Part Mol Spectrosc 47:1531–1539

Edwards HGM, Holder JM, Wynn-Williams DD (1998) Comparative FT-Raman spectroscopy of Xanthoria lichen-substratum systems from temperate and antarctic habitats. Soil Biol Biochem 30:1947–1953

Edwards HG, Newton EM, Wynn-Williams DD (2003a) Molecular structural studies of lichen substances II: atranorin, gyrophoric acid, fumarprotocetraric acid, rhizocarpic acid, calycin, pulvinic dilactone and usnic acid. J Mol Struct 651:27–37

Edwards HG, Newton EM, Wynn-Williams DD, Coombes SR (2003b) Molecular spectroscopic studies of lichen substances 1: parietin and emodin. J Mol Struct 648:49–59

Eisenreich W, Bacher A (2007) Advances of high-resolution NMR techniques in the structural and metabolic analysis of plant biochemistry. Phytochemistry 68:2799–2815

Eisenreich W, Knispel N, Beck A (2011) Advanced methods for the study of the chemistry and the metabolism of lichens. Phytochem Rev 10:445–456

Ejiri H, Sankawa U, Shibata S (1975) Graciliformin and its acetates in *Cladonia graciliformis*. Phytochemistry 14:277–279

Elix JA (1996) Biochemistry and secondary metabolites. Lichen Biol 1:154–180

Elix J, Norfolk S (1975) Synthesis of β-orcinol metadepsides. Aust J Chem 28:2035–2041

Elix JA (2014) A catalogue of standardized chromatographic data and biosynthetic relationships for lichen substances, 3rd edn. Published by the author, Canberra. http://www.anbg.gov.au/abrs/lichenlist/Chem%20Cat%203.pdf

Ertl L (1951) Über die lichtverhältnisse in laubflechten. Planta 39:245–270

Esquenazi E, Yang Y-L, Watrous J, Gerwick WH, Dorrestein PC (2009) Imaging mass spectrometry of natural products. Nat Prod Rep 26:1521–1534

Ewing DF (1979) 13C substituent effects in monosubstituted benzenes. Org Magn Reson 12:499–524

Exarchou V, Krucker M, van Beek TA, Vervoort J, Gerothanassis IP, Albert K (2005) LC-NMR coupling technology: recent advancements and applications in natural products analysis. Magn Reson Chem 43:681–687

Fahmy AR, Niederwieser A, Pataki G, Brenner M (1961) Dünnschicht-Chromatographie von Aminosäuren auf Kieselgel G. 2. Mitteilung. Eine Schnellmethode zur Trennung und zum qualitativen Nachweis von 22 Aminosäuren. Helv Chim Acta 44:2022–2026

Fan TW-M, Lorkiewicz PK, Sellers K, Moseley HNB, Higashi RM, Lane AN (2012) Stable isotope-resolved metabolomics and applications for drug development. Pharmacol Ther 133:366–391

Feige GB, Lumbsch HT, Huneck S, Elix JA (1993) Identification of lichen substances by a standardized high-performance liquid chromatographic method. J Chromatogr A 646:417–427

Feuerer T, Hawksworth DL (2007) Biodiversity of lichens, including a world-wide analysis of checklist data based on Takhtajan's floristic regions. Biodivers Conserv 16:85–98

Fisher RF (1979) Possible allelopathic effects of reindeer-moss (*Cladonia*) on jack pine and white spruce. For Sci 25:256–260

Freysdottir J, Omarsdottir S, Ingólfsdóttir K, Vikingsson A, Olafsdottir ES (2008) In vitro and in vivo immunomodulating effects of traditionally prepared extract and purified compounds from *Cetraria islandica*. Int Immunopharmacol 8:423–430

Friedman M (2004) Applications of the ninhydrin reaction for analysis of amino acids, peptides, and proteins to agricultural and biomedical sciences. J Agric Food Chem 52:385–406

Furuya T, Shibata S, Iizuka H (1966) Gas–liquid chromatography of anthraquinones. J Chromatogr A 21:116–118

Götz A, Eylert E, Eisenreich W, Goebel W (2010) Carbon metabolism of enterobacterial human pathogens growing in epithelial colorectal adenocarcinoma (Caco-2) Cells. PLoS ONE 5:e10586

Greer T, Sturm R, Li L (2011) Mass spectrometry imaging for drugs and metabolites. J Proteomics 74:2617–2631

Greving MP, Patti GJ, Siuzdak G (2011) Nanostructure-initiator mass spectrometry metabolite analysis and imaging. Anal Chem 83:2–7

Grigsby RD, Jamieson WD, McInnes AG, Maass WSG, Taylor A (1974) The mass spectra of derivatives of polyporic acid. Can J Chem 52:4117–4122

Grube M, Cardinale M, Berg G (2012) 17 Bacteria and the Lichen Symbiosis. Fungal Associations, Springer, pp 363–372

Halama P, Van Haluwin C (2004) Antifungal activity of lichen extracts and lichenic acids. Biocontrol 49:95–107

Hale ME (1983) The biology of lichens. Edward Arnold, London

Hauck M, Huneck S (2007a) Lichen substances affect metal adsorption in *Hypogymnia physodes*. J Chem Ecol 33:219–223

Hauck M, Huneck S (2007b) The putative role of fumarprotocetraric acid in the manganese tolerance of the lichen *Lecanora conizaeoides*. Lichenologist 39:301–304

Hauck M, Willenbruch K, Leuschner C (2009) Lichen substances prevent lichens from nutrient deficiency. J Chem Ecol 35:71–73

Heilman AS, Sharp AJ (1963) A probable antibiotic effect of some lichens on bryophytes. Rev Bryol Lichenol 32:215

Heinemann M, Sauer U (2010) Systems biology of microbial metabolism. Curr Opin Microbiol 13:337–343

Holder JM, Wynn-Williams D, Rull Perez F, Edwards HGM (2000) Raman spectroscopy of pigments and oxalates in situ within epilithic lichens: Acarospora from the Antarctic and Mediterranean. New Phytol 145:271–280

Holland PT, Wilkins AL (1979) Mass spectra of some naturally occurring stictane triterpenoids and their trimethylsilyl derivatives. Org Mass Spectrom 14:160–166

Holzmann G, Leuckert C (1990) Applications of negative fast atom bombardment and MS/MS to screening of lichen compounds. Phytochemistry 29:2277–2283

Honegger R (1986) Ultrastructural studies in lichens. New Phytol 103:797–808

Horhant D, Lamer A-CL, Boustie J, Uriac P, Gouault N (2007) Separation of a mixture of paraconic acids from *Cetraria islandica* (L.) Ach. employing a fluorous tag—catch and release strategy. Tetrahedron Lett 48:6031–6033

Huneck S (1972) Chemie der Flechteninhaltsstoffe, XCI. Chromonglucoside aus Flechten. J Für Prakt Chem 314:488–498

Huneck S (1976) Inhaltsstoffe von Pyxine coccifera. Phytochemistry 15:799–801

Huneck S, Himmelreich U (1995) Arthogalin, a cyclic depsipeptide from the lichen arthothelium galapagoense. Z Naturforschung Sect Ba J Chem Sci 50:1101–1103

Huneck S, Höfle G (1980) Structure of acaranoic and acarenoic acids. Phytochemistry 19:2713–2715

Huneck S, Schmidt J (1980) Lichen substances—126 mass spectroscopy of natural products—10. Comparative positive and negative ion mass spectroscopy of usnic acid and related compounds. Biol Mass Spectrom 7:301–308

Huneck S, Takeda R (1992) Zur Chemie der Proto-und allo-Proto-lichesterinsäure. Z Für Naturforschung B J Chem Sci 47:842–854

Huneck S, Yoshimura I (1996) Identification of lichen substances. Springer, Berlin

Huneck S, Djerassi C, Becher D, Barber M, Von Ardenne M, Steinfelder K, Tümmler R (1968) Flechteninhaltsstoffe—XXXI: Massenspektrometrie und ihre anwendung auf strukturelle und streochemische probleme—CXXIII Massenspektrometrie von depsiden, depsidonen, depsonen, dibenzofuranen und diphenylbutadienen mit positiven und negativen ionen. Tetrahedron 24:2707–2755

Huneck S, Tønsberg T, Bohlmann F (1986) (−)-Allopertusaric acid and (−)-dihydropertusaric acid from the lichen Pertusaria albescens. Phytochemistry 25:453–459

Huneck S, Feige GB, Lumbsch HT (1994) High performance liquid chromatographic analysis of aliphatic lichen acids. Phytochem Anal 5:57–60
Huovinen K (1987) A standard HPLC method for the analysis of aromatic lichen in progress and problems in lichenology in the eighties. Bibl Lichenol 25:457–466
Hyvärinen M, Koopmann R, Hormi O, Tuomi J (2000) Phenols in reproductive and somatic structures of lichens: a case of optimal defence? Oikos 91:371–375
Ikekawa N, Natori S, Itokawa H, Tobinaga S, Matsui M (1965) Gas chromatography of triterpenes. I. Ursanane, Oleanane, and Lupane Groups. Chem Pharm Bull (Tokyo) 13:316–319
Isenor M, Kaminskyj SG, Rodriguez RJ, Redman RS, Gough KM (2010) Characterization of mannitol in *Curvularia protuberata* hyphae by FTIR and Raman spectromicroscopy. Analyst 135:3249–3254
Jilkine K, Gough KM, Julian R, Kaminskyj SG (2008) A sensitive method for examining whole-cell biochemical composition in single cells of filamentous fungi using synchrotron FTIR spectromicroscopy. J Inorg Biochem 102:540–546
Jones DF, MacMillan J, Radley M (1963) Plant hormones-III. Identification of gibberellic acid in immature barley and immature grass. Phytochemistry 2:307–314
Jones DL, Clode PL, Kilburn MR, Stockdale EA, Murphy DV (2013) Competition between plant and bacterial cells at the microscale regulates the dynamics of nitrogen acquisition in wheat (*Triticum aestivum*). New Phytol 200:796–807
Jorge-Villar SE, Edwards HGM (2010) Lichen colonization of an active volcanic environment: a Raman spectroscopic study of extremophile biomolecular protective strategies. J Raman Spectrosc 41:63–67
Karakus B, Odabasoglu F, Cakir A, Halici Z, Bayir Y, Halici M, Aslan A, Suleyman H (2009) The effects of methanol extract of Lobaria pulmonaria, a lichen species, on indometacin-induced gastric mucosal damage, oxidative stress and neutrophil infiltration. Phytother Res 23:635–639
Kim HJ, Jang YP (2009) Direct analysis of curcumin in turmeric by DART-MS. Phytochem Anal 20:372–377
Kinraide WTB, Ahmadjian V (1970) The effects of usnic acid on the physiology of two cultured species of the lichen alga Trebouxia Puym. Lichenologist 4:234–247
Kirschbaum U, Wirth V (1997) Les lichens bio-indicateurs: les reconnaître, évaluer la qualité de l'air. E. Ulmer, Paris
Kutney JP, Sanchez IH, Yee TH (1974) Mass spectral fragmentation studies in usnic acid and related compounds. Org Mass Spectrom 8:129–146
Lane AL, Nyadong L, Galhena AS, Shearer TL, Stout EP, Parry RM, Kwasnik M, Wang MD, Hay ME, Fernandez FM (2009) Desorption electrospray ionization mass spectrometry reveals surface-mediated antifungal chemical defense of a tropical seaweed. Proc Natl Acad Sci 106:7314–7319
Lang Q, Wai CM (2001) Supercritical fluid extraction in herbal and natural product studies—a practical review. Talanta 53:771–782
Lawrey JD (1977) Adaptive significance of O-methylated lichen depsides and depsidones. Lichenologist 9:137–142
Le Pogam P, Legouin B, Le Lamer A-C, Boustie J, Rondeau D (2014) Analysis of the cyanolichen Lichina pygmaea metabolites using in situ DART-MS: from detection to thermochemistry of mycosporine serinol. J Mass Spectrom, in press
Lee YJ, Perdian DC, Song Z, Yeung ES, Nikolau BJ (2012) Use of mass spectrometry for imaging metabolites in plants: mass spectrometry for imaging metabolites in plants. Plant J 70:81–95
Letcher RM, Eggers SH (1967) Chemistry of lichen constituents. Part IV. Tetrahedron Lett 8:3541–3546
Li B, Knudsen C, Hansen NK, Jørgensen K, Kannangara R, Bak S, Takos A, Rook F, Hansen SH, Møller BL, Janfelt C, Bjarnholt N (2013) Visualizing metabolite distribution and enzymatic conversion in plant tissues by desorption electrospray ionization mass spectrometry imaging. Plant J 74:1059–1071
Liao C, Piercey-Normore MD, Sorensen JL, Gough K (2010) In situ imaging of usnic acid in selected *Cladonia* spp. by vibrational spectroscopy. Analyst 135:3242–3248
Lines CEM, Ratcliffe RG, Rees TAV, Southon TE (1989) A 13C NMR study of photosynthate transport and metabolism in the lichen Xanthoria calcicola Oxner. New Phytol 111:447–456
Lisboa BP, Diczfalusy E (1963) Colour reactions for the in situ characterisation of steroid oestrogens on thin-layer chromatograms. Acta Endocrinol (Copenh) 43:545–560
Lisickov K, Najdenova V, Zoltan D (2002) Application of supercritical CO_2 extraction for separation of natural antibiotics from lichens. Herba Pol 48:32–39
Luftmann H, Aranda M, Morlock GE (2007) Automated interface for hyphenation of planar chromatography with mass spectrometry. Rapid Commun Mass Spectrom 21:3772–3776
Lumbsch HT, Kashiwadani H, Streimann H (1993) A remarkable new species in the lichen genus Placopsis from Papua New Guinea (lichenized ascomycetes, Agyriaceae). Plant Syst Evol 185:285–292
Luo H, Yamamoto Y, Jeon H-S, Liu YP, Jung JS, Koh YJ, Hur J-S (2011) Production of anti-helicobacter pylori metabolite by the lichen-forming fungus nephromopsis pallescens. J Microbiol 49:66–70
Marante FT, Castellano AG, Rosas FE, Aguiar JQ, Barrera JB (2003) Identification and quantitation of allelochemicals from the lichen *Lethariella canariensis*: phytotoxicity and antioxidative activity. J Chem Ecol 29:2049–2071
Mathey A (1979) Contribution a l'étude de la famille des Trypéthéliacées (lichens pyrénomycètes). Nova Hedwigia 31:917–935
Mathey A (1981) LAMMA: new perspectives for lichenology? Fresenius. J Anal Chem 308:249–252
Mathey A, Steffan B, Steglich W (1980) 1, 2-Naphthochinon-Derivate aus Kulturen des Mycosymbionten der Flechte Trypethelium eluteriae (Trypetheliaceae). Liebigs Ann Chem 1980:779–785

Mathey A, Van Vaeck L, Steglich W (1987) Investigation of semi-thin cryosections of lichens by laser microprobe mass spectrometry. Anal Chim Acta 195:89–96

Mathey A, Van Roy W, Van Vaeck L, Eckhardt G, Steglich W (1994) In situ analysis of a new perylene quinone in lichens by Fourier-transform laser microprobe mass spectrometry with external source. Rapid Commun Mass Spectrom 8:46–52

Metz H (1961) Dünnschichtchromatographische schnellanalyse bei enzymatischen steroid-umsetzungen. Naturwissenschaften 48:569–570

Mietzsch E, Lumbsch HT, Elix JA (1994) WINTABOLITES (Mactabolites for Windows). Users Man Comput Program Univ Essen 2nd edn. p 54

Miglietta ML, Lamanna R (2006) 1H HR-MAS NMR of carotenoids in aqueous samples and raw vegetables. Magn Reson Chem 44:675–685

Millot M, Kaouadji M, Champavier Y, Gamond A, Simon A, Chulia AJ (2013) Usnic acid derivatives from Leprocaulon microscopicum. Phytochem Lett 6:31–35

Mitsuno M (1953) Paper chromatography of lichen substances. I. Pharm Bull 1:170–173

Molnár K, Farkas E (2010) Current results on biological activities of lichen secondary metabolites: a review. Z Naturforsch C 65:157–173

Morlock G, Schwack W (2010) Hyphenations in planar chromatography. J Chromatogr A 1217:6600–6609

Munier R, Macheboeuf M (1949) Microchromatographie de partage des alcaloïdes et de diverses bases azotées biologiques. Bull Soc Chim Biol (Paris) 31:1144–1162

Narui T, Sawada K, Takatsuki S, Okuyama T, Culberson CF, Culberson WL, Shibata S (1998) NMR assignments of depsides and tridepsides of the lichen family Umbilicariaceae. Phytochemistry 48:815–822

Nascimbene J, Tretiach M, Corana F, Lo Schiavo F, Kodnik D, Dainese M, Mannucci B (2014) Patterns of traffic polycyclic aromatic hydrocarbon pollution in mountain areas can be revealed by lichen biomonitoring: a case study in the Dolomites (Eastern Italian Alps). Sci Total Environ 475:90–96

Nash TH (2008) Lichen biology. Cambridge University Press, Leiden

Nemes P, Vertes A (2012) Ambient mass spectrometry for in vivo local analysis and in situ molecular tissue imaging. TrAC Trends Anal Chem 34:22–34

Nemes P, Barton AA, Vertes A (2009) Three-dimensional imaging of metabolites in tissues under ambient conditions by laser ablation electrospray ionization mass spectrometry. Anal Chem 81:6668–6675

Nguyen K-H, Chollet-Krugler M, Gouault N, Tomasi S (2013) UV-protectant metabolites from lichens and their symbiotic partners. Nat Prod Rep 30:1490

Nielsen KF, Smedsgaard J (2003) Fungal metabolite screening: database of 474 mycotoxins and fungal metabolites for dereplication by standardised liquid chromatography–UV–mass spectrometry methodology. J Chromatogr A 1002:111–136

Nimis PL, Skert N (2006) Lichen chemistry and selective grazing by the coleopteran Lasioderma serricorne. Environ Exp Bot 55:175–182

Nishikawa Y, Michishita K, Kurono G (1973) Studies on the water soluble constituents of lichens. I. Gas chromatographic analysis of low molecular weight carbohydrates. Chem Pharm Bull (Tokyo) 21:1014–1019

Nylander W (1866) Les lichens du Jardin de Luxembourg

Odabasoglu F, Cakir A, Suleyman H, Aslan A, Bayir Y, Halici M, Kazaz C (2006) Gastroprotective and antioxidant effects of usnic acid on indomethacin-induced gastric ulcer in rats. J Ethnopharmacol 103:59–65

Okuyama E, Umeyama K, Yamazaki M, Kinoshita Y, Yamamoto Y (1995) Usnic acid and diffractaic acid as analgesic and antipyretic components of Usnea diffracta. Planta Med 61:113–115

Oliver SG, Winson MK, Kell DB, Baganz F (1998) Systematic functional analysis of the yeast genome. Trends Biotechnol 16:373–378

Ozenda P, Clauzade G (1970) Les lichens: étude biologique et flore illustrée. Masson Paris

Parrot D, Jan S, Baert N, Guyot S, Tomasi S (2013) Comparative metabolite profiling and chemical study of Ramalina siliquosa complex using LC–ESI-MS/MS approach. Phytochemistry 89:114–124

Patton A, Chism P (1951) Quantitative paper chromatography of amino acids. An evaluation of techniques. Anal Chem 23:1683–1685

Pengsuparp T, Cai L, Constant H, Fong HH, Lin L-Z, Kinghorn AD, Pezzuto JM, Cordell GA, Ingolfsdöttir K, Wagner H (1995) Mechanistic evaluation of new plant-derived compounds that inhibit HIV-1 reverse transcriptase. J Nat Prod 58:1024–1031

Pernice M, Meibom A, Van Den Heuvel A, Kopp C, Domart-Coulon I, Hoegh-Guldberg O, Dove S (2012) A single-cell view of ammonium assimilation in coral–dinoflagellate symbiosis. ISME J 6:1314–1324

Polborn K, Steglich W, Connolly JD, Huneck S (1995) Structure of the macrocyclic bis-lactone lepranthin from the lichen arthonia impolita; an X-ray analysis. Z Für Naturforschung B J Chem Sci 50:1111–1114

Pöykkö H, Hyvärinen M, Bačkor M (2005) Removal of lichen secondary metabolites affects food choice and survival of lichenivorous moth larvae. Ecology 86:2623–2632

Pyatt FB (1967) The inhibition influence of *Peltigera canina* on the germination and subsequent growth of graminaceous seeds. Bryologist 70:326–329

Ramaut JL (1963a) Chromatographie sur couche mince des despsides et des depsidones. Bull Sociétés Chim Belg 72:316–321

Ramaut JL (1963b) Chromatographie en couche mince des depsidones du β orcinol. Bull Sociétés Chim Belg 72:97–101

Ranković B, Mi\vsić M, Sukdolak S (2008) The antimicrobial activity of substances derived from the lichens *Physcia aipolia*, *Umbilicaria polyphylla*, *Parmelia caperata* and *Hypogymnia physodes*. World J Microbiol Biotechnol 24:1239–1242

Rezanka T, Dembitsky V (1999a) Novel brominated lipidic compounds from lichens of Central Asia. Phytochemistry 51:963–968

Rezanka T, Dembitsky VM (1999b) Fatty acids of lichen species from Tian Shan Mountains. Folia Microbiol (Praha) 44:643–646

Rezanka T, Guschina IA (2001a) Glycoside esters from lichens of Central Asia. Phytochemistry 58:509–516

Rezanka T, Guschina IA (2001b) Macrolactone glycosides of three lichen acids from *Acarospora gobiensis*, a lichen of Central Asia. Phytochemistry 58:1281–1287

Rikkinen J (1995) What's behind the pretty colours? A study on the photobiology of lichens. Finnish Bryological Society

Roullier C, Chollet-Krugler M, Pferschy-Wenzig E-M, Maillard A, Rechberger GN, Legouin-Gargadennec B, Bauer R, Boustie J (2011) Characterization and identification of mycosporines-like compounds in cyanolichens. Isolation of mycosporine hydroxyglutamicol from *Nephroma laevigatum* Ach. Phytochemistry 72:1348–1357

Russo A, Piovano M, Lombardo L, Garbarino J, Cardile V (2008) Lichen metabolites prevent UV light and nitric oxide-mediated plasmid DNA damage and induce apoptosis in human melanoma cells. Life Sci 83:468–474

Santesson J (1967a) Chemical studies on lichens—III: the pigments of Thelocarpon epibolum, T. laureri and Ahlesia lichenicola. Phytochemistry 6:685–686

Santesson J (1967b) Chemical studies on lichens. 4. Thin layer chromatography of lichen substances. Acta Chem Scand 21:1162–1172

Santesson J (1969) Chemical studies on lichens. 10. Mass spectrometry on lichens. Ark För Chem 30:363–377

Schmeda-Hirschmann G, Tapia A, Lima B, Pertino M, Sortino M, Zacchino S, de Arias AR, Feresin GE (2008) A new antifungal and antiprotozoal depside from the Andean lichen Protousnea poeppigii. Phytother Res 22:349–355

Schmidt J, Huneck S, Franke P (1981) Lichen substances—128 Mass spectroscopy of natural products—12. Comparative positive and negative ion mass spectroscopy of nitrogen-containing and ring C cleaved usnic acid derivatives. Biol Mass Spectrom 8:293–300

Seco JM, Quinoa E, Riguera R (2004) The assignment of absolute configuration by NMR. Chem Rev 104:17–118

Sharma J, Hood LV (1965) Thin-layer solubilization chromatography: I Phenols. J Chromatogr A 17:307–315

Shibata S (2000) Yasuhiko Asahina (1880–1975) and his studies on lichenology and chemistry of lichen metabolites. Bryologist 103:710–719

Shibata S, Takito M, Tanaka O (1950) Paper chromatography of anthraquinone pigments. J Am Chem Soc 72:2789–2790

Shrestha G, St. Clair LL (2013) Lichens: a promising source of antibiotic and anticancer drugs. Phytochem Rev 12:229–244

Shukla V, Joshi GP, Rawat MSM (2010) Lichens as a potential natural source of bioactive compounds: a review. Phytochem Rev 9:303–314

Silverstein RM, Bassler GC, Morrill TC (1998) Spectroscopic identification of organic compounds, 5th edn. Wiley, New York

Stahl E, Kaltenbach U (1961) Anisaldehyde-sulphuric acid for sugars, steroids, and terpens. J Chromatogr 5:35–40

Stocker-Wörgötter E (2008) Metabolic diversity of lichen-forming ascomycetous fungi: culturing, polyketide and shikimate metabolite production, and PKS genes. Nat Prod Rep 25:188–200

Strack D, Feige GB, Kroll R (1979) Screening of aromatic secondary lichen substances by high-performance liquid-chromatography. Z Naturforschung CA J Biosci 34:695–698

Stuart B (2004) Infrared spectroscopy. Wiley Online Library

Stuelp PM, Carneiro Leão AMA, Gorin PAJ, Iacomini M (1999) The glucans of *Ramalina celastri*: relation with chemotypes of other lichens. Carbohydr Polym 40:101–106

Sundholm EG (1978) C-13 NMR-spectra of lichen xanthones—temperature-dependent collapse of long-range couplings to hydrogen-bonded hydroxyl protons. Acta Chem Scand 32B:177–181

Sundholm EG (1979) Syntheses and 13C NMR spectra of some 5-chloro-substituted lichen xanthones. Acta Chem Scand 33B:475–482

Sundholm EG, Huneck S (1980) C-13 NMR-spectra of lichen depsides, depsidones and depsones. 1. Compounds of the orcinol series. Chem Scr 16:197–200

Sundholm EG, Huneck S (1981) C-13 NMR-spectra of lichen depsides, depsidones and depsones. 2. Compounds of the beta-orcinol series. Chem Scr 18:233–236

Szeghalmi A, Kaminskyj S, Gough KM (2007) A synchrotron FTIR microspectroscopy investigation of fungal hyphae grown under optimal and stressed conditions. Anal Bioanal Chem 387:1779–1789

Tabacchi R, Allemand P, Tsoupras G (1991) Direct analysis of lichens by tandem mass spectrometry. Symbiosis 11:193–206

Takeda K, Hara S, Wada A, Matsumoto N (1963) A systematic, simultaneous analysis of steroid sapogenins by thin-layer chromatography. J Chromatogr A 11:562–564

Tay T, Turk AO, Yilmaz M, Turk H, Kivan\cc M (2004) Evaluation of the antimicrobial activity of the acetone extract of the lichen Ramalina farinacea and its (+)-usnic acid, norstictic acid, and protocetraric acid constituents. Z Naturforschung C J Biosci 59:384–388

Torres A, Hochberg M, Pergament I, Smoum R, Niddam V, Dembitsky VM, Temina M, Dor I, Lev O, Srebnik M (2004) A new UV-B absorbing mycosporine with photo protective activity from the lichenized ascomycete *Collema cristatum*. Eur J Biochem 271:780–784

Tschesche R, Wulff G (1961) Über Saponine der Spirostanolreihe, VII. Über Digalogenin, ein neues Sapogenin aus den Samen von Digitalis purpurea L. Chem Ber 94:2019–2026

Tschesche R, Lampert F, Snatzke G (1961) Über triterpene: VII. Dünnschicht-und ionenaustauscherpapierchromatographie von triterpenoiden. J Chromatogr A 5:217–224

Van Roy W, Mathey A, Van Vaeck L (1996) In-situ analysis of lichen pigments by Fourier transform laser microprobe mass spectrometry with external ion source. Rapid Commun Mass Spectrom 10:562–572

Vijayakumar CS, Viswanathan S, Kannappa Reddy M, Parvathavarthini S, Kundu AB, Sukumar E (2000) Anti-inflammatory activity of (+)-usnic acid. Fitoterapia 71:564–566

Wachtmeister CA (1952) Studies on the chemistry of lichens. 1. Separation of depside components by paper chromatography. Acta Chem Scand 6:818–825

Wachtmeister CA (1956) Identification of lichen acids by paper chromatography. Bot Not 109:313–324

Wang X-N, Zhang H-J, Ren D-M, Ji M, Yu W-T, Lou H-X (2009) Lobarialides A-C, antifungal triterpenoids from the lichen Lobaria kurokawae. Chem Biodivers 6:746–753

Wenkert E, Baddeley GV, Burfitt IR, Moreno LN (1978) Carbon-13 nuclear magnetic resonance spectroscopy of naturally-occurring substances-LVII triterpenes related to lupane and hopane. Org Magn Reson 11:337–343

Whiton JC, Lawrey JD (1984) Inhibition of crustose lichen spore germination by lichen acids. Bryologist 87:42–43

Wilkins AL, Elix JA, Whitton AA (1990) A one-and two-dimensional 1H and 13C NMR study of some lichen triterpenoids of the pyxinol group. Aust J Chem 43:411–417

Wink M, Heinen HJ, Vogt H, Schiebel HM (1984) Cellular localization of quinolizidine alkaloids by laser desorption mass spectrometry (LAMMA 1000). Plant Cell Rep 3:230–233

Wojciechowski ZA, Goad LJ, Goodwin TW (1973) Sterols of the lichen *Pseudevernia furfuracea*. Phytochemistry 12:1433–1436

Wolfender J-L, Ndjoko K, Hostettmann K (2001) The potential of LC-NMR in phytochemical analysis. Phytochem Anal 12:2–22

Woodruff HB, Munk ME (1977) A computerized infrared spectral interpreter as a tool in structure elucidation of natural products. J Org Chem 42:1761–1767

Wynn-Williams DD, Edwards HGM, Garcia-Pichel F (1999) Functional biomolecules of Antarctic stromatolitic and endolithic cyanobacterial communities. Eur J Phycol 34:381–391

Yamamoto Y, Miura Y, Kinoshita Y, Higuchi M, Yamada Y, Murakami A, Ohigashi H, Koshimizu K (1995) Screening of tissue cultures and thalli of lichens and some of their active constituents for inhibition of tumor promoter-induced Epstein-Barr virus activation. Chem Pharm Bull (Tokyo) 43:1388–1390

Yoshimura I, Kinoshita Y, Yamamoto Y, Huneck S, Yamada Y (1994) Analysis of secondary metabolites from lichen by high performance liquid chromatography with a photodiode array detector. Phytochem Anal 5:197–205

Zizovic I, Ivanovic J, Misic D, Stamenic M, Djordjevic S, Kukic-Markovic J, Petrovic SD (2012) SFE as a superior technique for isolation of extracts with strong antibacterial activities from lichen *Usnea barbata* L. J Supercrit Fluids 72:7–14

Index

D.K. Upreti et al. (eds.), *Recent Advances in Lichenology*,
DOI 10.1007/978-81-322-2181-4